Praise for *Unbound*

"Currier's seamless narrative recalls Jared Diamond's sprawling histories of human civilization, and like Diamond, Currier manages to be thorough in synthesizing a great deal of specialized knowledge . . . while telling a story that is gripping."

—*Library Journal*

"A breathtaking analysis of human technological, social, and cultural history . . . An original telling of the human story, beautifully written by an erudite anthropologist . . . *Unbound* should be on every educated person's reading list."

—Jack M. Potter, Professor of Anthropology Emeritus,
University of California at Berkeley

"An ambitious and fascinating account of the role of technologies in the evolution of the human species . . . All in all, a 'Jared Diamond-esque' tour de force."

—Richard Robbins, SUNY Distinguished Teaching Professor,
Department of Anthropology,
State University of New York at Plattsburgh

"*Unbound* is a fascinating, accurate, and highly readable account of human cultural progress from the earliest hominid toolmaking to the age of digital technology."

—William O. Beeman, professor and chair,
Department of Anthropology, University of Minnesota

"How has technology figured in bringing us to our present-day predicament? . . . Given anthropologist and author Richard Currier's broad knowledge and lucid prose, this innovative, broad-ranging, beautifully illustrated exploration promises—and deserves—to find wide use, both in the classroom and beyond."

—Robert Bates Graber, Professor Emeritus of Anthropology,
Truman State University

"*Unbound* is a fine book, written in clear, non-technical language. Richard Currier succeeds in analyzing and clarifying the ways that technologies have affected both social structure and communication throughout the history of cultures."

—Eugene A. Hammel, professor of Anthropology and Demography, emeritus, University of California at Berkeley

"Sweeping in scope, daring in proposition, *Unbound* looks backward to look forward at a future mediated—and threatened—by human technologies."

—Brandy Schillace, PhD, managing editor of *Culture, Medicine, and Psychiatry*

"*Unboun*d combines the best of a lifetime of discriminating, multidisciplinary scholarship with the storytelling abilities of great scholarly, humanistic writers like Desmond Morris, Carl Sagan, and historian James McPherson. Dr. Currier's highly accessible work is alive with the energy of discovery."

—William Manley, Technical Lead for Cultural Resource Programs Worldwide and Deputy Federal Preservation Officer, United States Navy

"A satisfying complete work beautifully written and presented . . . I have just put down what will be a classic and required reading in anthropology departments. *Unbound* is more: it is a highly readable, believable history of how Man became what s/he is . . . Rarely can one call a work of non-fiction a page-turner, but this one is . . . This is one of those satisfying, complete works that will appeal equally to the academic and the educated 'man in the street'. . . Read Darwin, then read Currier. We are going to hear much of this book."

—Robert Cooper, anthropologist and author of *Thailand: Culture Shock! A Survival Guide to Customs & Etiquette*

"*Unbound* is a fascinating overview of the influence of technology on human evolutionary history. It is an excellent example of being understood by a wide audience of educated readers."

—Anek R. Sankhyan, PhD, President, Palaeo Research Society, Ghumarwin, Himachal Pradesh, India

UNBOUND

UNBOUND

HOW EIGHT TECHNOLOGIES MADE US HUMAN AND BROUGHT OUR WORLD TO THE BRINK

RICHARD L. CURRIER, PHD

FOREWORD BY TOM GJELTEN

Arcade Publishing • New York

First Edition

Arcade Publishing books may be purchased in bulk at special discounts for sales promotion, corporate gifts, fund-raising, or educational purposes. Special editions can also be created to specifications. For details, contact the Special Sales Department, Arcade Publishing, 307 West 36th Street, 11th Floor, New York, NY 10018 or arcade@skyhorsepublishing.com.

Arcade Publishing® is a registered trademark of Skyhorse Publishing, Inc.®, a Delaware corporation.

Visit our website at www.arcadepub.com.
Visit the author's website at www.richardlcurrier.com.

10 9 8 7 6 5 4 3 2 1

Library of Congress Cataloging-in-Publication Data

Currier, Richard L.
Unbound : how eight technologies made us human and brought our world to the brink / Richard L Currier. - First edition.
pages cm
Includes bibliographical references and index.
ISBN 978-1-62872-522-3 (hardback) - ISBN 978-1-62872-776-0 (E-book) -
ISBN 978-1-62872-752-4 (paperback)
1. Technology and civilization-History. 2. Human evolution. 3. Population.
4. Environmental degradation. 5. Sustainable development. I. Title.
CB478.C87 2015
303.4-dc23
2015014024

Cover design by Anthony Morais
Cover photo: *Prometheus Carrying Fire* by Jan Cossiers, courtesy of Wikimedia Commons

Printed in the United States of America

CONTENTS

FOREWORD IX
by Tom Gjelten

INTRODUCTION XIII
The invention of eight key technologies during the five-million-year history of humanity freed our species from the natural constraints that govern all other forms of life.

1 THE PRIMATE BASELINE: TOOLS, TRADITIONS, MOTHERHOOD, WARFARE, AND THE HOMELAND 1
The tree-dwelling life of the primates foreshadowed not only the evolution of the human physical form but also human social life, culture, and technology.

2 THE TECHNOLOGY OF SPEARS AND DIGGING STICKS: UPRIGHT POSTURE AND BIPEDAL LOCOMOTION 29
Prehistoric apes learned to make wooden spears and digging sticks, began to walk and run on two legs, revolutionized sex, and evolved into early hominids.

3 THE TECHNOLOGY OF FIRE: COOKING, NAKEDNESS, AND STAYING UP LATE 55
As early hominids evolved into emerging humans, they learned to sleep with fire, live in caves, and cook food—and became naked in the process.

4 THE TECHNOLOGIES OF CLOTHING AND SHELTER: HATS, HUTS, TOGAS, AND TENTS 81
The emerging humans constructed dwellings to shelter themselves, fashioned clothing to protect their bodies, and established new homelands in the cold northern latitudes.

5 THE TECHNOLOGY OF SYMBOLIC COMMUNICATION: MUSIC, ART, LANGUAGE, AND ETHNICITY 106

Modern humans developed visual and verbal symbolism, invented traditions of music, art, and design, adopted distinct tribal and ethnic identities, and began to fuse into large-scale groups.

6 THE TECHNOLOGY OF AGRICULTURE: PERMANENT VILLAGES AND THE ACCUMULATION OF WEALTH 139

The technology of agriculture led to the creation of permanent human settlements, the accumulation of wealth, the formation of socioeconomic classes, the economic importance of children, the suppression of female sexuality, and the genesis of organized warfare.

7 THE TECHNOLOGIES OF INTERACTION: SHIPS, WRITING, THE WHEEL, AND THE BIRTH OF CIVILIZATION 167

Innovations in the technologies of human interaction—in travel, trade, and the written word—enabled agricultural people to create social relationships across previously unbridgeable spans of time and space, leading to the development of cities and the birth of urban civilization.

8 THE TECHNOLOGY OF PRECISION MACHINERY: CLOCKS, ENGINES, AND INDUSTRIAL SOCIETY 201

When the clockmakers of Medieval Europe created precision machinery, they unleashed a transformation that produced the industrial revolution, another explosive growth in the human population, and the exploitation of virtually the entire planet for the fulfillment of human needs.

9 THE TECHNOLOGY OF DIGITAL INFORMATION: THE WORLD WIDE WEB OF HUMAN INTERACTION 240

From the nature of work and human interaction to the design of everything humans build and manufacture, and even to the nature of history itself, the digital revolution will transform human life and society as profoundly as any technology has transformed it in the past.

10 OUR WORLD AT THE BRINK: IS HUMANITY DRIFTING TOWARD A PLANETARY CATASTROPHE? 266
The explosive growth of the human population and its voracious consumption of the earth's resources has led to rapid environmental disintegration and a growing risk of planetary catastrophe.

ACKNOWLEDGMENTS 301
NOTES 303
BIBLIOGRAPHY 325
INDEX 361

Foreword

I met Richard Currier in 1969, when I was an undergraduate anthropology major at the University of Minnesota. Cultural anthropology to me was a social science like no other, one that not only offered the clearest and most intriguing explanations for human life and behavior but also featured the most heroic researcher role imaginable. One could unlock the secrets of an exotic culture only by living in it wholeheartedly for months on end; a serious anthropologist was as much adventurer as scholar. I wanted to see what it was really like to do ethnography, and I was unwilling to wait until graduate school to find out. Richard was a newly minted PhD from Berkeley, just a few years older than I, and he was the only professor in the department who understood and welcomed my enthusiasm. When I got the opportunity that year to join a university study trip to Honduras, I came up with the crazy idea to break off from the group and settle for a few weeks by myself in a Black Carib village that was on our itinerary. Richard knew I was unprepared at that point to do serious field research but agreed nevertheless to advise me in any independent project I devised.

I was told that the Black Carib, or Garifuna, people were descended from African slaves who had escaped slavery and found refuge among the Carib Indians who inhabited islands across the Caribbean. In time, they adopted much of the Carib language and culture, which explains their name. Theirs was a fascinating and unique history, and the eight lazy weeks I spent in a Garifuna village on the north coast of Honduras were the highlight of my college years. I can't say I came up with a clear research focus. I went fishing each morning with the men of the village and spent the afternoons chatting up my neighbors. The "thesis" I wrote afterwards was more travel log than ethnography, and many college professors would have dismissed my experience as un-serious. Not Richard Currier. In a department staffed largely by self-important old-timers, he brought energy and intellectual curiosity to his teaching, routinely challenging conventional wisdom and orthodox scholarship and encouraging his students to do the same. Always generous with his time, he listened to my stories of the strange fishing customs I observed or the loose "marriage" arrangements among my neighbors or the rowdiness I saw at a wake and then helped me see the logic behind the behaviors. It was from Richard that I learned how to think about culture. While I eventually opted for a career in the news media, doing cultural anthropology is itself a form of reporting, and I could not have had a better intellectual preparation for journalism than I got under his tutelage at the University of Minnesota.

Richard Currier's sharp intellect and original thinking burn brightly in the pages of *Unbound.* He reveals the remarkable story of how we humans became "unbound" from the natural forces that limit the habitats and populations of other animals. We stand upright and walk on two legs because our prehistoric ancestors figured out how to use wooden spears and digging sticks. We developed far larger brains than those of our closest primate ancestors largely because we learned how to control fire. Cooking our food made it more digestible, so we were able to eat more in less time and have more opportunity for thinking. We learn in *Unbound* that the use of clothing and shelter enabled the hominids, unlike other primates, to migrate out of the tropics and settle in the cold northern latitudes. By inventing language, we made it possible to communicate the connections between actions and outcomes, thus making it possible for hunters and gatherers to become agriculturalists. At almost every step of our journey to civilization, the development of new technology—from huts to carts, from ceramics to looms—was key. The invention of technologies of interaction, such as writing and ships, was followed by the emergence of technologies of precision machinery, which led to the industrialization of society. Some of the linkages he explains are unexpected. The industrial revolution may be responsible for the custom of marrying for love. The mechanization of home life can be tied to the movement for gender equality.

But this is not a story of unending progress. Richard Currier warns that our use of technology to master and conquer the natural world has also given us the means to destroy it. In the final chapter of *Unbound*, he reminds us how human invention has brought our planet to the brink of environmental catastrophe by introducing the scourges of pollution, deforestation, mass extinction, and overpopulation. As a result of man's activity, the planet is entering a period of climate change that is likely to have devastating consequences. At the same time, we could ultimately face another ice age, which would bring an even greater threat to human civilization. And then we must consider the ultimate "doomsday machine," the development of a nuclear technology with the power to exterminate the entire human species in a matter of moments, with irreparable damage to the biosphere.

Unbound is a book deeply informed by science and the wisdom of a well-trained anthropologist, and yet Richard Currier is still the same popularizer of knowledge that I encountered as a college student more than forty years ago. Here follows three million years of human history, but the account of our evolution moves as quickly and effortlessly as a bedtime story.

Tom Gjelten

Introduction

"All truth passes through three stages: first, it is ridiculed; second, it is violently opposed; third, it is accepted as self-evident."

—Anonymous[1]

Sixty-five million years ago, an asteroid more than six miles wide, traveling at a speed of 67,000 miles per hour, slammed into the earth off the coast of southern Mexico with a force 500 million times greater than the atomic bomb that was dropped on Hiroshima. That event triggered major disruptions in the earth's climate, ultimately producing an environmental catastrophe that resulted in the extinction of the dinosaurs and 75 percent of all species of life on Earth.

We are now in the process of another mass extinction of plants and animals caused by human activity that may ultimately become as deadly as the previous five mass extinctions that have taken place in the earth's geologic history. A majority of biologists believe that more than half of all living things will become extinct within the next century or two, with unknown consequences for the future and for the dwindling number of species that will remain alive.

We humans are no longer a species of simple hunters and gatherers living within the constraints of a stable natural world. Instead, freed from many of our natural limitations by the relentless progress of technology, we have become the unbound masters of the biosphere.

Over the course of the last five million years, eight key technologies have profoundly altered the relationship of our species with the natural environment, liberating us from the natural forces that restrain the populations of all other living things. One by one, each of these technologies has initiated a major transformation, or metamorphosis, in human life and society. These metamorphoses have revolutionized the structure of our bodies, expanded the capabilities of our minds, and given birth to human societies of unparalleled size and power.

In the modern era, humanity has gained control over nearly all of the earth's natural environments and has essentially converted the entire planet into a vast production unit for its own exclusive benefit. In the process, the newly unbound human species has taken over much of the natural environment, polluted the earth's soils, oceans, and atmosphere, and brought our world to the brink of catastrophe.

Our species is unique among all earthly creatures in its ability to comprehend and plan for the long term. Yet we are still motivated by ancient animal instincts, including the drive to expand and multiply to the limits of the possible. Other living things are limited in their ability to reproduce by the relatively fixed nature of their relationship to the environment. But by allowing us to escape the bonds of our biological destiny, technology has made it possible for us to continue multiplying, even as we have moved the world ever closer to an uncertain and potentially disastrous future.

Five million years ago, the adoption of fabricated spears and digging sticks by our ape-like ancestors encouraged us to stand, walk, and run upright. This innovation eventually produced a radical restructuring of mammalian anatomy that freed the forelimbs from the duties of locomotion. With the free use of their powerful forelimbs and dexterous hands, our ancestors were able to control fire, fashion clothing, and build dwellings. These technologies liberated us from the need to live in the tropical environments where we originated and allowed us to populate the vast temperate regions of Europe and Asia.

One hundred thousand years ago or more, when we began to use verbal and visual symbols to communicate, we freed ourselves from the limitations of direct personal experience. We gained the ability to share information over space and time, enabling us to pool our knowledge with others and develop cultures that were passed down through the generations in the oral traditions of song, story, and mythology.

Ten thousand years ago, the technology of agriculture liberated us from the constant search for food that preoccupies every other animal species. In the process, we were no longer bound to the endless wandering that had always been our fate as hunters and gatherers. We began to grow our own food, live in villages, and accumulate both the material wealth and the knowledge and wisdom that we passed down to our descendants.

Five thousand years ago, we developed powerful new technologies of transportation and communication. These included large seagoing ships, wagons pulled by beasts of burden, and forms of writing that enabled us to record information for posterity and to communicate with others over vast distances. These technologies of interaction enabled us to build cities, create civilizations, and evolve increasingly sophisticated forms of art, science, commerce, warfare, and religion that soon lifted humanity into a new position of supremacy over all other forms of life.

Five hundred years ago, the precision instruments of clocks, sextants, compasses, microscopes, and telescopes freed us from the limitations of our unaided sensory organs. And scarcely more than two hundred years ago, the technology of reciprocating engines liberated us from our ancient dependence on the physical power of the human body and our beasts of burden. As a result, we have conquered the world with the powers of science and the machinery of industry, and we have created immense nations in which millions of people live and work together as members of a single human society.

An eighth metamorphosis is now under way, triggered by the key technology of digital information, which has made it possible for all human beings to visit and communicate with each other, anywhere on Earth. This has enabled us to create a global culture and society that transcends national boundaries. The challenge for humanity will be to embrace this global civilization without sacrificing either the individual liberties or the ethnic identities that we all need to realize our goals in life and belong to something larger than ourselves.

But before we begin the remarkable story of how technology has freed humanity from the bonds of its natural origins, I would like to define and clarify four essential concepts that I have used in this book in ways that are a bit out of the ordinary. These concepts are 1) the nature of technology in the broadest sense of the word; 2) my decision to use the term "hominids" instead of the currently more fashionable "hominins"; 3) the three distinct phases of human evolution as they unfolded over the past five million years; and 4) the essential difference between a revolution and a metamorphosis.

The Nature of Technology

In modern speech, we generally use the word "technology" when referring to the most complex machines, structures, tools, techniques, and processes of modern life—things like spacecraft, automation systems, chemical processes, computer networks, and electronic devices. But in this book I have used the word "technology" as it has been defined by anthropologists and primatologists, who encountered preindustrial technologies in the ancient societies of hunters and gatherers and in the primeval societies of wild chimpanzees. Thus, anthropologists have defined technology—in its widest and most inclusive sense—as *the deliberate modification of any natural object or substance with forethought to achieve a specific end or serve a specific purpose.* Anthropologists have always regarded the tools, weapons, garments, and dwellings of hunting and gathering societies as true technologies. This book faithfully follows this traditional view.

Unlike the very simple technologies of chimpanzees and other animals, most human technologies involve complex processes and multiple materials that are used together to achieve a specific end. The prehistoric bow and arrow, for example, was typically made of a stone arrowhead and bird feathers fastened to opposite ends of a wooden shaft with vegetable gum and bound together with animal sinews. Each of these materials was not only derived from a different source but also required its own process of extraction and preparation—

yet we typically regard the bow and arrow as a single technology. Each of the eight key technologies described in this book is actually a complex collection of things and processes. What ties each of them together as a single entity is the common purpose for which each was created and used.

Hominids, Hominins, or Homininas?

For the past 250 years, all of the fully upright bipedal primates in the human family tree were called hominids—a word derived from the Latin term *Hominidae*, originally defined by the eighteenth-century Swedish naturalist Carolus Linnaeus, who founded the modern scientific method for classification of species. For many decades, scientists and writers have used the term "hominid" to refer to all of the species, both prehistoric and modern, who walked and ran fully upright and whose arms and hands were free, uniquely among the higher animals, to make and carry things.

But the traditional meaning of the term "hominids" changed in the 1990s, when major revisions were made to the classification of the monkeys and apes that belong to the mammalian order called primates. Advances in DNA analysis in the 1990s made it possible to quantify the precise genetic distance between one species and another, and since the genetic distance between humans and the great apes—such as the chimpanzee and gorilla—turned out to be relatively small, the official classification was substantially revised.

Under the new classification, the *Pongidae*—the biological family that formerly included the chimpanzees, gorillas, and orangutans—was abolished, and all of these species were placed together with humans into the family *Hominidae*. Therefore, technically speaking, the term "hominids" no longer means "the family of modern and prehistoric humans" but now literally means "the family of modern and prehistoric humans, chimpanzees, gorillas, and orangutans."

Once the *Hominidae* or hominids were no longer restricted to two-legged animals, anthropologists and paleontologists began to use the term "hominins" to refer to modern and prehistoric humans. Unfortunately, however, "hominins" has exactly the same problem as "hominids," because the subfamily *Homininae* includes not only humans but also gorillas and chimpanzees—and the tribe *Hominini* includes not only humans but also chimpanzees.

Technically, therefore, neither "hominids" nor "hominins" refers exclusively to prehistoric and modern humans. In fact, the only remaining scientific term that refers exclusively to the upright, bipedal humans, both modern and prehistoric, is the subtribe *Hominina.* But still struggling with the change from "hominids" to "hominins," writers and scientists can be forgiven for their reluctance to transition yet again to the unused and little-known "homininas"—especially considering that the subtribe *Hominina* may be doomed to the same fate that has already befallen its predecessors. Some scientists have proposed that the chimpanzees be reclassified as a species of our own genus, *Homo.* If this should come to pass, even the *Hominina* will include the chimpanzee, a quadrupedal animal neither built for, nor capable of, true bipedal locomotion—and not by any stretch of the imagination a legitimate member of the human family.[2]

For all of these reasons, I have used the traditional term "hominid" throughout this book as the preferred term for *all prehistoric and modern bipedal species in the human family tree.* Unlike "hominins"—which has recently become the favored term in academic anthropology and paleontology—the term "hominids" has been part of the scientific lexicon for centuries. It long ago became firmly accepted in common usage, and it continues to be recognized and understood by all educated readers. Most importantly, it is no less appropriate than "hominins," given the current scientific definitions of the *Hominidae*, *Homininae*, and *Hominini*.

The Three Distinct Phases of Human Evolution

If we go back to its earliest beginnings, we can see that human evolution has unfolded in three distinct phases. The hominids of each phase possessed a characteristic anatomy, a characteristic range of brain sizes, a characteristic collection of tools and weapons, and a distinct geographical distribution. The species that typify these three phases can be easily identified as falling into three groups, which I will refer to in this book as "early hominids," "emerging humans," and "modern humans."

The first phase—that of the early hominids—began several million years ago, when a population of prehistoric apes gradually evolved the ability to stand, walk, and run fully upright. The famous fossil remains of Lucy, one of the most ancient of these early hominids, was an *Australopithecus afarensis,* and at least five other species are currently recognized by paleontologists.

The early hominids made crude Oldowan stone tools but left no evidence that they used fire or lived in caves. Although they stood, walked, and ran fully upright, they retained the long arms, curved finger bones, long toes, and narrow shoulders typical of their tree-dwelling ancestors. The persistence of these ape-like characteristics in the fully upright early hominids is compelling evidence that they continued to climb high into the trees to sleep at night to avoid the large predators, especially the big cats, that were their most dangerous natural enemies.

Although they were inventive and resourceful, it is unlikely that the early hominids were significantly more intelligent than the great apes. Compared to the modern human brain's average size of roughly 1,400 cubic centimeters (cc), the brains of the early hominids were only slightly larger than the 375 cc brain of a typical chimpanzee—and their brains never expanded in any meaningful way during the millions of years that they inhabited tropical Africa.

The long history of the early hominids should be viewed as the successful adaptation of upright posture and bipedal locomotion by primates with the brain power of very intelligent apes. These creatures made spears and digging sticks, successfully hunted and killed other animals, defended themselves against their natural enemies, and flourished for about four million years—a period of time roughly eight hundred times longer than the entire history of urban civilization that began in ancient Mesopotamia five thousand years ago.

Beginning sometime after two million years ago, a population of more highly evolved hominids, with significantly larger brains, began to appear on the African continent. Over the course of the next million or so years, these emerging humans, with their superior Acheulean stone tools and technologies, gradually outcompeted and replaced the more primitive early hominids. By approximately one million years ago, all evidence of the early hominids had disappeared from the fossil record. They had apparently become extinct.

The emerging humans were larger and taller, with the broad shoulders and narrow waist that characterizes modern human populations. Moreover, the bones of their fingers were straight, not curved, their arms were shorter, and their toes were short and stubby. This indicates that the emerging humans were no longer adapted to climbing into the trees to sleep at night. The emerging humans became cave dwellers, and they developed a different strategy—the use of fire—to protect themselves from the large and dangerous predators in their environment.

Homo erectus, the most important and most successful of the emerging humans, migrated out of Africa to inhabit the tropical environments of South and East Asia and eventually settled across all the colder northern latitudes of Eurasia, from the British Isles to China. The brains of *Homo erectus*, which in the earliest finds averaged about 650 cc, grew in size until they reached 1,250 cc—almost within the normal range of modern human beings. It was *Homo erectus*, the "upright man," who decisively crossed the gulf between humanity and the rest of the animal kingdom.

Finally, beginning approximately 250,000 years ago, the first modern humans began to appear in Africa, sporting giant brains of 1,300 and 1,400 cc—roughly triple the size of the early hominid brains. *Homo sapiens*, the "thinking man," spread throughout the African continent, while other populations of modern humans migrated into Europe and Asia. Their descendants included the Neandertals who hunted the wooly mammoth and the wooly rhinoceros during the last ice ages and the anatomically modern humans who made the famous cave paintings of prehistoric France and Spain.

Between twenty-five thousand and fifteen thousand years ago, some of the tribes of anatomically modern humans who were living in Siberia crossed into Alaska, rapidly spread throughout all of North and South America, and completed the human conquest of all the earth's continents.

Early hominids, emerging humans, and modern humans were thus the three dominant populations during each of the major phases in the evolution of humanity. The early hominids survived for well over four million years, the emerging humans for nearly two million years. We modern humans, with our "superior" brains, have inhabited this earth for at most a quarter of a million years—one-eighth as long as the emerging humans and one-sixteenth as long as the early hominids. Modern humans have a long way to go before equaling the longevity of our most ancient ancestors.

Eight Metamorphoses, Many Revolutions

Although numerous revolutions have come and gone in the course of human history, humanity has experienced only seven fundamental transformations, or metamorphoses. Some of these metamorphoses have been called revolutions (the metamorphosis of agriculture is often called the Neolithic revolution, and the metamorphosis of science and industry is commonly known as the industrial revolution).

But the word "revolution" is also used to describe any sudden and sweeping change in a particular political power structure or in a particular realm of culture, including science, technology, and art. A metamorphosis, however, describes *a sweeping change in every aspect of culture and society:* diet, habitat, social relationships, economic behavior, group size, technology, evolutionary pressures, and even human anatomy itself. There have been thousands of revolutions in the course of humanity's evolution and history, but there have been only a few genuine metamorphoses.

The first three metamorphoses occurred literally millions of years ago, among populations of prehistoric apes, early hominids, and emerging humans. These metamorphoses led to the fabrication of lethal weapons, the development of full upright posture and bipedal locomotion, the expansion and intensification of sexual behavior, the control of fire, the fabrication of clothing and dwellings, and the creation of that uniquely human innovation, the nuclear family.

The next three metamorphoses occurred thousands of years ago, among populations of biologically modern humans. These transformations led to the invention of language and symbolic communication, the emergence of tribal and ethnic identities, the domestication of plants and animals, the birth of civilizations, and a massive increase in the earth's human population.

The seventh metamorphosis, which produced the industrial revolution, occurred only a few centuries ago and is well-documented by a wealth of historical sources. This technological transformation so radically increased the ability of humans to feed and protect their offspring that human overpopulation has now become the primary threat to the earth's environment.

At the present time, an eighth metamorphosis is under way, set in motion by the key technology of digital communications. For the first time in human history, it has become possible for anyone on Earth to interact with almost anyone else on Earth, quickly and affordably. Human society will be transformed as much by this latest metamorphosis as it was by the seven key technologies of the past and the seven metamorphoses they unleashed.

May the Fittest Hypotheses Survive

My goal in writing this book has been to identify the fundamental biological and cultural transformations—and the technologies which triggered them—by which, step by step, the human species has arrived at our present exalted yet precarious state of being. In the process, I have attempted to make sense out of several distinctive anatomical characteristics of hominids that exist nowhere else in the animal kingdom and for which a clear evolutionary advantage has not always been apparent.

Why did our ancestors adopt upright posture in the first place? Why did we lose the formidable dental weaponry we inherited from our primate ancestors and become unable to defend ourselves without fabricated weapons? Why are human females the only mammals whose breasts swell and become permanently enlarged at sexual maturity, regardless of whether they are pregnant or nursing? Why is our sexual behavior more or less continuous, rather than coordinated with periods of fertility, as it is with every other species? How did we become the only animal on Earth that is attracted to fire rather than repelled by it? Why did we lose the natural protective coat of fur possessed by all other primates and become naked? Why did humans, the descendants of a group of mammals that evolved to live in the treetops, become adapted to living not just *on* the ground but in some cases actually *beneath* the ground, in caves both natural and artificial? And how do all of these uniquely human characteristics fit together as a coherent whole?

When Galileo proposed that the earth revolved around the sun, he was denounced by the papacy and the astronomers of his day, convicted of heresy, and condemned to house arrest for the rest of his life. When Darwin proposed that the human species had evolved from an ape-like ancestor, he was greeted with skepticism, scorn, and ridicule by the learned scholars of his time.

In modern times, scholars and scientists have been no less prone to dismissing an unorthodox explanation, if it happens to challenge the assumptions of conventional scientific wisdom. Full upright posture and bipedal locomotion turned out to be millions of years older than was originally believed. The use of fire is hundreds of thousands of years older than was originally believed. Chimpanzees make and use a variety of tools, a capability once considered the exclusive province of humans. And the astonishing realism of prehistoric art turned out to be tens of thousands of years older than the early paleontologists were prepared to believe.

Readers familiar with the study of human evolution will find that some of the explanations I have offered concerning human origins run counter to orthodox scientific thinking. In itself, this should trouble no one, since orthodox scientific thinking about human origins and human evolution has changed many times, and in some cases what was heresy in one generation has become orthodoxy in the next.

Most of the facts and theories about human evolution presented in this book are consistent with current scientific thinking. When they are not, I have attempted to show why I believe an alternative explanation is called for. Some of the hypotheses I have proposed in this book may be unorthodox, but in my view they best fit all the facts as we know them. It will be up to others to evaluate their validity and to determine their fitness for survival within the scientific understanding of human origins.

UNBOUND

CHAPTER 1

The Primate Baseline

Tools, Traditions, Motherhood, Warfare, and the Homeland

> *"We must . . . acknowledge . . . that man, with all his noble qualities . . . still bears in his frame the indelible stamp of his lowly origin."*
>
> —Charles Darwin, *The Descent of Man*

When Charles Darwin first advanced the idea that the human species had evolved from an ape-like ancestor, he was greeted with howls of indignation from the clergy, the public, and many of the learned scholars of his time. In spite of all the obvious similarities between the human body and the bodies of monkeys and apes, Darwin was portrayed as a heretic whose theories contradicted not only the biblical story of creation but also the widely held belief that the human species was far too unique to have sprung from such a "lowly origin" (see Figure 1.1). By the time of his death, however, Darwin's views on evolution had become widely accepted, and since his time, the evolutionary connection between humans and prehistoric apes has been demonstrated by paleontologists and geneticists with a thoroughness and precision that Darwin himself could never have imagined.

Human beings may be related to apes and monkeys in the sense that we share a common ancestry, but humans are unique in ways that unequivocally separate us, not only from primates, but also from all other forms of life. Most of this book explores the technology-driven changes that gradually transformed us into much more than just another animal species. But in order to make sense out of the strange complexities of human society and culture, we have to begin by understanding the primate baseline—the anatomy and behavior of monkeys and apes. These were the genetic starting points, the natural raw materials, out of which the unique anatomy and behavior of human beings evolved. By understanding the nature of these evolutionary building blocks, we can more fully appreciate how far we have come—and how far we have yet to go.

The stamp of our primate ancestry is obvious in every aspect of human anatomy. The human hand evolved from the need to grasp the branches of trees with a powerful, secure grip. The human shoulder, which allows the arm to rotate into a fully vertical position, evolved from the need to hang by the arms from overhead branches. The human foot, beautifully adapted to walking and running on two legs over the open ground, was originally a grasping "hand" designed for climbing trees.

Apart from its comparatively small mouth and high forehead, the human face is a typical primate face. It is hairless around the eyes, ears, nose, and mouth, with prominent eyebrows.[1] The nose is short, because the sense of smell is not important for survival in the trees. And both eyes face forward for binocular vision, because this type of vision enables primates to judge distances in three-dimensional space and is thus vital for moving quickly and easily through the trees.

Even the human voice evolved from the ancient primate need to communicate with friends, relatives, or rivals who may be hidden in the dense tropical foliage. In fact, all primates are vocal, and most species make a variety of sounds, each with its own special meaning. The gibbons of Southeast Asia even invent their own songs, which they sing in the forest every day before dawn. And while our bodies are no longer capable of climbing trees to dizzying heights with the ease of apes and monkeys, we still find a singular pleasure in being perched in high places with commanding views.

In addition to providing us with the basic features of our distinctive human physical characteristics, the influence of our primate ancestry can be clearly seen in many basic elements of human behavior. Like most primates, we are a social, group-living species. We mature slowly and remain dependent on our mothers for the first several years of our lives, and we form intense bonds with our mothers, siblings, and mates that often last for our entire lives. We organize ourselves into social hierarchies, and within these hierarchies we compete with our siblings, classmates, and coworkers who are similar in rank to ourselves. At the same time, we defer to our parents, teachers, and bosses who outrank us, and we expect deference from our children, students, and employees whom we outrank.

Even the much-vaunted human ability to create and pass on distinctive cultures exists in a rudimentary form among many other higher animals, including whales, elephants, and even prairie dogs. Modern field studies by primatologists have established beyond question that monkey and ape societies are also capable of creating and maintaining the basic building blocks of culture, in which customs and traditions are invented by individuals, passed to other members of the group through imitation and practice, and handed down to succeeding generations from parents to offspring. Lastly, the use of tools and weapons, once considered the defining difference between humans and all other animals, has been identified unequivocally as part of the normal behavior of our closest genetic relative, the common chimpanzee.

Group Solidarity and the Homeland

Primates are highly social animals, and for the most part they spend their days in the companionship of other members of their group, sharing a common territory or home range from which other groups are excluded. While a group of primates willingly shares its homeland with its own members, it will aggressively drive away other members of its own species who belong to groups from other territories, and it will defend its own territory against neighboring groups—just as we do now and the hunter-gatherers did before us (see Figure 1.2).

For every primate group there is a "we" and a "they"—the insiders who belong to one's own group and live in one's own territory versus the outsiders who belong to other groups and live in alien territories. And every primate group defends its homeland against rival groups with noisy, hostile, and sometimes violent confrontations that often take place at the borders where adjacent territories meet. The members of two different groups of monkeys or apes scream at each other, make threatening gestures, break branches, throw things, and generally attempt to intimidate the other side.

Primates also distinguish between two fundamentally different types of ownership: communal property and personal property. The territory and natural resources of a group's homeland—including sleeping trees, fruit trees, honeycombs, birds' nests, drinking places, and so on—are generally considered to be the communal property of the group as a whole, and any member of the group has a right to use them. But when a particular piece of fruit is picked, a tasty insect is captured, or a nest of branches is constructed in a sleeping tree, that nest or morsel of food becomes the property of the individual who gathered or built it—and it is rarely shared with others.

Human societies all recognize these two types of property in the personal possessions that belong only to individuals versus the public territory shared by all members of the community (which in our society includes streets, roads, parks, and other public spaces). To these, humans have added a third type of property, the family possessions (especially food and dwellings) that are shared by the members of the family but not by the society as a whole.

Primate groups vary in size from a handful to 150 or more individuals. This size range is exactly the same as that found among the nomadic bands of human hunters and gatherers studied by anthropologists. The smallest primate groups consist of little more than a mother and her offspring, but most primate species live in larger groups that include adults, juveniles, and infants of both sexes. Some groups are little more than harems, in which several females live with a single dominant alpha male. This pattern is typical of gorillas, langurs, howler monkeys, and baboons.[2] Still other species, such as rhesus monkeys and chimpanzees, live in groups with multiple males and females. These species typically have strong and lasting bonds with their mothers, but their sexual relationships are, from our human point of view, promiscuous—very casual and neither permanent nor exclusive.

Almost all primate groups consist of a complex web of relationships that binds the members of the group together with four distinct types of social bonds that are also fundamental building blocks in all human societies. These are: 1) the maternal relationship between mother and offspring; 2) the social hierarchies that bind individuals together in relationships of dominance and submission; 3) the friendships and alliances that can form between any two individuals; and 4) the sexual relationships that are formed and maintained between adult males and females.

Primate Motherhood

The maternal bond tends to be stronger and more intense among mammals than other animals simply due to the physical and emotional attachments formed during the weeks, months, or years that every female mammal spends nursing her young. And because primates are specially adapted to living in the trees, the maternal bond is more powerful and lasting among them than it is among any other group of mammals. Almost alone among the many species of placental mammals, primate mothers must physically carry their offspring with them, wherever they go, throughout the first months or years of life.

The reasons for this extraordinary maternal burden are easily identified. Since primates are adapted for a life in the trees—and since they must be constantly on the move to search for seasonal tree-borne foods—primates cannot construct permanent nests or burrows. This means that—unlike burrowing animals such as mice, rabbits, or foxes—they cannot hide their young from danger until they are old enough to be on their own. Moreover, a single stumble or fall from the treetops could easily be fatal to the immature primate.

It requires literally years of development before a young ape or monkey can safely travel through the treetops on its own, and until then it is dependent on its mother to provide safe transport from place to place. This is very different from terrestrial animals, whose young can harmlessly stumble and fall over and over as they learn to walk and run. For these reasons, the intensity, duration, and life-or-death significance of intimate physical contact between the primate mother and her offspring dwarfs that of any other higher animal.

In infancy, a monkey or ape will cling to the fur on its mother's body with all four limbs, riding upside down under her belly almost continuously during the first few weeks or months of life. As it grows larger and stronger, the baby primate will begin to move about cautiously on its own, but it rushes back to its mother at the first sign of danger. As it passes out of infancy, the juvenile ape or monkey gradually makes the transition from riding upside-down on its mother's belly to riding right-side-up on her back or shoulders—and this will continue for months or even years before it is old enough to give up this constant need for maternal contact.

The bond that develops between the primate offspring and its mother during these initial months and years of intimate physical contact typically lasts for life (see Figure 1.3). It is not surprising, therefore, that the maternal bond is central in the social life of all species of primates, while the paternal bond varies, depending on the species, from great importance to complete irrelevance.

The already powerful maternal bond typical of all primates became even more intense as four-legged, tree-dwelling primates evolved into two-legged, ground-dwelling humans. Human offspring mature more slowly than those of any other primate, and the period of maternal dependence is correspondingly longer. In the societies of monkeys and apes, adult females gain status and prestige in the group when their offspring are born. Likewise, the unique burdens and responsibilities of motherhood are recognized, valued, and celebrated in every human society by a wealth of cultural traditions that honor the special, life-long relationship between the human mother and her children. Humans are, however, unique in the strong bonds that typically develop between fathers and their offspring, a revolutionary development among group-living primates.

Primate Sexual Relationships

Stable sexual bonds and exclusive sexual relationships, including cooperation between males and females in the rearing of their offspring, appeared long ago in the history of life on Earth. Such relationships can be found among animals as primitive as fish, and they are nearly universal among birds, some of whom, such as geese, mate for life. Among mammals, however, sexual relationships are often neither stable nor exclusive, and they tend to vary greatly in character and importance from one species to another.

Sexual relationships among goats and sheep, for example, are usually limited to a few brief acts of copulation. The alpha males in these species, who have exclusive sexual rights over their herd of females, are far too busy mating and defending their rights to help any of their dozens of mates with the task of rearing their offspring. This is a typical pattern among animals that live in herds, such as the grazing herbivores.

At the other end of this continuum, the titi monkeys of South America form monogamous, often exclusive sexual relationships, and pairs of titi monkeys are rarely out of each other's sight. When at rest, mated titis often sit next to each other with their tails intertwined, and they often seem to be more disturbed by expressions of distress from their mates than by expressions of distress from their offspring. It is notable that among titis, both sexes participate actively in caring for the young. In fact, after the first three months of infancy, the male titi may carry the offspring as much as 90 percent of the time. On the surface, this seems to parallel a pattern common to many human societies, but unlike the titi, different human societies vary tremendously in their attitudes toward the care of the young by males.

The various species of apes and monkeys exhibit many different patterns of sexual relationships, and not surprisingly the sexual relationships typical of humans are very different from those of any monkey or ape. Yet for all their differences, human sexual relationships contain many of the elements found in the typical relationships of the non-human primates, including the tendency of both humans and other primate males to compete for access to sexually active females. But among the most important differences between humans and all other primates is that women are the only female primates capable of being sexually active more or less continuously from puberty to old age.

When female monkeys and apes ovulate, they go into heat—technically the state of estrus—for approximately five to seven days about once each month, and females are sexually active only during these relatively brief estrus periods when they are ovulating and fertile. Estrus in most primates is characterized by a sudden and intense appetite for sex, accompanied by a conspicuous swelling of the female genitalia. But when estrus subsides, sexual relations cease until ovulation occurs again.

Among all non-human primates, females who are immature, pregnant, nursing, or have become infertile with age do not ovulate, do not experience estrus periods, and with rare exceptions do not engage in sexual intercourse at these times. Although different primate species vary greatly in the age of sexual maturation, in the frequency and duration of estrus, and in the intensity of sexual interactions, all non-human primate species—and indeed all higher animals—exhibit some version of this ancient sexual pattern.

The only exception to this rule occurs among the bonobo, or pygmy chimpanzee, in which many kinds of sexual play and genital stimulation take place daily between individuals of the same and opposite sexes and even among individuals of very different ages. But most of this sexual behavior does not include actual intercourse and seems to have little or no connection to either procreation or to forming bonds between males and females. Instead, sexual behavior among bonobos seems to function as a means of resolving conflict and reducing tension between individuals.

It is notable that of all the primates, the bonobo, by far the most sexually active of all primates (followed closely by the common chimpanzee), is generally considered to be most closely related to humans genetically. Bonobos have extremely long estrus periods (thirty days, versus approximately five to seven days for most primates), and estrus among bonobos occurs about once every forty-five days, leading to considerably more opportunities for sexual intercourse than is typical for other non-human primates. Even the hypersexual bonobos, however, do not continue having sexual intercourse during pregnancy, during nursing, or after menopause, as most humans do.

It is also noteworthy that neither bonobos nor common chimpanzees are sexually possessive or sexually exclusive. This is a striking contrast with humans, who are intensely competitive over sexual partnerships and who tend to be extremely possessive toward their mates. The significance of this radical departure from typical primate behavior will be explored in detail in the next chapter.

While intermittent sexual relations governed by the hormonal cycles of ovulation and estrus are universal among apes and monkeys, the patterns of mating and sexual bonding among the various primate species vary greatly between one species and another.

Monogamy, the most common form of sexual bonding among humans, is nonexistent among 97 percent of all mammalian species and is rare among apes and monkeys. Among the orangutans of Southeast Asia, a single adult male will establish a sexual relationship with two or more adult females who live apart from each other. Each female orangutan and her immature offspring typically live by themselves in the rainforest, where they are visited periodically by the adult male that fathered these offspring. At other times, this same adult male may also visit his other mates and their offspring living in nearby areas.

The gibbons and the closely related siamangs of Southeast Asia also live in isolated nuclear families, where male gibbons play an active role in caring for the offspring. The Southeast Asian rainforest is characterized by widely dispersed food sources, and this is thought to be the reason for the lack of large groups in this habitat. Groups this small, essentially consisting of a single nuclear family, however, tend to be rare in the primate world.

Two other types of sexual bonding among primates are much more common than the nuclear family group. The first is the harem system, in which a single dominant or alpha male has exclusive sexual rights over a specific group of females. The second is the multi-male, multi-female system, in which sexual relationships are transitory and sexual behavior is generally considered promiscuous.

The harem system is typical of gorillas, patas monkeys, langurs, and most baboons. It is no coincidence that these species are all largely terrestrial and spend most of the daylight hours on the ground foraging for food. In these species, the females form a stable community, together with a dominant adult male who protects them and who has exclusive sexual rights to any of them that come into estrus. At the same time that a single adult male monopolizes several females, most of the immature and older adult males are condemned to a form of "bachelorhood" in which they live alone, or with shifting groups of other bachelors, each waiting for the opportunity (which for most of them never arrives) to acquire a harem of his own. The alpha male must therefore always be on his guard, ready to defend his position against the many would-be challengers lurking nearby in other parts of the homeland.

As time passes and the dominant male ages, he is increasingly challenged and ultimately replaced by a younger and stronger male who had previously been living as a bachelor. A series of increasingly vicious confrontations will usually end with the defeat and departure of the original alpha male, who, if he is lucky enough to survive the final battles, will live out the rest of his life in the twilight of aging bachelorhood. Meanwhile, the females remain together as a group and accept the dominance of the new alpha male.

Upon assuming ownership of the harem, the new alpha male often kills the offspring that are still in infancy. Since these infants were fathered by the previous male, this behavior probably evolved to ensure that the new male will be able to invest his time and energy protecting and defending the survival of those offspring that are carrying his own genes rather than the genes of his predecessor. And once ovulation is no longer suppressed by the hormones released during nursing, the now childless females begin ovulating again, come into estrus, become sexually active again, and eventually conceive new offspring fathered by the new alpha male.

Given the pattern of violent competition among adult males of these species to possess their own harems, natural selection among such species tends to favor males that are larger, more aggressive, more possessive, and more dangerous. This constant selection in favor of superior fighting abilities and possessiveness toward females promotes the qualities that make it more likely that the males of such species will defend their harems and offspring from terrestrial predators. If a primate group wanders far out into open country and cannot easily flee to the safety of the trees, it is the males who must confront and defend the group from the lions, wild dogs, hyenas, and other predators that hunt in the open countryside.

The multi-male, multi-female system is probably the most common type of primate group. It is characterized by males and females of all ages living together within a single group structure. Chimpanzees and bonobos, as well as savannah baboons, macaques, and many other Old World monkeys, all live in multi-male, multi-female groups. Sexual bonds in such groups are not exclusive, and sexual possessiveness is either greatly diminished or nonexistent. Both males and females have sexual relations with a variety of partners. Among the promiscuous species, females in estrus typically have sexual relations with several different males, often one right after another, while an adult male will readily have sex with any estrus female who will accept his advances.

Yet even among the multi-male, multi-female species, a sexually active female and a particular male often form a more permanent bond known as a consort pair. The consort pair tends to separate themselves from the group, engage in frequent sex and mutual grooming, and even share food—a behavior that is otherwise exceedingly rare among primates. In fact, primate mothers have been known to snatch food away from their own offspring and eat it themselves.

The consort pair relationship tends to be transitory, however, typically lasting only for the few days during each monthly cycle that the female ape or monkey remains sexually active. Yet the consort relationship seems to have provided the biological starting point for the evolution of the human nuclear family—and a strong, stable nuclear family is an indispensable feature of all successful human cultures.

Like all other primates, humans have an intense and abiding interest in sex. But the continuous sexual relationship typical of humans goes far beyond the norm for any other primate—or for that matter, any other animal. The origins and implications of this greatly expanded sexuality will be discussed in detail in the next chapter. But what is most striking about the patterns of human sexual bonding is that it does not conform to any of the various types that have just been described. Instead, it includes elements of all of them—monogamy, polygamy, and promiscuity—to one degree or another.

Monogamy is by far the most common sexual bond among humans, but in the majority of cases it coexists in the same societies with forms of polygamy similar to the harems of monkeys and apes. Primate groups that form harems are often compared to human societies that practice polygyny—the type of polygamy in which a single man is allowed to marry and father children by more than one woman. It has often been noted that there are (or at least used to be) many more cultures and societies that allowed polygyny than there were cultures and societies in which it was prohibited. But there are also important differences between the harems of non-human primates and the human practice of polygyny.

The first difference is that, since female apes and monkeys have sex only during ovulation and estrus, it is rare for more than one female in a group to be sexually active at any given time—except in the case of a takeover of a group by a new alpha male and the wholesale slaughter of infants. Female humans, however, tend to be sexually active most of the time (except immediately before and after childbirth—and, in many societies, during menstruation).

The effect of this continuous sexual availability is a constant undercurrent of competition and jealousy among human co-wives for the sexual attention and favors of their shared husband. Anthropologists who study such societies typically report that friction between co-wives is the bane of their polygynous marriages. In fact, the most productive hunters and most prosperous farmers who practice polygyny in these societies do so more for the economic advantages of having more women to work for them and more children to support them in their old age than for any expectation of sexual nirvana.

The second important difference is that even in societies that permit polygyny, the vast majority of marriages are monogamous, because most men simply cannot afford to provide for more than one wife. By contrast, in the single-male, multi-female societies of non-human primates, essentially all of the adult females belong to one of the harems of females who mate with their currently reigning alpha male on the relatively rare occasions when they are ovulating.

Finally, the human tendency toward promiscuity often overwhelms even the strictest cultural prohibitions, and in some of the sexually permissive cultures studied by anthropologists, certain kinds of sexual possessiveness were frowned upon. In fact, in some cases, the sharing of sexual partners was not only tolerated but was actually considered good form. The Inuit Eskimos of the Arctic were famous for sharing their wives and unmarried daughters with other males who came to stay as visitors or guests. To the Inuit, sexual possessiveness was an undesirable trait, and an Inuit man would no sooner deny his guest the pleasure of his wife than he would deny him the hospitality of food or shelter.

Primate Friendships

Animal friendships exist in many different forms. In primates, these can range from casual acquaintances with familiar members of the group to special "political" alliances among adults who may be sticking together for mutual defense or who may be competing with other individuals and alliances for status and power. Bonds of friendship are found to some extent in most higher animals, and primates are notable in the important role played by bonds of friendship in the course of their daily lives.

Lastly, there is a special bond that can develop between any two individuals who are usually (but not always) members of the same group or species. Most birds are monogamous during the nesting season, when both males and females engage in the arduous tasks of raising the next generation, including building the nest, hatching the eggs, and feeding and protecting the young. In fact, many species of birds, such as geese, mate for life.

Elephants, dolphins, wolves, lions, and many other mammals may develop intense and lifelong bonds with members of their own species. Many domesticated animals, including dogs, cats, horses, pigs, parrots, and geese, are capable of forming strong attachments with other species—including human beings—sometimes for reasons that defy logic. It is tempting to think that only human beings fall in love, but the intense affection and attraction for another individual—an attraction that often rises to the level of an obsession with the other individual—occurs among many animal species and is entirely consistent with our definition of "love."

Primate Social Hierarchies

Higher animals that live in groups typically form social hierarchies in which every individual occupies a specific place in a pecking order, with the most dominant individuals at the top and the most submissive individuals at the bottom. The highest-ranking chickens can peck any of the chickens below them in the hierarchy, while lower-ranking chickens are pecked by those above them and in turn can peck those below them. Unable to defend themselves or attack others, the lowest-ranking chickens are often pecked to death. While we humans may regard this behavior as cruel, the ability to form social hierarchies based on differences of dominance and submission among individuals is actually essential to the preservation of peace and solidarity within the group.

Social hierarchies promote social stability in a very simple way: they ensure that the members of a group will not engage in repeated battles as they compete for food, mates, sleeping places, and other desirable things. Hostile confrontations between individuals of similar rank may occur, but they will typically continue only until one individual wins and the other loses. The winner becomes dominant over the loser, who becomes and remains subordinate in the relationship. From that time forward, the subordinate individual will normally yield to the dominant individual whenever they both desire the same thing, whether it is an attractive member of the opposite sex or a favorite item of food tossed on the ground between them. The original hostilities between them have been resolved by a stable arrangement. There will be no further confrontations, battles, or life-threatening injuries.

All primates—humans included—tend to act in submissive ways toward their elders and in dominant ways toward their juniors. At the same time, they tend to fight and compete for dominance with those of similar status, especially members of their own age group. These struggles for dominance are particularly intense in adolescence and young adulthood, when the social hierarchies of the newly mature generation are being established. The most dangerous of these struggles for dominance occurs when the position of an aging alpha male is challenged by a younger and stronger individual, temporarily upsetting and destabilizing the known and familiar social hierarchy that has been in effect for months or years up to that time.

While on the surface it may seem odd to describe hierarchical behaviors such as dominance and submission as a type of bonding, the fact is that these hierarchical relationships tend to be exceedingly stable and long-lasting. Social hierarchies are common among group-living animal species because they minimize hostilities among the members of the group as well as provide a mechanism that allows the entire group to take concerted action when necessary.

Just as monkeys and apes compete for dominance and social rank within their particular hierarchies, we humans also compete for dominance and social rank within our own hierarchies of art, science, technology, business, government, fashion, entertainment, and sports. But while the players change, the play remains the same: a stable social hierarchy in which each individual has a particular rank, in which life-or-death battles are infrequent, and in which winners and losers know their status, and stay in their places, but nevertheless remain watchful for opportunities to move up in the hierarchies to which they belong.

The Fission-Fusion Society

One particularly distinctive type of social structure that is typical of certain primate species—a type that is particularly common in human societies and that has been greatly elaborated by humans—is called the fission-fusion society. In these species, the active group tends to vary in size and composition from hour to hour, day to day, and season to season. The fission-fusion society breaks apart (the fission phase) into individuals and smaller groups that disperse throughout the home territory at certain times of day, or in certain seasons of the year, and then gathers together (the fusion phase) in a single location at other times to form one very large group. This type of society is typically found among three types of primates: 1) monkeys that have adapted to a terrestrial way of life; 2) the apes mostly closely related to hominids, especially the common chimpanzee and the bonobo; and 3) human beings.

In a fission-fusion society, the individual members of the group often spend their daylight hours scattered throughout the territory of their homeland, each individual or small group setting off in pursuit of some particular kind of plant or animal food in the morning. Later, all return to the home base as the day comes to an end, gathering for safety in the sleeping trees, cliffs, or (in the case of humans) campsites and dwellings. In spite of their many differences in social organization, all human societies exhibit the same type of fission-fusion behavior. The persistence of this pattern is quite evident in our own society, in which family members typically disperse to their separate destinations during the day and then gather at the end of the day to eat and sleep together in their single shared dwelling.

Perhaps because the composition of the various groups fluctuates constantly—each consisting of a different mix of individuals appropriate to the time of day or season of the year—species that practice this fission-fusion pattern are also more likely to interact with and even mingle with other groups from other home ranges. Chimpanzees are well known for the periodic occasions when two separate and independent groups undergo a kind of temporary fusion and mingle as a single group for hours at a time. The members of the two groups interact with great excitement, the males of each group putting on displays of dominance: hooting, screaming, and charging about, sometimes breaking off branches and brandishing them ferociously during their performance. After a while, the excitement dies down, the members of the different groups separate, the original groups reassemble, and the two groups go their separate ways.

These dominance displays function not only as a warning to other males but also as a means of attracting sexually active females from the other group, in the hopes of having sexual liaisons with one of them. When they do occur, these liaisons may persist even after the original groups re-form and depart to different destinations, resulting in the younger, lower-ranking females leaving the group into which they were born to join their new mates in a new group. This is a fundamental primate mating practice called "exogamy" that will be discussed later in this chapter.

The Inuit or Eskimos, who lived north of the Arctic Circle, typically spent the brief Arctic summer foraging for birds' eggs, berries, and small game in small family groups scattered widely throughout the mushy tundra. But in the winter, the hunt for large game such as caribou, seal, and walrus often required the cooperation of several families. During this time of year, the Inuit would build their igloos close together in a sheltered location, forming winter camps that included dozens or scores of people who often got together to gossip, share experiences, recount the traditional folktales, and practice the traditional ceremonies of their ancestors.

The !Kung Bushmen, hunters and gatherers of equatorial Africa's Kalahari Desert, typically scattered into widely dispersed family groups during the rainy season, when water was plentiful and easy to find. But during the dry season they would gather into camps consisting of dozens of families, settling close to each other in the vicinity of the few permanent water holes, where they would pass the time interacting with members of other families, while they waited for the rains to come again.

In our modern society, extended families often live scattered over large geographical areas, yet they will often go to great lengths to gather together periodically for the ceremonial occasions that are enshrined in our own cultural traditions, such as holidays, birthdays, weddings, anniversaries, and funerals. Modern civilization has created a complex variety of overlapping and interlocking groups related to many different spheres of human life, all made possible by the great human elaboration of the fission-fusion society.

We are members of our own nuclear families of parents and children and also of the larger extended families that include our grandparents, aunts, uncles, cousins, nieces, nephews, and grandchildren. We belong to several other groups defined by the communities in which we live, the schools we attend, the trades and professions we pursue, and the organizations we work for. We are born into a family with a specific ethnic identity (or mix of ethnic identities), and each of these ethnicities is associated with a specific geographic region, language, history, religion, diet, value system, and mode of dress.[3]

Finally, we express our individuality and satisfy our unique personal interests by purposefully joining one or more of the thousands of voluntary associations that have proliferated in modern times. These include religious groups, political parties, charitable organizations, social clubs, and the many professional associations that have sprung up in science, medicine, technology, commerce, sports, entertainment, the arts, and every other area of human endeavor.

Each of these many human groups has certain requirements for membership and certain obligations demanded of its members. Human society as we know it would not exist were it not for the innate primate passion for group identity and solidarity combined with the flexibility of the fission-fusion society. Our instinct for group identity allows us to form groups with enough coherence to stick together with a feeling of solidarity and to work together to achieve common goals. Our instinct for both fissioning and fusing allows many groups to proliferate within a single society, each defined in a different way, and each meeting a different set of social needs.

As we trace the evolution of modern humans and the development of increasingly complex human societies, we will see how the fusion phase of the fission-fusion cycle has allowed our species to forge new forms of solidarity within increasingly larger human groups. Modern nations and political movements include thousands or millions of members, most of whom have never met and never will meet, yet they collectively identify themselves as belonging to the same group and living within a single circle of trust.

The history of our species has, in fact, been marked by ever-larger patterns of fusion. These began to occur when hunter-gatherers developed tribal identities; it expanded again when members of different tribes began living together in villages, towns, and city-states, and it finally expanded to its present size when many thousands of city-states were forged into approximately two hundred nation-states, in which the members of our group typically number in the millions and tens of millions.

Primate Exogamy and the Incest Taboo

In spite of the group solidarity typical of primates—and the antagonistic relationships primate groups have with each other—most primates interbreed on a regular basis with members of rival groups. Upon reaching puberty, the adolescent primate typically leaves its mother's group and actively seeks out friendships and sexual relationships with members of another group—and ultimately to gain acceptance as a permanent member. This is called exogamy—meaning "marriage outside of the group"—and among most primate species it is the males who leave to join another group, a phenomenon known as male exogamy.

Thus, the most common type of primate group is composed of a core of females who are related to each other as mothers, sisters, and daughters. These females have spent their whole lives together, are familiar with each other's idiosyncrasies, have foraged for food together, shared some of their babies with each other, and have long since formed a strong and stable social hierarchy that promotes cooperation and peaceful relations within the group.

Some species of primates, however—most notably the chimpanzees—practice just the reverse: female exogamy. Chimpanzee males remain in the group to which they were born and have lifelong bonds not only with their mothers but also with their fathers, brothers, and sons. Female chimpanzees, however, gradually wander away from their home group upon reaching maturity, seeking to establish sexual relationships and other friendships with the members of a completely different group. If they are persistent, they will ultimately win acceptance by the adult males and females of the new group. Eventually, they will become pregnant and bear offspring themselves, and while their daughters will grow up and move away, their sons will remain with them for the rest of their lives.

The universal primate practice of exogamy is reflected in the social structure of all human societies, although it takes many different forms, including both male and female exogamy. The traditional peasant societies of China and India practiced a form of strict village exogamy based on residence. In these agricultural villages, all the land and dwellings were owned by males and inherited strictly in the male line. Thus, the men never moved from the village of their birth, while the women—whose property was limited to portable possessions such as clothing, jewelry, and furniture—left their home village forever upon marrying and spent the rest of their lives in their husband's village. But in the traditional island villages of Polynesia, the land and dwellings were owned by women and inherited strictly in the female line—and thus it was the males who moved away from their home villages when they married to spend the rest of their lives as inhabitants of their wife's village.

In addition to these forms of exogamy based on residence, all human cultures require a form of exogamy that applies to kinship relations, namely the universal human prohibition against incest. Incest taboos require men and women to choose partners from outside of their own kinship group. In our society, the kinship group is defined very narrowly as the nuclear family of parents and children—and to a somewhat lesser extent, the extended family of uncles, aunts, and cousins. But in tribal cultures where kinship relationships are more important and more complex than they are in our society, incest prohibitions are often based on clan affiliations that can define hundreds or even thousands of potential relationships as incestuous and therefore completely off limits.

Exogamy benefits primate societies in two ways. First, it ensures that there is constant genetic mixing among groups who live in adjacent territories, even if the groups are hostile to each other. This tends to minimize inbreeding, with its deleterious effects. Second, it ensures that within each group there are many adults that were born in other groups and already have relationships with their friends and family members who are members of those other groups.

These links between different and often competing groups help to minimize hostilities and violence between one group and another. In many tribal and peasant societies, certain clans and villages traditionally exchanged partners with certain other clans and villages. This custom, often observed for generations, tended to produce a web of social bonds between the groups that would bind them together, minimizing hostilities between them and providing them with valuable allies in case of attack from enemy groups.

Primate Hunting and Warfare

Monkeys and apes were once regarded as peaceful gatherers pursuing a vegetarian way of life supplemented by the occasional insect, bird's egg, or small reptile. But in recent years, field studies of primates in their natural habitats revealed that a number of them—including vervets, macaques, mandrills, baboons, orangutans, and chimpanzees—actively hunt and kill other warm-blooded animals, including other primates.

Chimpanzees are probably the most effective hunters among all the non-human primates. Chimpanzees have been observed hunting and killing at least nineteen different species of mammals, including wild pigs, antelopes, and several types of monkeys (their favorite prey). One chimpanzee group was found to have routinely killed more than 10 percent of the red colobus monkeys living in its territory every year—a kill rate equivalent to that of hyenas and lions. The hunt is usually carried out by a group of adult males, who cooperate in pursuing and corralling their prey until they can get close enough to kill it. This is basically the same method used by human hunters the world over, with one important difference: the chimpanzee must kill its prey by biting it with its long, razor-sharp canine teeth and mutilating it with its bare hands, while the human kills its prey with fabricated weapons that can inflict lethal injuries from a distance, avoiding the dangers of close combat.

Until recently, it was believed that, of all the primates, only humans engaged in lethal violence against members of their own species. Behaviors such as murder, warfare, and cannibalism were once believed to be unknown among the largely vegetarian (and presumably more peaceable) societies of apes and monkeys. But field studies of primates in their natural habitat have completely disproved this earlier view. Over a period of ten years, a particularly aggressive group of chimpanzees in Uganda were observed killing eighteen members of rival groups, a kill rate several times higher than the average for human hunter-gatherers. Numerous other studies have documented many instances of murder, infanticide, and cannibalism among a wide variety of primate species.

But of all the violent behaviors that were once believed unique to humans, perhaps none was more unexpected than the discovery that certain groups of chimpanzees are capable of sustained campaigns of violence against neighboring groups with the ultimate goal of occupying portions of their territory and claiming it as their own. This form of "warfare" is strikingly similar to the raiding parties carried out by hunter-gatherers and agricultural villagers all over the world.

One particularly large and aggressive group of chimpanzees lives at Ngogo in the Kibale National Park in Uganda. This group was studied by a team of American primatologists from 1999 to 2008, who discovered that the adult males of the Ngogo group were engaging in systematic raids in the territories of neighboring groups, periodically ambushing and killing other chimpanzees that belonged to these groups.

Chimpanzees typically move about in their own home range in noisy and rambunctious gangs, in which individuals may be scattered about in the forest within earshot of each other. But the behavior of the Ngogo chimpanzees changed dramatically when they embarked upon a raid into another group's home territory.

About once every ten to fourteen days, up to twenty adult males from Ngogo would set out to conduct a raid on one of their neighbors, falling silent and forming a single file as they crossed the border between their neighbors' territory and their own. As they penetrated deeper into "enemy territory," they would carefully scan the treetops for signs of the enemy and react nervously to the slightest sound.

If they encountered a group of defenders that outnumbered them, the Ngogo males would break ranks and flee back to their own territory. But if they chanced upon an unlucky male alone in the forest, they would pursue him, hold him down, and beat him to death. If a female was ambushed, she was usually released, but her offspring were usually killed and eaten. The majority of the Ngogo killings were directed at one particular group that lived to the northeast, and after several years of attacks, the Ngogo group moved into and appropriated a large portion of their neighbors' territory, expanding the area of their own home range by 22 percent.

The practice of infanticide is also common among many species of primates, occurring most often when a new alpha male wrests control of a harem of females from the alpha male that previously owned it. The new male will systematically kill all of the infants sired by his predecessor, which causes the now childless females to become fertile again and triggers the resumption of estrus and sexual activity. By mating with his new harem, the new alpha male can sire his own offspring, increasing the number of his own descendants and thereby maximizing the spread of his own genes.

Primate Tools and Weapons

In October of 1960, the primatologist Jane Goodall was in the first year of her historic research into the behavior of wild chimpanzees, living in an observation camp in a protected reserve on the shores of Lake Tanganyika in East Africa. One rainy October morning, wet and exhausted from hours of searching fruitlessly through the rain-soaked valleys for a chimpanzee to observe, she suddenly saw a movement in the long grass and focused her binoculars on the spot. Recognizing one of the adult males from the group she was studying, she cautiously approached.

The adult male was sitting next to a termite nest, repeatedly inserting a long grass stem into the entrance holes to the nest. After each insertion, he would withdraw the stem, now covered by clinging termites, lick the termites off the stem, and eat them. If the end of the stem became bent, he would bite it off to make a new end, or he would discard the stem and pick another. He feasted on the termites for an hour and then wandered away.

Goodall set up an observation post near the termite nest and soon was able to observe other members of the group fishing for termites, using not only the grass stems close at hand but also vines and twigs brought from several yards away, which they deliberately modified by stripping off the leaves (see Figure 1.4).

Before Goodall's research, the conventional wisdom generally accepted by behavioral scientists was that humans were the only species capable of deliberately fashioning tools from natural materials and using them for specific purposes. This was, in fact, one of the principal criteria used to distinguish humans from all other animals. But by 1973, Jane Goodall had recorded thirteen different types of tools made and used by the chimpanzees she was studying, and since that time more than twenty-five distinct types of tools, each deliberately fashioned from natural materials and each used for a specific purpose, have been identified by primatologists studying chimpanzees in the wild.

In addition to the probes they make for termite fishing, chimpanzees select and clean twigs and small sticks for a variety of uses, including gathering honey, extracting the edible portions of nuts from their shells, and teasing out the marrow from the bones of their prey. A primitive hammer and anvil made from stones or wood is used by one group for cracking kola nuts, and a pounding pestle made from the trunk of a palm tree is used for deepening holes in trees. Large, flat leaves are picked for use as mats for sitting on wet ground and as disposable "hats" when it rains, and smaller leaves are chewed into a soggy mass that the chimpanzees use as a sponge for collecting water and cleaning wounds.

Chimpanzee technology also includes weapons made from the natural materials in their environment. When engaging in their periodic threat displays, male chimps will break off branches and wildly brandish them about as they dash screaming through the forest. When they are attacking others, chimps will gather fruit, sticks, and even stones and throw them at their adversaries, and the members of at least one group of chimpanzees in West Africa have been observed making wooden "spears." Selecting a suitable branch about three or four feet long, they strip off the leaves and twigs, sharpen one end with their teeth, and use the spear to stab and kill bush babies, a small, primitive species of primate that typically sleeps in the cavities of hollow trees.

Technology gave humans a new kind of power over nature, but this power did not begin with the industrial revolution, the rise of civilizations, the development of agriculture, or even the invention of stone tools. The richness and variety of chimpanzee technologies indicates that the human use of technology began with prehistoric apes that were ancestral to both humans and chimpanzees.

Technology as a force in human evolution appeared millions of years ago, with the invention of the first primitive weapons—not by humans, but by prehistoric apes—who, by adopting wooden spears and digging sticks, set themselves upon the evolutionary path that eventually produced human beings. The evidence and rationale for this proposition will be presented in detail in the next chapter.

Primate Customs and Traditions, the Building Blocks of Culture

In the early 1950s, a group of primatologists from Kyoto University began a long-term study of a colony of about one hundred wild Japanese macaques that had been living on Koshima, a tiny islet off the coast of southern Japan. To better observe the monkeys' social behavior, the scientists would throw piles of sweet potatoes on the beach every time they visited the little island. The macaques would emerge from the forest to gather up the sweet potatoes, but before they would begin eating them, they would laboriously brush the sand off the potatoes with their hands, as is typical of this species.

But one day a young female named Imo, only eighteen months old, discovered that washing her sweet potatoes in a nearby stream was a much faster and easier way to clean them. Soon, Imo's mother began washing her potatoes instead of brushing them, and over the next few years, potato-washing spread among Imo's playmates and their mothers, until by 1958 the entire colony was washing their sweet potatoes either in the stream or the ocean. This newly discovered behavior persisted as a tradition in the Koshima colony and has been handed down through the generations ever since.

Later, the scientists began leaving wheat on the sandy beach, and the wheat and the sand became mixed together. Imo discovered that she could separate them by casting a handful of sandy wheat into the sea, whereupon the sand would sink, leaving the wheat floating on the surface to be scooped up. Other members of the colony began to adopt this method, and wheat-washing joined potato-washing as a local custom found only in this one colony. Since then, primatologists have observed other such feeding customs in individual colonies of macaques all over Japan, including washing apples, digging hidden peanuts out of the sand, unwrapping caramels, and even eating dead fish.

The Japanese macaque inhabits a wide variety of environments, from the subtropical jungles of Koshima to the snowy alpine forests of Honshu. It makes perfect sense that such a species would have the ability to adapt to different environments by discovering new foods, inventing new ways to eat them, and passing these customs along to other members of the group—and, via their own offspring, to future generations. In this way, the discoveries and inventions of innovative individuals such as Imo can not only benefit other members of the group, but will also be preserved over time, providing future generations, which will live in the same habitat, with ready-made solutions for the food-getting challenges posed by that particular environment.

In fact, similar customs and traditions—the building blocks of culture—have been observed among many other primate species. A group of chimpanzees in Guinea, West Africa, has learned to crack open very hard-shelled kola and panda nuts using the "hammer and anvil" technique, and this novel technique has been transmitted from one generation of chimpanzees to the next. Many other examples of distinct customs and traditions have been observed in wild chimpanzee groups, including various kinds of tool use, grooming techniques, and greeting behaviors such as hand-clasping, in which the members of some groups raise their arms and clasp each other's hands and wrists when they meet. Significantly, none of these behaviors are universal among chimpanzees but instead are found only in specific groups. This tells us that the behavior is transmitted by learning and not by genetic factors.

We take it for granted that every human culture has its own distinctive customs and traditions, handed down from one generation to the next. What we rarely appreciate is that the ability to invent new behaviors—and, through imitation, to pass them on as the customs and traditions of a specific group—is not at all limited to humans. Similar elements of animal culture have also been observed in the wild among whales, elephants, birds, and even rodents. There can be little doubt that customs and traditions among group-living animals, the foundations of human culture, first appeared millions of years before humans walked the earth.

This is not to say that we humans are not unique. Our mental and physical uniqueness is profound and undeniable. But the elements from which this uniqueness evolved unquestionably appeared long ago in the behavior of monkeys and apes. The physical similarities between humans and other primates are obvious, while the behavioral similarities are not always as apparent. Group solidarity and the concept of a homeland; the social bonds forged by motherhood, sex, friendship, and social hierarchies; the flexibility of the fission-fusion society; the advantages of exogamy; and the rudiments of hunting, warfare, tools, weapons, customs, and traditions all exist among non-human primates. These are the genetic building blocks of human behavior. Without them, human society would have never evolved, and the world we live in would have never come into being.

ଔଌ

CHAPTER 2

The Technology of Spears and Digging Sticks

Upright Posture and Bipedal Locomotion

> *"[O]ne day, about noon . . . I was exceedingly surprised with the print of a man's naked foot on the shore, which was very plain to be seen in the sand."*
>
> —Daniel Defoe, *Robinson Crusoe*

One evening in 1976, two scientists went out for a stroll after a day's work in a paleontological site 3.6 million years old near the African village of Laetoli in present-day Tanzania. The scientists were amusing themselves by throwing chunks of elephant dung at each other, when one of them slipped and fell face down on a layer of rock that had begun as volcanic mud millions of years ago but that had long since hardened into a kind of natural cement. There, inches from his face, was the unmistakable impression of fossilized raindrops.

Further investigation revealed that the volcanic mud also bore the imprint of numerous fossilized animal tracks. Careful excavation of this volcanic layer over many months revealed the tracks of numerous prehistoric animals ranging in size from elephants to mice. Finally, after two years of painstaking excavations and the discovery of hundreds of animal tracks, the archeologists at Laetoli discovered one of the most important finds in the history of human paleontology: an eighty-foot trail of footprints, made by two individuals, an adult and a child, walking together across the volcanic mud more than three million years ago (see Figure 2.1).[1]

The Laetoli footprints provided direct, indisputable evidence that full bipedal locomotion—the ability to stand, walk, and run for long periods of time and cover long distances using only the hind legs—had emerged millions of years earlier than had once been assumed. Furthermore, the footprints confirmed that paleontologists had indeed drawn valid conclusions about the behavior of prehistoric hominids[2] (the traditional scientific term for all prehistoric and modern humans), although their conclusions had until then been based only on the indirect evidence of human anatomy.

Absence of Evidence Is Not Evidence of Absence

The fossil remains of early hominids showed that the pelvis and leg bones of these ancient species were all very similar to their equivalents in the anatomy of modern humans. This suggested that all of these early hominids had been capable of at least some form of bipedal locomotion. But until the discovery of the footprints at Laetoli, there had been a great deal of scientific controversy about whether the earliest hominids had actually walked fully upright in the manner of modern humans. This controversy was fueled by the nearly complete absence of fossil foot bones from early hominid finds and by the fact that the oldest fossil hominid footprints found prior to 1978 were all less than one hundred thousand years old, by which time the human body had already evolved into its fully modern form.

While many thousands of fossilized footprints had been identified from time periods millions of years ago, all of these footprints had been made by other animals, not by hominids. In fact, dozens of species of dinosaurs—ranging in size from creatures larger than elephants to some as small as mice—had left thousands of footprints in the fossil record from time periods dating from 75 to 250 million years ago.

Many of these dinosaurs had left behind more fossil footprints than the remains of their actual skeletons.

Was the anatomical evidence from the skull, spine, and pelvis alone sufficient to conclude that the early hominids had really walked upright, in spite of the puzzling absence of any clearly bipedal footprints in the fossil record? These doubts were erased once and for all by the discovery of the Laetoli footprints. These footprints confirmed the old maxim that "absence of evidence is not evidence of absence"—in other words, that the absence of physical evidence of some past event is not, in itself, proof that the event did not actually occur. This is especially likely when the evidence in question has been destroyed by the passage of time.

The study of human evolution has been plagued by numerous occasions in which scientists have seriously underestimated the true antiquity of some important evolutionary milestones. For example, in 1948 a leading anthropology textbook stated that while occasionally "someone has claimed . . . [an age older than one million years] for this or that human fossil . . . there is as yet no single find of anything in man's direct line of ancestry that . . . dates back as far as that."[3] Yet paleontologists now agree that human evolution began at least *five million* years ago.

Similarly, the practice of tool-making and the development of cultures and traditions were once believed to be the exclusive province of humanity. Yet the work of Goodall and others has now demonstrated beyond doubt that other primate species exhibit clear evidence of both tool-making and culture—and it is therefore safe to assume that both of these capabilities existed among prehistoric species that lived well before the appearance of the earliest hominids.

Finally, until the end of the twentieth century, virtually all of the most ancient remains of wooden artifacts came from desert environments and were less than fifteen thousand years old. (The sole exception was a single spear point found in 1911 in Clacton, England, in a deposit roughly four hundred thousand years old—but this was a controversial find that was not universally accepted.) Yet in 1997, a set of finely crafted wooden throwing-spears, or javelins, positively dated at four hundred thousand years old, were found at a prehistoric site in Schöningen, Germany, proving that prehistoric woodworking was vastly more ancient than had generally been believed.

"Lucy" and the Earliest Hominids

In 1974, a relatively complete fossilized skeleton of an upright female *Australopithecus* was unearthed in Ethiopia. This find, which was more than three million years old, achieved instant fame as Lucy, the oldest reasonably complete skeleton found to that date that belonged unmistakably to the human lineage. Lucy was a member of the species *Australopithecus afarensis,* one of the early hominids that inhabited prehistoric Africa. Her discovery caused a sensation, because her pelvis, leg bones, and knee joints were strikingly similar to our own, indicating that her species was fully upright and capable of full bipedal locomotion.

The early hominids such as Lucy were burly little creatures who stood, walked, and ran upright easily over the ground as we do. At a distance, you might have mistaken them for very small people, but on approaching them more closely you would notice that their heads were ape-like—with low foreheads, prominent brow ridges, and protruding jaws—their bodies were probably covered with fur, their arms seemed rather long for their bodies, and their toes seemed rather long for their feet. The females stood about three and one-half feet tall and weighed about seventy-five pounds, while the males stood slightly over four feet tall and weighed about ninety-five pounds. They made tools and weapons, hunted and butchered wild game, and lived in family groups that were probably similar to our own.

Although the australopithecines like Lucy had the straight, downward-facing big toe of the true hominid, their narrow shoulders and their long, curved fingers and toes provide clear evidence that they spent at least part of their lives in the trees. Yet the chemical analysis of their bones—as well as the structure of their molar teeth—indicates that their diet consisted mainly of foods found not in the trees but on the ground. If they no longer subsisted mainly on the leaves, fruits, and nuts found in the trees, why did they retain many critical features of their tree-climbing anatomy? The most likely answer is that the trees provided the safest place to sleep at night, when the early hominids were particularly vulnerable to attack by the large predators in their environment.

Lucy was about three feet six inches tall, weighed about seventy pounds, and had a brain only slightly larger than the brain of a chimpanzee. Above the waist, she looked like an ape. Below the waist, however, she looked like a human, except that her entire body was probably covered with hair. Lucy's pelvis was already well on its way to evolving into the rigid, doughnut-shaped structure typical of modern humans, a shape that is distinctly different from the more loosely structured and more tubular-shaped pelvis of apes and monkeys.

Lucy's leg bones were long and straight, like ours, with locking knee joints that enabled her to stand for long periods without straining her leg muscles, and the fossil feet that had been recovered from other *Australopithecus* finds were also like ours, with a well-defined arch to provide propulsion when walking. All of Lucy's toes faced in the same direction, indicating that—unlike an ape's foot with its opposable big toe—the feet of the early hominids were no longer adapted for grasping tree branches, but instead, like the feet of modern humans, had become adapted for traction and forward motion while walking on the ground. In addition to all these important anatomical changes, Lucy and her kind had completely lost the formidable dental weaponry—the long, sharp canine teeth—that are possessed by the males (and to a lesser extent, the females) of all other primate species.

The loss of the primates' only biological weapons—in a species that spent most of its time living on the ground among dangerous predators such as lions, leopards, hyenas, and wild dogs—suggests that Lucy and her contemporaries, unable to defend themselves without biological weapons, would have long since become extinct unless they had access to some other type of lethal weapons. Yet no remains of such weaponry had ever been found from this very early time period until the discovery of stone artifacts 3.3 million years old from the site of Lomekwi 3 in Kenya in 2011.[4] Until then, the earliest stone tools ever found were all dated at more than a million years after Lucy's time. How can we account for this discrepancy?

Considering the well-documented tool-making habits of wild chimpanzees, the first tools and weapons were probably not made of stone at all, but instead fashioned from branches, twigs, leaves, and other highly perishable materials, none of which could have survived for more than a few dozen years—let alone millions of years—in the warm, moist, tropical climate of prehistoric Africa. In this case, the absence of evidence is entirely to be expected.

Furthermore, the anatomical changes required in order for upright posture and bipedal locomotion to have evolved in the first place were tremendous, radical, and truly revolutionary. Changes of such magnitude do not happen unless there are clear survival advantages for the organism that adopts them, and they must produce increased rates of survival as these changes take effect. And yet, when the early hominids evolved upright posture and bipedal locomotion, they had to contend with some very definite *disadvantages* brought about by these changes. We will review the most important of these disadvantages later in this chapter.

The Radical Redesign of the Mammalian Body Plan

The full upright posture and true bipedal locomotion achieved by Lucy and her kind required a wholesale transformation of the anatomy characteristic not only of all the primates but also of all mammals. This redesign included major changes to the skull, spine, pelvis, legs, feet, and other musculoskeletal structures. For the first time, the weight of the entire upper body had to be supported by the hind legs alone, an arrangement that had never before existed among any of the placental mammals.[5] What survival advantage was powerful enough to completely reorganize the way the primate body stood, walked, and ran?

Let's begin with the head. The place where the skull is attached to the spinal column, called the foramen magnum, had to move from the back of the skull, where it is located in all other animals, to the bottom of the skull, where it is located in humans. Otherwise, the face would tend to be pointed upward rather than forward. The spine is supported in all other mammals at both ends, with the forelegs and shoulders supporting the weight of the head and forequarters while the hind legs and pelvis support the weight of the tail and hindquarters. But in a two-legged mammal, the spine and pelvis must support the weight of the entire upper body, including the head and neck, forearms and shoulders, chest, and abdomen—in short, everything except the hind legs themselves. To accomplish this feat, the spine had to be greatly strengthened and bent into an S curve, to bring the center of gravity of the entire body directly over the pelvis.

In order to support the weight of the entire upper body, the pelvis also had to undergo a major restructuring. Since the longer and more flexible pelvic bones typical of all other primates were not designed to support the weight of the upper body for extended periods of time, the pelvic bones had to become shorter, thicker, and fused together into a single rigid, doughnut-shaped structure. At the same time, the bones of the hind legs became long and straight, with the ability to lock at the knee joints. When other primates stand on their hind legs, their knees are bent. The evolution of locking knee joints enabled the bipedal hominids to stand fully upright for long periods of time without putting constant strain on their leg muscles.

Finally, the grasping feet and opposable big toes typical of all other primates had to be completely transformed. Very early in human evolution, the "thumb" of the foot actually began to migrate from the side of the foot to the front. If you look at the thumb on the rear foot of a chimp or gorilla, you will see that it is located on the side of the foot and is *opposable* to the other toes, similar to the opposable thumb in the human hand. This is necessary for grasping with the hind legs while climbing about in tree branches, but it is poorly suited to walking on the flat surface of the ground. So this grasping thumb migrated from its original position opposite the other four toes to a new position in which it is lined up with the other toes, allowing all five toes to face in the same direction. At this point, the ape's thumb had become the human's big toe.

At the same time, the palm of the foot lengthened, eventually developing a distinct arch, and the "fingers" on the hind feet became shorter, eventually turning into the four stubby little toes of our own species that are no longer capable of grasping much of anything but are perfectly capable of supporting their new function of walking and running for long distances over the open ground.

When an animal evolves a radical head-to-toe redesign of the basic body plan that served its ancestors well for tens of millions of years, especially in a short space of geological time, we can assume that powerful evolutionary forces are at work. But scientists have struggled to reach consensus about what these evolutionary forces were. In fact, a number of different theories have been proposed to explain precisely why our ancestors embarked upon the evolution of their unique and unprecedented form of upright posture and bipedal locomotion.

I will argue in this chapter that it was the technology of wooden spears and digging sticks—an innovation that must have begun with prehistoric apes ancestral to the hominids—that provided the survival advantages that were great enough to produce an evolution into full upright posture and true bipedal locomotion. This line of thinking was originally suggested by Charles Darwin and, as we will see, it has been supported since then by several contemporary anthropologists. But first, let us review the various competing theories and briefly note the weaknesses in each of them.

The early hominids lived during a period in African prehistory when the forests were shrinking and the grasslands were expanding. For a long time, it was believed that bipedal locomotion evolved primarily as an adaptation to the drier savannah environment—characterized by grasslands dotted with scattered trees—that was spreading throughout equatorial Africa. It was assumed that walking and running on two legs had enabled Lucy's ancestors to leave their original forest habitat, see more clearly over the tops of the grasses, and cover longer and longer distances over the open ground.[6] This is known as the "savannah-based hypothesis." There are, however, some major problems with this theory.

First, all other grassland species—including both herbivores such as antelopes and zebras as well as carnivores such as lions and hyenas—are strictly quadrupedal. (In fact, all land mammals except hominids are quadrupedal.) Second, of all the other primates who also made the transition to a terrestrial way of life (these include baboons, mandrills, vervet monkeys, and patas monkeys), none of them adopted upright posture. Instead, they all adapted quite successfully a grassland existence while remaining firmly planted on all fours.

Third, the earliest anatomical evidence of bipedal locomotion comes from the fossil remains of an ape-like creature called *Ardipithecus*, who appeared at least 4.5 million years ago—more than a million years before Lucy was born. In a major setback for the savannah-based hypothesis, it has been firmly established that *Ardipithecus* lived and died in a forest environment and never inhabited the grasslands at all.

The fossil remains of *Ardipithecus* were first discovered in the Awash Valley of Ethiopia in 1992.[7] Its feet are clearly those of an arboreal animal, with long, flexible digits and an opposable big toe that was adapted for grasping tree branches. Yet while *Ardipithecus* had not yet lost the ability to climb trees like an ape, the characteristics of its pelvis and leg bones showed that, long before Lucy and the other early hominids roamed the African savannah, the evolution of upright posture was already well under way.

Due to the problems of the savannah-based theory, a variety of alternative and often competing theories have been proposed in recent years to explain the evolution of upright posture and bipedal locomotion. Each of these theories has its scientific supporters and detractors, and many of them describe certain evolutionary mechanisms that doubtless played a role in the evolution of bipedalism. But none of these mechanisms seems to involve the kind of life-or-death survival advantages that would have been necessary for the radical restructuring of primate anatomy that bipedalism required.

The "provisioning hypothesis" proposes that hunting males needed free hands in order to carry their prey back to the home base for the females and offspring.[8] However, this does not account for the fact that other predatory mammals bring meat back for their offspring by simply carrying it in their mouths. While this behavior might have required an evolutionary adjustment (since primates rarely use their mouths for carrying things), it would hardly have required a complete restructuring of primate anatomy.

The "thermal loading hypothesis" proposes that by standing upright, the early hominids exposed a smaller area of their bodies to the fierce tropical sun and were thus able to remain cooler while exerting themselves out in the open.[9] But in addition to the fact that this does not explain why no other tropical animals have ever adopted a similar strategy, we should recall that bipedalism first began with *Ardipithecus,* who lived largely in the shade of the tropical forests.

The "warning display hypothesis" proposes that individuals (especially males) that could threaten others by frequently standing erect were more dominant and thus sexually outcompeted individuals who stood erect less frequently.[10] However, this hypothesis fails to account for the fact that gorillas, bears, and other species that also threaten their rivals by standing erect have never adopted bipedalism.

The "postural feeding hypothesis" proposes that bipedalism enabled prehistoric apes to stand upright while gathering low-hanging fruit with both hands, rather than hanging on to a branch with one hand and gathering fruit with the other.[11] However, it seems unlikely that this would confer a sufficiently powerful survival advantage to produce the massive anatomical changes that bipedalism entailed. Furthermore, the essence of bipedalism is not merely standing upright but walking and running on two legs. Apes, monkeys, bears, beavers, prairie dogs, meerkats, and many other mammals can easily *stand* upright, but hominids are specially adapted for bipedal *locomotion.*

The primary weakness of these theories is that none of them explains why the large canine teeth of our primate ancestors disappeared at the same time as bipedal locomotion evolved. Instead, the loss of large canines among the hominids is typically explained by the "diminished male rivalry hypothesis," which proposes that large canines disappeared in order to prevent adult male hominids from engaging in deadly combat, because this might have interfered with their ability to cooperate and become more effective hunters.[12]

However, lions, wolves, hyenas, chimpanzees, and many other species that engage in cooperative hunting do not have diminished biological weapons. Besides, male humans are entirely capable of *both* close cooperation *and* deadly conflict with each other, and the absence of large canines has not prevented male humans from engaging in deadly conflict throughout all of human history. Instead of using their teeth to kill each other, they just use their lethal weapons.

So, the question remains. How could any of the early hominids have survived for millions of years in an environment full of large and dangerous predators without any effective way of defending themselves?

The Disappearing Canine Teeth

The canine teeth of *Ardipithecus ramidus*—the earliest species to display skeletal evidence of bipedal locomotion—had already become significantly diminished compared to the canines of its ancestors. By the time of Lucy and the early hominids, the large, deadly, weapons-grade canine teeth possessed by all other primate species had completely disappeared, never to be seen again among the hominids.

In fact, one of the members of the team that investigated the *Ardipithecus ramidus* site observed that male canine reduction was well under way by six million years ago.[13] The loss of weapons-grade canine teeth began millions of years before Lucy appeared, and by the time of the early hominids these large canine teeth—the only biological weapon the primates ever possessed during their entire fifty-million-year history—had completely disappeared.

No monkey, ape, or hominid has claws, horns, antlers, hooves, or tusks. But they all have formidable canine teeth that can inflict painful and even lethal damage to an adversary. Yet the *Australopithecus* fossils have only small, dull, flat-wearing canine teeth that could not possibly have competed with the dangerous biological weapons of their natural enemies (see Figure 2.2).

Furthermore, apes and monkeys all have superior tree-climbing abilities that enable them to climb high into the treetops, far beyond the reach of their natural enemies like the big cats. Yet the long, straight legs and shrunken toes of the australopithecines and the other hominid species who followed them would have seriously reduced their ability to climb out of danger, making them that much more vulnerable to their natural enemies.

A species' natural defenses disappear only when they are no longer needed for survival. Although primates can certainly throw stones, a stone heavy enough to cause real damage to a large predator would be too heavy to throw any distance with any great accuracy. Besides, how many stones can a hominid throw in the time it takes for a lion to charge and attack? Heavy clubs made from large branches can crush the skull of an adversary, but only at close range. If you get close enough to a predator to hit it over the head with a club, it will probably be close enough to maul you with its claws, and—unless you have dispatched it with a single blow—it will already have its fangs at your throat. How did the early hominids compete so successfully against "nature red in tooth and claw" with neither the teeth nor the claws to fight back, when all they had to fight with was their "bare hands"?

The most likely answer is that the hands of the early hominids were not bare. They must have been carrying weapons. And the weapons they fashioned to attack both predators and prey must have been so superior to the formidable canine teeth of their primate ancestors that having effective dental weaponry was no longer necessary.[14] As we saw in chapter 1, the use of technology is not limited to hominids. Certain wild chimpanzees have even been observed making primitive spears that they use for killing their prey. It is therefore most likely that when prehistoric apes began to make spears, they unleashed the evolutionary forces that eventually produced an ape with full upright posture and bipedal locomotion.

Apes are most likely to walk on their hind legs when they are carrying things in their hands (see Figure 2.3). The enhanced ability to walk and stand on the hind legs would have become a powerful evolutionary advantage for those prehistoric apes that adopted a revolutionary weapon—a long, sharpened stick, the first primitive spear—that could be used to attack another animal from a distance.

The spear would have completely changed the game of hunting and defense, because long spears can be used with lethal force while hunters remain safely out of reach of the biological weapons of their prey. If a group of hunters can impale an animal with spears that are several feet long, it can be killed without ever coming close enough to use its biological weapons on its attackers.

In *The Descent of Man,* Darwin wrote that "The early forefathers of man were . . . probably furnished with great canine teeth; but as they gradually acquired the habit of using stones, clubs, or other weapons, for fighting with their enemies or rivals, they would use their jaws and teeth less and less. In this case, the jaws, together with the teeth, would become reduced in size."[15]

In his classic work, *The Stages of Human Evolution*, the physical anthropologist C. Loring Brace wrote that "we can guess that the wielding of a pointed stick was the crucial element that led to the change in selective forces that produced a tool-dependent biped . . . we could add . . . that the digging stick redirected is a more effective defensive weapon than even the formidable canine teeth of the average male baboon. After all, to bring canine teeth into effective use, the baboon literally has to come to grips with its adversary, and if that happens to be 200 pounds of hungry leopard, the chances are poor that . . . the . . . baboon can get away unscathed."[16]

The most recent expression of this line of thought was formulated by the anthropologist Robert Bates Graber, who wrote in 2000 that "the earliest stone tools, like those of modern chimps, no doubt were of material softer than stone . . . . Indeed it is quite possible that sharpened sticks—relied on heavily, for survival, by known foraging people as digging sticks and as spears, but unknown among monkeys and apes—may have been the tool that tipped natural selection decisively toward upright posture."[17]

The ability to carry and wield a spear long enough to attack and kill another animal from beyond the range of its biological weaponry would have had powerful survival value. Once the ancestors of the hominids had adopted the technology of the spear, those individuals who could stand firmly on their hind limbs—while jabbing and thrusting a spear with their forelimbs—would have had a clear advantage over their competitors. The longer these individuals could remain standing, the farther they could walk and run on two legs, and the larger and heavier weapons they could carry with them, the more effectively they could defend their mates and offspring against attacks from would-be predators—and the more meat they could bring back to share with the other members of the group. For all of these reasons, the survival value of fabricated spears would have been sufficient to bring about the major anatomical changes required for a four-footed animal to evolve into a two-footed animal.

Behaviors of the sort that might have led to spear use have been observed among chimpanzees and gorillas, which, as noted in chapter 1, commonly break off tree branches and brandish them about during their episodes of threat displays. The soft wood of a fresh branch can easily be sharpened by rubbing it against an outcropping of rock or against the rough bark of certain tropical trees. And wild chimpanzees in Senegal, West Africa, have been observed to fashion wooden spears by stripping the branches and bark from a straight stick, sharpening one end with their teeth, and using them to kill bush babies by stabbing them in the tree hollows where they sleep.

But if large numbers of prehistoric species made and used spears and digging sticks for millions of years, why have no remains of these wooden tools and weapons ever been found in archeological sites dating from these time periods? This is a classic case where "the absence of evidence is not evidence of absence."

For a long time, the oldest human artifacts made of wood ever found were only a few thousand years old. Archeologists disagreed sharply as to how old such artifacts would turn out to be, but for a long time the prevailing view was that the manufacture of lethal weapons from wood—and the use of such weapons in cooperative hunting of large game—did not begin until the appearance of anatomically modern humans roughly fifty thousand years ago.

It therefore came as a shock to the scientific world when, in 1997, several finely made and perfectly balanced wooden javelins or throwing-spears approximately four hundred thousand years old were recovered from an ancient peat bog in Schöningen, Germany. Due to the highly acid nature and lack of oxygen in these bogs, the wood had been literally "pickled"—protecting them from bacterial decay. The Schöningen javelins were made by *Homo erectus,* an emerging human, who apparently used them for hunting wild horses long before modern humans appeared.

The Schöningen javelins were made from the fire-hardened wood of the yew tree, and their manufacture required not only cutting down a good-sized tree but carving away the softer wood on the outside of the tree trunk to expose the harder heartwood within. It was then necessary to harden the business end of the javelin in a fire. This was a delicate operation, and it had to be performed without charring the wood.

It was also necessary to make the business end of the javelin thicker and heavier than the throwing end, just as modern javelins are made today. Remarkably, the center of gravity of the Schöningen javelins is located exactly one-third of the way back from the point of the spear—the optimum balance-point for throwing and in fact identical to the balance-point of the modern javelins used to this day. This sophisticated weapon—showing clear evidence of advanced planning and sophisticated technical knowledge in woodworking—was made by the emerging human, *Homo erectus,* who lived long before the Neandertals and whose brain was considerably smaller than our own. This sophisticated knowledge of woodworking, involving a complex, multistep process, would not have suddenly emerged full-blown from the mind of *Homo erectus.* Instead, it would have been the result of thousands of generations of slowly accumulating knowledge, stretching back beyond the appearance of these emerging humans to the dawn of the earliest hominids.

The loss of dental weaponry among all prehistoric bipeds proved to be complete. None of the early hominids ever regained the large and dangerous canines of their prehistoric ancestors. When the technology of making and using spears was adopted by an ancient population of prehistoric apes, it became a strategy for hunting and defense that effectively trumped the biological weaponry of both the predators and prey living in those prehistoric environments.

Once the hind legs were capable of assuming full responsibility for locomotion, the forelegs were free to serve other purposes. They could be used to make tools and weapons, and they could also be used to carry these tools and weapons from place to place. The bipedal hominids were able to use their free hands and arms to transport relatively heavy burdens—such as the spoils of the hunt or a trove of ripe fruit—from one place to another, making it possible to bring food that was acquired in distant locations back to a communal home base, as the provisioning hypothesis suggested. And the same basic technology—in the form of a digging stick—would have also been used to gather foods that are buried in the earth. In hunting and gathering societies, women routinely used such sticks to dig up roots, bulbs, tubers, termite nests, and the burrows of small animals, as well as to knock down nuts and fruits from the ends of tree branches that were too slender to climb.

In the final analysis, it appears that the technology of spears and digging sticks not only stimulated the evolution of bipedalism but also was ultimately responsible for the evolution of humanity itself. However outlandish this claim may seem, the human body's unique, much loved, and much celebrated physical form most likely owes its origin to a group of ancient apes—long since lost in the mists of prehistory—that first mastered the extraordinary offensive, defensive, and food-gathering possibilities of the long, sharp stick.

But the transformation to upright posture and bipedal locomotion also placed some important limitations on the ability of prehistoric hominid females to hunt large game. It may have been these very limitations that resulted in the appearance of a unique sexual division of labor among humans, a division of labor that is found in no other animal species.

Hunting and the Maternal Burdens of Female Hominids

Once the early hominids adopted a technology that included tools and weapons made of wood, they began to use their lethal weapons for hunting and killing other animals for meat. In doing so, they created a uniquely human ecological adaptation—a way of life known as hunting and gathering—that has been practiced by every member of the human lineage until humans began adopting the technology of agriculture roughly eleven thousand years ago. But unlike the females of every other predatory species, the demands of the hunt conflicted with the heavy burdens of motherhood among the hominids.

In all other predatory species, females participate equally with males in all aspects of the hunt, and in some species—notably including that preeminent carnivore, the African lion—the females outshine the males in the quality and quantity of their kills. Female lions can do this because their offspring are safely hidden in nests and burrows, where they will not interfere with the hunt or be in danger of injury at the scene of a deadly attack. The same is true of all other female predators, such as tigers, leopards, wolves, bears, foxes, weasels, killer whales, porpoises, seals, eagles, hawks, owls, and falcons—the list goes on.

Female hominids must always keep their offspring under close supervision, and they can hardly be expected to use sharpened sticks as weapons while carrying babies in their arms. Thus, female hominids would have used their long, sharp wooden implements as digging sticks to unearth the edible roots and tubers that are largely unavailable to apes and monkeys. In fact, it is possible that sharpened sticks were originally invented by females to gather underground foods and were only later adapted for hunting by males. Either way, this primordial technology allowed both sexes to significantly expand the range of foods available to them. And as these changes became firmly established, the hominids adopted another highly unusual type of food-getting behavior: both males and females brought their different types of food to their mutual home base at the end of the day, where the females shared the fruits of their labors and the males shared the spoils of the hunt.

Hominids are the only animal species in which the males are predators, the females are foragers, and both sexes regularly share the different foods they obtain. Moreover, this unique pattern of adult food-sharing between the sexes is intimately bound up with the hominids' open-ended, nearly continuous sexual availability. In order to understand how this extremely unusual pattern evolved, it is important to appreciate the immense maternal burdens that are involved in hominid reproduction. Not only are these maternal burdens the most severe of all primate species, they are also the heaviest maternal burdens among all female mammals.

When you touch the palm of a newborn infant with your finger, it will grasp your finger with surprising strength. This grasping reflex, which disappears in the first weeks of life, is the remnant of a powerful instinct we inherited from our primate ancestors. Its original purpose was to ensure that every primate infant would cling to the fur of its mother with unremitting tenacity, since continual physical contact with its mother was its only place of safety and vital to its very survival. A primate infant clings to its mother's body with all four limbs during the first weeks of life, riding upside down beneath her like a sloth dangling from a tree limb. Even after infancy has passed, the baby primate continues to ride on its mother's back or shoulders for months—or in some cases even years—before it finally learns to move about safely on its own.

The maternal bond is greatest in primates because primates are specialized for a life in the trees, in which their offspring are constantly at risk of falling to their deaths. From the moment of birth, primate mothers must be in close physical contact with their offspring wherever they go. This is a sharp contrast to most terrestrial mammals, whose infants are either hidden away in burrows (such as rabbits and foxes) or are able to walk on the day they are born (such as horses and elephants). The grasping reflex of the primate infant allows the female monkey or ape to use all four limbs to move about and gather food. She doesn't have to hold the infant, because the infant holds on by itself. She can climb, jump, pick fruit, run from her enemies, and swing from the branches, using all four of her powerful limbs, secure in the knowledge that her baby is clinging to her like a barnacle and will never let go.

The primate offspring thus lives in intimate physical contact with its mother for the first months or years of its life. And the bond that develops between primate mother and child as a result of this continual physical contact is not only powerful but also unique to the primates. Scientists who study the behavior of apes and monkeys in the wild have observed many lifelong relationships between mothers and offspring—especially among our closest relatives, the chimpanzees—and such relationships are rare or nonexistent in other animal species.

But the offspring of the early hominids were not able to cling to their mothers' fur in the same way. The opposable big toe had rotated downward for more efficient bipedal locomotion, and the feet of early hominid infants were no longer capable of securely grasping their mothers' fur. With only two grasping hands instead of four, early hominid infants became increasingly unable to hold on to their mothers by themselves and instead needed constant support from their mothers' arms.

Upright posture also added significantly to the burden of pregnancy. Pregnancy in a four-footed animal adds weight, but it does not change the mother's center of gravity. But in an upright hominid, pregnancy causes the body's center of gravity to be shifted forward, requiring the mother to compensate by leaning backwards more and more as the pregnancy progresses.

Predatory hunting is relatively rare among monkeys and apes, but it does occur among chimpanzees and baboons, where it is typically carried out by groups of cooperating males. All of these factors suggest that predatory hunting among the early hominids was pursued mainly by adult males, while adult females pursued the more traditional primate strategy of foraging for fruits, berries, roots, tubers, seeds, eggs, and insects. Yet hominid mothers and their offspring needed the highly nourishing, high-protein diet of the carnivore as much as the male hunters themselves. This means that the female hominids who gained access to the spoils of the hunt were inevitably better nourished—and their offspring more likely to survive—than the females who did not have access to those spoils.

While monkeys and apes do not normally share food—even with their offspring—they will typically share food with their sexual partners. Thus, in order for both hominid sexes to take maximum advantage of this unusual division of labor, a pattern of sexual behavior evolved that is found nowhere else in the animal kingdom. The most remarkable aspect of this unique sexual adaptation is that the majority of all sexual encounters—including sexual intercourse—occur when the female is not ovulating and there is no possibility of conceiving offspring.

Given this simple fact, it is apparent that by far the greater part of human sexual behavior fulfills a purpose other than that of reproduction. And while we cannot observe prehistoric hominids engaging in sexual behavior, even the earliest hominids had probably already evolved the nearly continuous sexual behavior characteristic of modern humans as a way of maintaining strong bonds—that included food-sharing—between males and females.

Almost Endless Estrus: The Hominids Revolutionize Sex

Most primates tend to be very selfish about food. They typically prefer to eat in solitude, and they often impassively ignore others, even their own offspring, who may beg and plead for a morsel. A notable exception, however, occurs when a female comes into estrus or "heat" and forms a type of sexual relationship called a consort pair with an adult male. The newly joined consort pair will separate themselves from the group and spend most of their time together, having sex, grooming each other, and sharing food. But when estrus disappears, the female loses interest in sex, the male loses interest in the female, and the former consorts go their separate ways. Sexual activity, consort pairing, and food-sharing will not occur again until the new offspring has been born, has survived infancy, and is old enough to be weaned.

This is not a problem for monkeys and apes, whose largely vegetarian diet is only occasionally augmented by meat, and whose tightly clinging babies allow their mothers nearly complete freedom of movement. But since adult female hominids were not generally free to hunt, this problem was solved when the outward manifestations of estrus were suppressed and an expanded receptivity to sex was substituted in its place. Instead of the hormone-driven but fleeting promiscuity of estrus periods, there evolved something new: a more generalized appetite for sex that spilled far beyond the old boundaries of ovulation and fertility to fill as much of the monthly calendar as possible and that continued even during pregnancy and nursing.

Although the human fertility cycle still exhibits the monthly time-clock typical of primates, female humans are the only female mammals who do not have clearly defined periods of sexual receptivity. Human females do not experience hormone-induced periods of uncontrollable sexual desire, nor do they exhibit the massive genital swellings timed to coincide with ovulation that occur among non-human primates. Human females are free to become sexually aroused during almost any part of the reproductive cycle, and they will engage in sexual intercourse not only when they are infertile but also during pregnancy, nursing, and after menopause. And with this massive increase in female sexual receptivity, the pattern of sexual behavior among male hominids also underwent significant evolutionary change.

Male hominids have developed a pattern of sexual behavior that also appears to be unique among group-living primates. The typical adult male hominid is strongly bonded to a single female consort who is sexually available most of the time—the sexual pattern we call monogamy. And to accommodate the tremendous increase in female sexual behavior, the typical male hominid has vastly increased the duration of his sexual activities. Unlike apes or monkeys, who have intercourse for only a few seconds at a time, most male humans have intercourse for several *minutes* before experiencing orgasm. In fact, the typical act of intercourse among humans lasts approximately fifty times longer than it does among other primates.

Finally, although the typical pattern among monkeys and apes is that the most dominant males enjoy a virtual monopoly over sexual access to females, dominant male humans enjoy a sexual advantage but hardly a monopoly. Since sexually available females are not the rarity in human groups that they are among monkeys and apes, most male humans, even if they are not the most dominant members of their group, still have regular sex partners and enjoy active sex lives.

Sexual and Maternal Bonds: The Foundations of the Human Family

When female hominids abandoned the estrus cycle to form permanent sexual relationships with individual males, a powerful new sexual bond was added to the ancient primate bond between mother and offspring. For the first time, a group-living primate species began to practice monogamy, and nuclear families became clearly defined structures within the larger social structure of the group.[18]

As the object of both maternal and sexual bonds, the female hominid became the emotional anchor of a social institution that had never before existed among group-living primates: the permanent nuclear family of mother, father, and offspring. In this unique adaptation, a female lives intimately together with a single male and their offspring for years at a time, binding them together into a basic building block of society even as they remain firmly integrated within the larger society of the nomadic band.

The hominid pattern of monogamy also created a new role in primate society: the role of the father, bonded to a single female and her offspring. In this way, the human family became an effective way of distributing the resources and channeling the energies of a hunting and gathering species. Since sexual access was no longer a scarce resource, this arrangement also minimized conflict and competition among males, allowing them to form stable, cooperative alliances with other males that would increase their power and effectiveness as hunters and warriors.

At some point in the history of human evolution, another very curious and completely unique change occurred in the anatomy and neurobiology of sex: the female breasts became linked with sexual feelings and behavior—a linkage that does not appear to exist among any other species of mammal. The nipples of the female human are neurologically wired as erogenous zones, and women the world over have noted that even the suckling of an infant at the breast can easily trigger feelings of sexual arousal.

Moreover, touching, fondling, and kissing the breasts is an important element in human sexual foreplay. While we may take these facts for granted, it is important to note that the mammary glands, whose principle function is to nourish immature offspring, play no such role in the sexual behavior of lions, tigers, dogs, sheep, goats, or cattle—or, for that matter, in the sexual behavior of any other primate.

It is notable that in all other mammals, the breasts swell and become enlarged only during the later stages of pregnancy, which reflects their primary function of delivering milk to the newborn. But in humans, the breasts swell and become enlarged at the time of puberty, usually well before a female human is capable of conceiving offspring. It is no accident that the timing of breast enlargement occurs at precisely the point in the female life cycle when she is approaching sexual maturity. Only in our species does the female breast have this dual role both as a source of nourishment for offspring and as a source of sexual attraction for the opposite sex. What is the purpose of this curious dual role, and why did it evolve only in humans? Unfortunately, the two most popular theories that propose to explain this curious phenomenon both have serious weaknesses.

It has been proposed that in an upright animal, the swollen breasts mimic the buttocks of our quadrupedal ancestors and therefore evolved to replace the buttocks as a sexual signal. But while it is true that the buttocks have a certain sexual appeal, standing upright hardly conceals them or renders them irrelevant as objects of desire.

Another theory holds that since the female sex organs became essentially hidden with the evolution of upright posture, the breasts took over the role of sexual signaling that the swollen sex organs had played during the estrus periods of the quadrupedal monkeys and apes.

But because the female breasts are permanently distended, they cannot serve as visual signals that ovulation is occurring and conception is possible. And human females in virtually every society go to great pains to *conceal* their sex organs from the sight of men. The fact that a woman's sex organs are not prominently displayed hardly makes them less of an object of interest on the part of the typical adult male.

A more straightforward explanation is that the female breasts evolved into organs of sexual significance as a result of two important changes in hominid behavior. The first was the disappearance of estrus periods and its replacement by continuous sexual availability in females. The permanent enlargement of the breasts at the time of sexual maturity thus became the visual signal of a woman's continual sexual readiness. The second change occurred when, over the course of hominid evolution, the male primate's relationship with its mother—a nurturing relationship of lasting affection and protectiveness—may have become associated, on a deep neurological level, with the male hominid's relationship with his mate.

Male apes and monkeys typically display a level of affection and protectiveness toward their mothers that they rarely display toward their female sex partners. The permanently distended breasts of sexually mature females may therefore have evolved as a strategy for redirecting the adult males' maternal affections into feelings of caring for their sex partners. At the very least, these feelings would have translated into greater opportunities for food-sharing between mates as well as more diligent protection against threats from predators and other hominids. All this would have enhanced the longevity and reproductive success of those hominid females whose breasts became permanently distended at puberty.

More than thirty-five thousand years ago, prehistoric people were carving "Venus figurines" out of stone and ivory, rendered with enormous breasts, buttocks, abdomens, and vulvas. These figurines are among the oldest surviving representations of the human figure, and we will examine them in detail in chapter 5. The sexual significance of the female breast is clearly an ancient phenomenon in human history, yet the evolution of the permanently distended female breast in our species remains a mystery that neuroscience and evolutionary psychology may eventually succeed in unraveling. But whatever its origin, its function as a linkage between maternal and sexual feelings can hardly be ignored. And this linkage is one of the many elements in the unique network of relationships that bind all humans together into the universal, permanent groups we call families.

The special web of feelings and relationships that naturally develop among males and females, parents and children, and siblings who live together for years and become bonded for life is a uniquely hominid innovation. The hominid family is more than a strategy for survival; it is the bedrock of human society.

The evolution of the human family created a web of enduring personal relationships that bound nuclear families together into larger extended families. As the early hominids evolved into modern humans, these extended family systems gave rise to complex kinship systems, rules of marriage, descent, and inheritance, and the hereditary transfer of wealth and power from one generation to the next. The tribal clans and royal dynasties that gave structure and stability to hunter-gatherer and civilized societies alike throughout most of human history could never have existed were it not for the deep emotional bonds forged in the crucible of the family.

When the ancestors of the hominids began to make, carry, and use spears and digging sticks in their daily lives, they set in motion a cascade of events that culminated in the evolution of an animal with a radically new physical form, an adaptation to the environment that required unprecedented cooperation between males and females, a huge expansion of sexual behavior, and the emergence of family ties that would provide the building blocks for the larger and more advanced societies of modern humans.

The technology of spears and digging sticks transformed humanity because fabricated tools and weapons were superior to their biological counterparts. And the superiority of technology over biology allowed the hominids to begin their long evolutionary journey toward dominance over all other forms of life. As we shall see in the next chapter, the power of the human lineage dramatically increased again during the second major transformation, when a very special population of hominids mastered the technology of fire and unleashed yet another game-changing strategy in the struggle for survival.

CHAPTER 3

The Technology of Fire

Cooking, Nakedness, and Staying Up Late

> *"Prometheus . . . went up to heaven, and lighted his torch at the chariot of the sun, and brought down fire to man. With this gift man was more than a match for all other animals."*
>
> —Thomas Bulfinch, *Stories of Gods and Heroes*

One Saturday afternoon in November 1924, Professor Raymond Dart was dressing for a wedding reception when two boxes of fossils from a limestone mine were delivered to his house in Johannesburg, South Africa. Opening the first box, he saw nothing of great interest. But when he opened the second box, as he later recalled, "a thrill of excitement shot through me." Dart, only thirty-two at the time, was a native Australian who had studied anatomy in Australia and England and had been sent to the fledgling University of Witwatersrand in Johannesburg to establish a credible department of anatomy there. Earlier that year, he had heard that fossilized baboon skulls had been found at a limestone mine in the remote village of Taung, and he asked that any new fossils be sent directly to him.

The contents of the second box included an endocast—a replica of the inside of a skull, created when limestone gradually replaces the soft brain tissues, roughly duplicating the shape of an ancient creature's original brain. "I knew at a glance," he continued, "that what lay in my hands was no ordinary anthropoidal brain. Here . . . was the replica of a brain three times as large as that of a baboon and considerably bigger than that of an adult chimpanzee." But where was the rest of that ancient skull? Where was the face that fit the brain? Dart searched feverishly through the boxes, and he soon found a large chunk of stone with a bowl-shaped depression in it. The endocast fit perfectly into the depression. Without doubt, the face of this creature was somewhere in the stone.

"I stood in the shade," said Professor Dart, "holding the brain as greedily as any miser hugs his gold, my mind racing ahead. Here I was certain was one of the most significant finds ever made in the history of anthropology." But just then, Dart felt a tug at his sleeve. It was the bridegroom, imploring him to finish dressing; the bridal car was expected at any moment. Reluctantly loading the rocks back into their boxes, Dart put the endocast and the large rock with the bowl-shaped depression aside and locked his precious stones in the wardrobe.

For weeks, Dart carefully chipped away at the large lump of limestone, certain that the face of this large-brained creature was embedded in it. He used his wife's sharpened knitting needles to pry the soft limestone away from the skull inside. Finally, two days before Christmas, the juvenile face of an ancient hominid could be clearly seen emerging from the rock. "I doubt," he wrote, "if there was any parent prouder of his offspring than I was of my Taung baby on that Christmas of 1924" (see Figure 3.1).

Dart christened his discovery *Australopithecus africanus* or "the southern ape of Africa," and this name has survived unchanged in the study of human paleontology ever since. The "Taung Child" was not only the first of the early hominids to be identified by science, but it also provided the first solid evidence that humanity did not evolve in Europe or Asia—which was the fashionable theory in the scientific circles of that time—but in Africa, as Charles Darwin had predicted more than fifty years earlier in *The Descent of Man*.

Dart published his findings in a now-classic article in the British scientific journal *Nature* in February of 1925, where he made a key anatomical observation. He pointed out that the foramen magnum—the point of attachment between the skull and the spinal column—was located at the bottom of the Taung skull, as it is in bipedal humans, rather than at the back, as it is in the quadrupedal apes. Dart believed that the position of the foramen magnum proved that the juvenile from Taung had stood and walked upright. But Dart's belief that *Australopithecus* was a very ancient type of early hominid was rejected by the scientific authorities in Europe, who concluded that Dart's fossil was most likely the abnormal remains of an extinct ape.

It was an article of faith among the paleontologists of those days that the ancestors of prehistoric humans had first developed large brains, and then only later—presumably as a result of their increasing intelligence—had lost their ape-like characteristics. This theory had been greatly reinforced in 1912, when Charles Dawson, an amateur British archeologist, announced that the remains of an individual with an ape-like face and a large brain had been found by a workman in a gravel pit in Piltdown, England. Before long, "Piltdown Man," with its ape-like face and its human-like brain, was being hailed by many in the scientific community as the "missing link" between humans and the apes, while Dart's discovery, with its human-like face and ape-like brain, simply did not fit the scientists' expectations.

Much to the embarrassment of the scientists who had accepted Dawson's find as the true missing link, however, Piltdown Man was eventually exposed as a hoax—but not until 1943, nearly forty years after its "discovery." Piltdown Man turned out to be a deliberate fabrication, composed of a medieval human skull, an old orangutan jawbone, and some chimpanzee teeth—all deliberately stained to give the appearance of great age. Charles Dawson died in 1916, and to this day the perpetrator of the Piltdown hoax has never been positively identified.

Deeply discouraged by the cool reception that his find received at the hands of European paleontologists, Professor Dart abandoned his investigations into human prehistory for the next twenty years and focused on his original mission of building up the Department of Anatomy at the University of Witwatersrand. But he continued to enjoy the support of paleontologists in his home country, and his students and colleagues continued to unearth a growing collection of ancient fossil remains from a number of prehistoric sites in South Africa.

Around the time of Dart's publication of the "Taung Child," a local schoolteacher had sent him fossil materials that had been unearthed at Makapansgat, the site of another South African limestone quarry. This quarry, located in an area of numerous ancient caves, has yielded a rich harvest of prehistoric human fossils over the years. Finally, at the end of World War II, encouraged by the increasing acceptance of *Australopithecus africanus*—and by the growing body of corroborating evidence that was being unearthed by his colleagues—Dart returned to the field in 1947 and began a long series of excavations in the Makapansgat caves. The numerous blackened bones that he found there led him to conclude that early hominids living in Makapansgat had built fires and roasted the flesh of their prey more than two million years ago. He named the early hominid fossils found at Makapansgat *Australopithecus prometheus* ("the southern ape that tamed fire").

But Dart was in for another disappointment. Chemical analysis revealed that the bones from Makapansgat had been blackened not by fire but rather by the chemical action of manganese dioxide leaching into the deposits. Moreover, Dart's *Australopithecus prometheus* proved not to be a new species at all but rather the fossil remains of *Australopithecus africanus* individuals, members of the same species as the Taung Child. With this turn of events, the true antiquity of the hominids' first use of fire was thrown into doubt, and the entire subject of how and when fire was first used by the hominids—one of the longest-running controversies in human paleontology—has remained unresolved for decades.

Yet, in the end, Dart's conclusion proved to be much closer to the truth than it might have originally seemed. In 1989, two South African paleontologists, Charles K. Brain and Andrew Sillen, published the results of an exhaustive study of blackened bones from Swartkrans Cave, about 150 miles southeast of Makapansgat, proving conclusively that bones of game animals had been repeatedly burned in campfires in the cave as long as 1.5 million years ago.[1] And although it seems likely that these campfires were set by the emerging human *Homo erectus*, so far the only hominid fossils that have been found in the same geological stratum as these burned bones are those of the early hominid *Paranthropus robustus*, a not-too-distant relative of Dart's signature discovery, the early hominid *Australopithecus africanus.*

Over the years, paleontologists' estimates of when our hominid ancestors first began to use and control fire have been all over the map. Some have argued that the control and use of fire as a regular aspect of human life did not begin until about 130,000 years ago, while others have proposed that the earliest hominids began using fire as early as seven million years ago.[2] This is a staggering fifty-fold difference in the estimated age of fire use.

But in recent years, evidence has been accumulating that indicates fire was used by *Homo erectus* at least as early as 1.5 million years ago, and evidence of fire associated with the remains of *Homo erectus* has now been found in the sites of Koobi Fora and Chesowanja in East Africa and in Wonderwerk Cave in South Africa. It now seems certain that the technology of fire was first adopted sometime between two million and 1.75 million years ago, when the early hominids, who had already lived in Africa for at least three million years, found themselves competing with a more advanced hominid, the emerging human, *Homo erectus.*

Early Hominids and Emerging Humans

You may recall that Lucy, the Australopithecine—as well as the other early hominids, which first appeared more than four million years ago—were small in comparison to ourselves. They stood between three and one-half and four feet tall and weighed between seventy-five and ninety-five pounds. Although they walked and ran fully erect on two legs, their bodies were, in many respects, much like the prehistoric apes from which they had descended. Their arms were long, the bones of their fingers were curved, their toes were long, and their torsos were pear-shaped, with narrow shoulders, a thick waist, and wide, flaring hips—all of which indicated that they continued to spend a significant part of their lives in the trees.

The savannah country of prehistoric Africa in which these early hominids flourished was inhabited by a number of large and dangerous predators—including leopards, cheetahs, lions, tigers, and hyenas—that could have easily preyed on defenseless two-legged creatures. However—as I argued in the previous chapter—armed with long spears and other fabricated weapons, the early hominids must have been able to defend themselves and their offspring against such predators during the daytime, which their long history of survival proves beyond doubt.

But even armed with lethal weapons, these early hominids would have been particularly vulnerable at night, when the big cats, with their superior night vision and heightened sense of smell, would have been able to approach them in the darkness and attack before being detected. In fact, many of the African caves that have been excavated from between four and two million years ago show clear evidence that lions and leopards living in the caves were killing and eating the early hominids on a regular basis.

Since the early hominids made tools and weapons, killed game, ate meat, and were clearly adapted to walking and running over the open ground, the most likely explanation for the persistence of their many ape-like anatomical characteristics was that they needed to retreat into the safety of the trees at night as their best defense against the large and dangerous predators that inhabited prehistoric Africa.[3] But beginning somewhat before two million years ago, a newer, larger, and more advanced hominid—an emerging human—begins to appear in the fossil record, and it was this creature who bridged the gap between the early hominids like *Australopithecus* and the large-brained modern humans like the Neandertals and ourselves.

Numerous fossil remains of emerging humans have been unearthed from many different archeological sites throughout both Africa and Eurasia, and the paleontologists who have unearthed these remains have tended to classify their various finds into numerous distinct species of the genus *Homo.* But some of these species are represented only by very fragmentary remains and many others are so similar to each other that it is doubtful that they are really distinct species at all.[4] Although there is a great deal of controversy about which of these finds actually represent distinct species, there is general agreement that—while they were primitive by our standards—the anatomy of the emerging humans was so similar to the anatomy of modern people that they should all be included in the genus *Homo*, to which all modern humans belong.

The most ancient of these emerging humans were two species, *Homo habilis* and *Homo ergaster*, which appeared in East Africa while the early hominids, such as the australopithecines, were still the dominant bipeds of that region. *Homo habilis* (the "skillful" or "handy" man) first appears in the fossil record roughly 2.3 million years ago, long before the early hominids became extinct. This species had the smallest brain of any of the emerging humans, averaging 500–600 cc—although it was larger than the brains of the early hominids, which averaged between 400 and 500 cc. *Homo habilis* also retained both the pear-shaped torso and the very small stature of the early hominids, although it made stone tools that were superior to those of the early hominids. These included a hand-held "hammer"—a round, fist-sized stone that was used to crack open the bones of its prey and that may have also been used to "tenderize" fresh meat by pounding it, rendering it easier to digest.

Homo ergaster (the "workmanlike man") appeared somewhat later, around 1.8 million years ago, and was the first hominid to fashion the "Acheulean" hand axes that represented a superior level of workmanship over the crude Oldowan pebble tools of the early hominids. The brain of *Homo ergaster* represents a huge increase in size over the hominids that preceded it, averaging around 850 cc. *Homo ergaster* was also much taller than *Homo habilis*, with the barrel-shaped chest and narrow waist that is typical of modern humans. Paleontologists are still unsure whether *Homo ergaster* was really a unique species or whether it was simply an early form of that preeminent emerging human, *Homo erectus.*

The most important of the emerging humans—whose existence as a separate species is not in doubt—was *Homo erectus* (the "erect man"), whose remains have been found throughout Africa, Asia, and Europe beginning nearly two million years ago and continuing until at least two hundred thousand years ago.[5] *Homo erectus* was much larger than the early hominids, almost as tall as modern people are today, and its anatomical characteristics all indicate that the emerging humans had completely abandoned the ancient primate habit of sleeping at night in the safety of the trees.

Homo erectus was one of the most intelligent of the emerging humans, and its brain—already significantly larger than the brains of either the apes or the australopithecines—continued to grow larger throughout its long and successful history. *Homo erectus* left behind large numbers of beautifully made, sophisticated Acheulean stone tools—notably an iconic, sharply pointed hand ax, shaped like a large teardrop (see Figure 3.2). The Acheulean hand ax marks the crossing of an especially significant threshold in human evolution, because its manufacture required a considerable amount of planning, foresight, and manual dexterity.

Perhaps the most notable characteristic of the anatomy of *Homo erectus*, in addition to its greater size, taller stature, and significantly larger brain, was that it no longer retained any of the ape-like physical characteristics associated with life in the trees. The toes of *Homo erectus* had shrunken to the point where they offered little or no help in climbing tree trunks, and the bones of its fingers had become straight, like our own finger bones, having lost the characteristic curved shape of the tree-dweller. Finally, *Homo erectus*'s upper torso was barrel-chested, with wide shoulders and a relatively narrow waist—features possessed by none of the tree-dwelling apes.

These changes indicated that *Homo erectus* was not only living on the ground during the daytime but was also sleeping on the ground at night. And the only reasonable way these emerging humans could have prevented being attacked by predators while sleeping on the ground would have been to set fires, sleep close to them, and keep them burning until dawn. This line of thought was spelled out in detail by the

Harvard anthropologist Richard Wrangham in *Catching Fire: How Cooking Made Us Human.*[6]

The Genesis of the "Cave Man"

Many of the behaviors that were once considered unique to humans have eventually been found to exist—although often in rudimentary form—among other animal species. These include, in addition to the ability to make and use tools, the ability to communicate complex information and the ability of group-living species to adopt new behaviors and pass them down to succeeding generations as cultural traditions. But of all the species in the animal kingdom, only the hominids have ever exhibited even the most rudimentary ability to make, control, or employ the use of fire as a regular feature of its daily life.

If it were not for their unique bipedal anatomy, hominids would have never been able to control and use fire. When a quadrupedal animal carries things, it typically carries them in its mouth, and since to carry a burning branch in the mouth would risk burning their mouths and faces or filling their eyes and noses with smoke, quadrupedal animals do not carry burning branches from one place to another. Apes and monkeys do have flexible forelimbs with grasping hands, and they are capable of walking for short distances on their hind legs while carrying things with their forelimbs, but no species of ape or monkey has ever been observed either using or controlling fire in the wild.[7] After all, fire is destructive to living tissue, and all animals instinctively fear and avoid it. All animals, that is, except hominids.

A hominid that can carry a spear in its hands for miles can also carry a burning branch for miles. Long before hominids learned how to make fire, they must have learned how to capture it, transport it, bring it to their home bases, and keep it going for days and weeks at a time. Lightning strikes, and the burning trees and grasses that can result from lightning strikes, were common in prehistoric Africa, and these were most likely the source of the fires that were brought home and tended by the emerging humans almost two million years ago.

Some of the earliest evidence of the controlled use of fire by *Homo erectus* a million years ago or older comes from caves in South and East Africa. Many of these large, dry caves were the favorite dwellings of some of the most dangerous of prehistoric African predators, including lions, leopards, bears, and snakes. But none of these animals would remain for long in a cave that was filled with fire and smoke. When *Homo erectus* learned to build fires inside of caves, it was able to drive out other animals, move into these choice dwelling places, and use them as a home base for weeks and months at a time.

By contrast, when the remains of early hominids such as *Australopithecus* are found in caves, it is usually evident that their bodies had been dragged into these caves by their natural enemies, especially the big cats, that preyed on them. Without fire, no hominid would ever have been able to make its home in a cave for very long. Yet in many of the earliest archeological sites that show evidence of the use of fire by emerging humans, there is no sign that either meat or vegetable foods had been cooked there. The archeologists who excavated these sites have concluded that fires were kept burning in these caves primarily as a source of light and as a form of protection from predators. With their new-found ability to control fire, the emerging humans were freed from the necessity of sleeping in the trees at night. For the first time, the hominids became fully terrestrial beings.

The ability to control fire conferred numerous benefits to the emerging humans. Fire not only drove large and dangerous predators out of the caves—it also drove out the insects, reptiles, and disease-carrying vermin that also inhabited the caves. And by setting fire to an area covered in dense brush, the harmful insects, reptiles, and poisonous snakes that lived in these habitats could be killed or driven out of fairly large areas.

Setting fire to the thickets of brush in the savannah environment also tended to clear the land, promoting the growth of tender new grasses and attracting the grazing animals, such as antelopes, that were among the hominids' favorite prey. In fact, many of the hunting and gathering societies that survived into modern times on the plains of

Africa, Asia, and the American West had been deliberately setting fires for as long as any of them could remember—both to promote the growth of new grasses and to drive herds of game animals toward their camps, where they could be more easily trapped and killed.

Fire provided a controllable source of heat that not only rendered the cold, damp interiors of caves more habitable but also allowed the emerging humans to migrate into colder climates. In fact, early in its long and remarkable history, *Homo erectus* migrated out of Africa and appeared in many locations in Europe and Asia more than 1.5 million years ago—becoming the first of the hominids to inhabit the Eurasian continent. Lastly, fire allowed the emerging humans to develop advanced tool-making techniques. These included the deliberate tempering or hardening of wooden tools and weapons, a technique that was well advanced by four hundred thousand years ago, as evidenced by the wooden throwing-spears from Schöningen described in the previous chapter.

Staying Up Late

Most primates are diurnal creatures, active by day, inactive by night. When darkness falls, apes and monkeys typically fall silent, stop moving about, and prepare for sleep. This behavior, common to nearly all non-human primates,[8] is regulated by a hormonal response to the level of light that reaches the eye. Melatonin—a hormone secreted by the tiny pineal gland, which is located at the base of the brain—is released when the level of light in the environment begins to drop with the approach of sunset.[9] The higher the concentration of melatonin in the bloodstream, the sleepier the individual becomes. Conversely, when the sun rises and the environment becomes flooded with light, the production of melatonin by the pineal gland drops off, and the melatonin remaining in the blood is removed from the body by the kidneys.

When the emerging humans began to use fire to ward off predators after sunset, they created an artificial source of light that had the effect of suppressing the production of melatonin and delaying the onset of sleep. As a result, the adoption of fire freed the hominids from the ancient limitations of the twelve-hour tropical day and extended their normal waking hours well into the night. This is the point in human evolution when our ancestors began staying up late—extending the time of wakefulness well beyond the average period of daylight that governs the circadian rhythms of all other tropical animals.

Staying up late gave the emerging humans unique new advantages. It provided additional time for eating, for making things, and for social life and communication. The human predilection to eat after the hours of darkness, to work on weaving, tool-making, and other tasks by the firelight, and to gossip, relate the events of the day, and recite folk tales and myths, represented unique behaviors that no other species could duplicate. All of these behaviors were made possible when the emerging humans first began building fires at night and staying up late, while lesser creatures slept or kept their distance.

The Raw and the Cooked[10]

While the earliest use of fire may have been to protect the hominids at night against attacks by predators, the regular building and maintaining of fires in the hominids' home bases would eventually find uses far beyond simple protection. And without doubt the most important of these uses—with consequences that reached far beyond the needs of the moment to profoundly shape the future of human evolution itself—was the discovery of cooking.

In a scientific paper published in the *Proceedings of the National Academy of Sciences* in 2012, an international team of scientists reported that microscopic analysis of deposits from Wonderwerk Cave in South Africa confirms that hominids were making and tending fires deep inside this cave more than one million years ago.[11] And of the hundreds of bones and bone fragments that were recovered from the site, as many as 80 percent of them showed evidence of having been burned in these fires. In other words, they were cooking their meat.

In *Catching Fire,* Richard Wrangham argued that the invention of cooking was ultimately responsible for most of the evolutionary achievements that led to the appearance of modern humans, including the process by which the brains of modern humans achieved their present immense size. Wrangham described at least five important advantages that the cooked food eaten exclusively by the emerging humans enjoyed over the raw food eaten by the early hominids and all other animals.

First, the cooking of animal flesh breaks down the fibrous collagen in muscle cells, rendering them into a protein-rich gelatin that requires far less time and energy to digest.

Second, cooking has the effect of disinfecting meat and rendering it fit for consumption. Some anthropologists have argued that early hominids derived much of the meat in their diets by scavenging the kills of larger predators such as lions and leopards. But carrion, especially in the tropics, is rapidly infested with harmful bacteria, and the prehistoric apes that were the ancestors of the early hominids, with their large intestines designed for the slow, laborious process of digesting leaves and other rough plant foods, would have been particularly susceptible to the effects of food-borne illnesses. Chimpanzees that will readily eat fresh meat will avoid and reject meat that has begun to decay.

Bacterial infestations by pathogens such as botulism, anthrax, salmonella, and E. coli, as well as by fungi and viruses, would have caused sickness and death among even the hardiest individuals, since these organisms would have multiplied to dangerous and potentially lethal levels in the long intestines of prehistoric apes. But cooking destroys the parasites and pathogens that multiply rapidly in an animal's carcass within a few hours of its death. And the disinfecting action of cooking allows cooked meat to be stored over much longer time periods than raw meat. Cooking would have kept the spoils of the hunt in edible condition for much longer time periods, making it possible for a group of hominids to feast for days on the remains of a large game animal.

Third, cooking of vegetable foods breaks down the cell walls of plants and turns indigestible cellulose into digestible starches and sugars. In fact, many of the vegetable foods that hominids have depended upon—in particular, the roots and tubers that they could unearth with their digging sticks—contain toxins that make them not only unpalatable but in many cases actually indigestible. Cooking breaks down and eliminates these toxins, opening up a whole range of vegetable foods for human consumption that would otherwise have been off limits. In fact, experiments with chimpanzees, gorillas, and orangutans have found that the great apes generally prefer their vegetables cooked instead of raw.[12]

Fourth, cooked food requires dramatically less time for chewing than raw food. The raw food diet of chimpanzees requires them to chew their food for hours on end. In fact, it has been estimated that chimpanzees spend roughly 50 percent of their waking hours simply chewing their food. By contrast, humans spend roughly 5 percent of their waking hours chewing—only one-tenth as much time as chimpanzees—and as a result, humans can devote far more of their waking hours to other pursuits, including hunting, making tools and weapons, and sharing information with other members of the group.

Fifth—and most significantly—the invention of cooking actually made it possible for the human brain to undergo its extraordinary expansion during the age of the emerging humans. While the brain of a chimpanzee averages about 400 cc and the brain of the early hominids averaged about 500 cc, the hominid brain *nearly tripled in size in the past two million years* and now averages between 1,300 and 1,500 cc. Without these immense brains, it would not have been possible for modern humans to create the unique language, culture, and technological capabilities of which they ultimately became capable. And the greatest part of this tremendous expansion in the size of the hominid brain was achieved by the emerging humans.

No known human society, however primitive or technologically unsophisticated, has ever been found to subsist entirely on a diet of raw food. In fact, people in modern society who have adopted a raw food diet must spend an extraordinary amount of time chewing their food, they have less energy, lose weight, and are hungry virtually all the time.[13] Cooking food, it seems, is an essential part of every human diet. In fact, the invention of cooking seems to have been an essential prerequisite for our species to have successfully evolved the enormous brains that have made us human and that distinguish us so utterly from every other animal species.

The Expensive Human Brain

The reasons that a diet of cooked food is necessary to support the energy needs of the modern human brain are summarized by the "expensive-tissue hypothesis" set forth in 1995 by the anthropologists Leslie Aiello and Peter Wheeler,[14] who began their now-classic study with the established fact that the brain is an "expensive" tissue in terms of its size and energy requirements.

Aiello and Wheeler observed that the size of the human brain is more than four times as large, relative to its body weight, as the brain of the average mammal. The human brain weighs about three pounds—about 2 percent of the average adult's total body weight. Yet when it is active, the human brain can consume as much as 20 percent of the body's available energy—roughly ten times as much energy, pound for pound, as is consumed by the human body as a whole.

Furthermore, the brain is not the only "expensive tissue" in the body. Other tissues with similarly high energy demands include the heart, liver, kidneys, and digestive organs. Taken together, the brain and these vital organs make up less than 7 percent of the body's weight, yet, when the body is in a resting state, they consume an astounding 60 to 70 percent of its available energy. The central question that Aiello and Wheeler attempted to answer is: How can the human body supply such substantial quantities of energy to its massive brain without reducing the energy requirements of other parts of the body?

Reducing the size of the muscles that constitute much of the more "inexpensive" tissues of the body would be entirely impractical. This is not only because these tissues normally consume only about one-third of the body's available energy but also because, in order to compensate for the increased energy demands of the expanded human brain, 70 percent of the muscles of the human body would have to be eliminated. This would make it difficult, if not impossible, for humans in their natural habitats to acquire the food they need to supply this energy in the first place.

Reducing either the size or the activity of the heart to any significant extent would reduce the flow of blood to levels that would be especially dangerous to the brain, which requires a steady and copious supply of blood. When the blood circulation falls significantly, the brain ceases to function effectively, ultimately resulting in the loss of consciousness.

Reducing the size or activity of the kidneys would seriously compromise one of their most important functions. The kidneys consume most of their energy when they are concentrating the urine by removing its precious water content and returning this water content to the bloodstream. Any reduction in this function could result in a dangerous level of dehydration, especially during the strenuous exercise involved in hunting and gathering in hot weather.

Reducing the size or activity of the liver would not only compromise the ability of this critical organ to cleanse the blood of toxins and various waste products, but it would also deprive the brain of its primary source of energy. The fuel that powers the activities of the brain is a large sugar molecule known as glycogen, and the body's supply of available glycogen is manufactured in the liver.

This leaves the digestive organs as the only candidates available for reduction in size and energy requirements. It is not surprising, then, to find that the human digestive organs—particularly the stomach and intestines—are the smallest, relative to body weight, of all the primates. In fact, the hominid fossil record contains clear evidence that the digestive organs underwent a significant reduction in size when the early hominids evolved into emerging humans that used fire.

The early hominids had rib cages that were wide and flaring toward the bottom, as well as wider and more flaring pelvic bones. These features indicate that the abdomens of these creatures were relatively large, similar to the abdomens of the great apes—the orangutans, gorillas, and chimpanzees. But with the appearance of *Homo ergaster* and *Homo erectus,* the rib cage became significantly narrower at the bottom and the pelvis became smaller in diameter. Both of these features suggest that *Homo erectus* had evolved the smaller, more compact abdomen that is characteristic of modern humans—and that would have been the result of a significant reduction in the size of the digestive organs.

Along with this reduction in gut size, the many fossils of emerging humans that have been excavated over the years show a profound and steady increase in the size of the brain, from roughly 600 cc with the appearance of *Homo habilis* nearly two million years ago to roughly 1,300 cc in the most modern forms of *Homo erectus*, which lived until about 250,000 years ago. This massive increase in the size of the brain in the space of two million years is unprecedented in the evolution of life on Earth. No other creature has done it, and the expensive-tissue hypothesis suggests that only a diet of cooked food—and the major reduction in the size of the digestive system that cooked food made possible—would have enabled this human ancestor to support as "expensive" an organ as the modern human brain.[15]

The cooking hypothesis is further corroborated by another major anatomical change that occurred with the appearance of *Homo erectus*: the teeth and jaws of the emerging humans became significantly reduced in size. You will recall that the chimpanzee, which lives entirely on raw foods, must spend roughly 50 percent of its waking hours chewing, while the modern human, which lives largely on a diet of cooked foods, can accomplish all of the chewing necessary for its nourishment during 5 percent of its waking hours. As you can imagine, the teeth and jaw of the chimpanzee are by necessity much larger than those of the human.

Not surprisingly, the fossil record also shows that the teeth and jaws of the australopithecines and other early hominids are also very large, while those of the emerging humans are considerably smaller. Cooked food not only requires less time to chew but also can be chewed sufficiently with much smaller teeth and jaws. All of this is evidence that *Homo erectus* had mastered the use of fire early in its history and developed a lifestyle in which cooked food had become the mainstay of its diet.

The many different kinds of evidence from archeological sites in prehistoric Africa all point to the conclusion that *Homo erectus,* the emerging human, was the first hominid to master the use of fire. Physical and chemical evidence shows that fires were burned for long periods of time deep inside caves inhabited by emerging humans. Physical and chemical evidence also shows that most of the bones found in some of these caves had been burned, indicating that at some point not long after the mastery of fire had been achieved, the emerging humans were cooking their meat.

The anatomical evidence from *Homo erectus* fossils shows that these emerging humans were no longer physically capable of climbing high into the trees at night for safety. And the anatomical evidence further shows that the emerging humans had much smaller digestive systems, much smaller jaws and teeth, and much larger brains than any of their predecessors. All things considered, *Homo erectus,* the emerging human, seems to have learned to use and control fire—and to have made it a fundamental component of its natural way of life—more than 1.5 million years ago.

The Naked Primate

When the emerging humans adopted fire, they became—and have remained—the world's only completely naked primate.

Sleeping with fire is one of the most universal of all human behaviors. It has been found among every hunting and gathering society that has been studied by anthropologists. And it was an especially common practice among hunters and gatherers to sleep within inches of a campfire that was kept going all night. Needless to say, this behavior would have been quite impossible if our bodies were still covered with a thick layer of fur, because, with the flames of a campfire only inches away, we would have quite literally set ourselves on fire.

None of the ways that humans interact with fire—carrying burning torches from one place to another, blowing on hot embers to light kindling, roasting meat and vegetable foods, feeding a campfire with fresh firewood, and holding the arms and legs out over a fire to warm them—would be possible if we had retained the long hairy fur that covers the bodies of all other primate species. Try to imagine tending a roaring campfire while wearing a thick, long-sleeved fur coat, and you will understand how difficult it would be to handle fire safely with thick fur all over your arms, legs, and torso. This is doubtless why the chimpanzee may learn to casually smoke a cigarette or cigar—and in some cases even operate a cigarette lighter—but will not, unlike humans, sit close to a roaring fire.

When the emerging humans began to use fire for light, warmth, and protection, those individuals and groups with the least amount of hair on their bodies would have been the most successful at handling fire, and it was doubtless at this point in human evolution that our ancestors grew naked. But losing all of the millions of hair follicles that primates have had all over their bodies for 55 million years would have required a number of significant genetic changes. Such changes did, in fact, take place when the ancestors of whales and dolphins adopted an aquatic existence and evolved into the physical forms of fishes. But the emerging humans achieved essentially the same result by retaining virtually all of their hair follicles while reducing the output of these follicles. As a result, humans have only a very short and fine coat of nearly invisible vellus hair that, while it can hardly be seen, nevertheless continues to cover almost the entire human body.[16]

The loss of a coat of thick fur had other advantages. Perhaps the most important of these was that it enabled the hominids to cool their bodies more efficiently by sweating. In fact, humans sweat more readily and more profusely than any other primate. The hominids' ability to sweat profusely made it possible for them to travel for long distances, when moving camp or pursuing game, without becoming overheated. While humans have about a thousand hair follicles per square inch—roughly the same number as chimpanzees—the human body has an average of 650 sweat glands per square inch. This is ten times as many sweat glands as is typical of non-human primate species. Finally, the loss of thick fur helped the hominids to control the populations of insect parasites that would have multiplied ferociously in the bedding of a species that spent night after night sleeping in the same place.

Out of Africa

In 1887, a young Dutch doctor by the name of Eugène Dubois resigned his coveted position as a lecturer in anatomy at the University of Amsterdam and—to the dismay of his horrified colleagues—joined the Dutch Army and requested that he be posted to the Dutch East Indies, where he hoped to find the fabled missing link between the humans and the apes. Dubois was fascinated by the theories of human evolution that were the talk of the European scientific world in those days, and he was particularly excited when two nearly complete Neandertal skeletons were unearthed in Belgium in the previous year, along with numerous stone tools. So with his young wife and baby in tow, Dubois left Europe behind, arriving on the island of Sumatra in December of 1887.

With two engineers and several dozen laborers that had been provided to him by the Dutch government, Dubois began to excavate a number of sites in Sumatra, but the remains he found were of comparatively recent origin. Moreover, conditions in Sumatra were difficult, to say the least. Several of his laborers became ill; others ran away. One of his engineers turned out to be useless; the other one died. But Dubois pressed on, and in 1890, after three years in Sumatra with little to show for it, he convinced the authorities to transfer him to the island of Java, where he hoped to have better luck.

In Java, Dubois's luck turned. In October of 1891, Dubois and his workers unearthed an intact skullcap from a deposit at least seven hundred thousand years old, which during the previous year had yielded part of a jaw and a molar tooth. And in August of the following year, the thigh bone of a fully erect hominid was found only a few yards away. Dubois was convinced that these were the remains of the "missing link," and he named his new-found hominid *Pithecanthropus erectus*, "the erect ape-man" (see Figure 3.3).

In 1894, Dubois published the results of his investigations, and the following year he returned to Europe to lecture and promote his find. But as with so many others who came before and after him, Dubois's discoveries generated mostly skepticism and controversy. In the end, only a few scientists accepted Dubois's contention that his "Java Man" was a prehistoric human ancestral to ourselves. The experience of being rejected by his colleagues so embittered Dubois that by the year 1900 he had locked his specimens away, refusing to discuss them or allow anyone to examine them. They would not been seen again for the next twenty-three years.

Dubois's specimens are now universally recognized as the fossil remains of *Homo erectus,* the first of the emerging humans to be excavated and described—and vastly older than the Neandertal fossils that had been found in Europe, which were less than one hundred thousand years old. At the time of its discovery, Java Man was not only the first fossil hominid to be found in Asia but was also the oldest fossil hominid ever found. Dubois died in 1940, unaware of the true antiquity of his discovery. The most recent analyses of similar *Homo erectus* fossils from Java have dated them as being more than 1.5 million years old, twice as old as Dubois had imagined.

The australopithecines and other early hominids lived out their entire history on the African continent. No remains of any of the early hominids have ever been found in either Europe or Asia. And although the many species of early hominids adapted to a variety of different habitats, including grasslands, gallery forests, marshes, river banks, and deserts, all of these habitats lay in the tropical latitudes of Africa. But *Homo erectus* and other emerging humans migrated out of Africa at least 1.8 million years ago, and evidence of their habitation from these ancient time periods has been found throughout Eurasia—in Britain, Spain, France, southern Russia, Pakistan, China, and Indonesia.

Lacking fire, the early hominids were dependent on their sleeping trees to provide protection at night, and they were unable to venture far from the mountains and river valleys, where trees grew in abundance. But changes in geology and climate that began roughly eight million years ago resulted in a huge expansion of grasslands throughout Asia and Africa, and by three million years ago, the savannah habitats of the two continents had merged into a vast belt of grasslands that stretched uninterrupted from West Africa to northern China.[17] And thus the emerging humans, with fire to protect them, were able to venture farther and farther into these immense grassland habitats, ultimately establishing new ways of life many thousands of miles from their African homeland.

Is the "Cave Man" Really Extinct?

In 1758, more than a century before the publication of Darwin's *Origin of Species*, the great Swedish naturalist Carolus Linnaeus published the tenth edition of his pioneering work, *Systema Naturae*, the product of twenty-five years of painstaking labor in which he had classified all known living things based on their perceived relationship to each other. Linnaeus's system was so useful and elegant that it was quickly adopted by the biological science of its day, and it has remained the basis of scientific classification of species ever since.

Linnaeus himself invented the term *Homo sapiens*, which has remained the proper scientific name for our species, and he considered the anatomical similarities between humans, monkeys, and apes to be so self-evident that he grouped them all together into a single order of mammals that he called the "primates." These were further subdivided into five distinct families, and various types of humans—both real and mythical—were placed in the family *Hominidae* (see Figure 3.4).

Anticipating the existence of the rumored "cave man," the tenth edition of *Systema Naturae* recognized two living species of humans: *Homo sapiens* or "thinking man" and *Homo troglodytes* or "cave-dwelling man." Linnaeus's "cave man" was not regarded as an extinct species of prehistoric humans (these were not to be discovered for another hundred years) but instead was believed to be a living population of diminutive people, living in caves hidden deep in the jungles of Southeast Asia. The evidence for the existence of *Homo troglodytes* was an account provided more than a hundred years earlier by a medical doctor named Bontius of the Dutch East India Company, who had returned from Southeast Asia with a detailed drawing and description of these mysterious creatures.

In time, it was discovered that Bontius's description was not of a cave man but rather of an orangutan—an ape that has never lived in caves and that in fact spends most of its life high in the treetops of the rainforest. But in a curious twist of fate, when the common chimpanzee was first identified in 1776 by the German physiologist Johann Friedrich Blumenbach, he named this species *Pan troglodytes* (the "cave-dwelling satyr"). Although it was eventually determined that chimpanzees were not the oversexed semi-human creatures living in caves imagined by Blumenbach, the name stuck and remains the correct scientific term for the common chimpanzee to this day.

But the persistent idea that primitive human-like creatures had once lived in caves was unexpectedly confirmed by the discovery in 1865 of the fossilized skeleton of a most unusual type of human, quite different from ourselves, that had lived forty thousand years ago in a cave in the Neander Valley (in German, *Neandertal*), near the city of Dusseldorf. Neandertal Man, as this species came to be known, flourished during the ice ages in Europe and Asia but disappeared about forty thousand years ago. Originally considered "sub-human," the Neandertals were eventually recognized as being so close to modern humans in brain size and other characteristics that they are now considered a subspecies of *Homo sapiens* and have been assigned the scientific name of *Homo sapiens neanderthalensis*. (The scientific name for our own subspecies is *Homo sapiens sapiens*.)

Since the discovery of the Neandertals, paleontologists have unearthed the fossilized remains of numerous ancient hominids from caves in Europe, Asia, and Africa, pushing the origins of full upright posture and bipedal locomotion ever farther into the past. Eugène Dubois first discovered the remains of *Homo erectus* in the caves of Java in the late nineteenth century, Raymond Dart discovered the remains of *Australopithecus* in South African caves in the 1920s, and other *Homo erectus* fossils were found in the 1920s with the discovery of the remains of Peking Man in a series of caves near Beijing, China.

Evidence of the hominid occupation of caves began with the mastery of the technology of fire by the emerging humans and continues to the present day. The Cro-Magnons famously painted lifelike pictures of animals on the walls of caves in France and Spain more than twenty-five thousand years ago, the inhabitants of ancient Israel left the Dead Sea Scrolls in caves, the Pueblo Indians built their picturesque cliff dwellings a thousand years ago by excavating the vertical sides of naturally occurring cliffs in the American Southwest, and about three thousand Gitano people, the Romani or gypsies of Spain, still live in a complex of caves near the present-day city of Granada. Finally, millions of people continue to live today in the *yaodongs* or "cave houses" of northern China, where entire towns are carved into hillsides or in pits dug in the soft earth.

Think about it. As the universal human practice of living in enclosed spaces with solid walls and roofs and small openings for entering and leaving clearly demonstrates, it is in our human natures to be cave dwellers. We humans, the descendants of 55 million years of tree-dwelling ancestors, who lived and died in the open air of the forest, are no longer able to live out in the open. Instead, we crave "a roof over our heads" in the form of a secure canopy that not only protects us from the wind, rain, and snow but also forms a closed, cave-like environment that traps the heat from our fires and protects us from attack by predators.

It no longer matters that the lions, tigers, leopards, hyenas, bears, and wolves of prehistoric times have long ago been driven from almost all of the places where humans live today. Our hairless bodies demand protection from the elements, and our ancestors long ago gave up the practice of sleeping out in the open. The tents and huts of the hunter-gatherers may have given way to the houses and apartments of the modern world, but all of these dwellings have retained the basic elements of the primeval cave dwelling: a solid roof overhead and around the sides, with openings to let in light and to provide a way of entering and leaving our cave-like homes.

In fact, if you look objectively at the architecture of a modern town or city, you can easily see that it is composed of two types of structures: 1) free-standing buildings and houses that are basically cave-like structures sitting on the ground next to other cave-like structures and 2) high-rise buildings that are essentially tall, cliff-like structures honeycombed with cave-like apartments. These apartments are equipped with openings in their sides to let in light and air (windows) plus larger openings (doors) that lead, through a network of passageways and stairways, to other doorways that open to the outside world.

The very fact that most humans feel little or no discomfort when they look out from windows and balconies that may be hundreds of feet above the ground is testimony to humanity's ancient tree-dwelling ancestry. And the fact that we do not feel trapped but instead feel safe and secure when we are shut inside a closed cave-like structure, with minimal openings for light, air, and egress, is a testament to our species' evolutionary history of living in caves—at first, in naturally occurring caves and later in caves of our own design and manufacture. The caveman is not extinct but is alive and well and living today in the houses and apartments of the modern world.

♦

When the emerging humans adopted the use of fire, they released a vast new set of human potentials, finally liberating themselves from the arboreal lifeways of their ancestors and becoming fully terrestrial, protected by their fires and their weapons from the predators that shared their many and varied habitats. For the first time in their multimillion-year history, hominids were able to make their homes in caves, protected from the elements and warmed by their campfires. They were also liberated from the nearly inflexible rhythm of the twelve-hour tropical day and night. Fire allowed them to stay awake far into the night, making things by the light it produced and communicating with each other about the events of the day. Finally, fire greatly expanded the hominid diet, allowing them to cook the seeds, nuts, roots, and tubers they gathered, freeing them to make edible that which was previously inedible due to its hardness, bitter taste, and dangerous toxins.

Fire also enabled the hominids to cook the meat that they acquired by hunting—and by stealing from the kills of other predators—making safe and easily digestible that which was previously unsafe and difficult to digest. And by adapting to a diet of cooked food, the emerging humans found a way to vastly reduce the amount of time and energy that was required to chew and digest their daily food intake. In the process, they made it possible to support their increasingly larger and more "expensive" brains. The technology of fire was, in fact, the singular achievement by which the emerging humans crossed—decisively and irrevocably—the immense gulf that has separated humanity from the rest of the animal kingdom ever since.

In the next chapter, we will examine the true antiquity of clothing and shelter. And we will consider the evidence that prehistoric humans have been keeping warm and dry in their artificial caves and body coverings, not for thousands of years—as is widely assumed—but rather for *hundreds of thousands* of years. For it was the technologies of clothing and shelter that made it possible for tropical hominids to take up residence in the temperate zones of the northern latitudes, to survive their bitter winters, and ultimately to populate virtually every terrestrial environment on the planet Earth.

CHAPTER 4

The Technologies of Clothing and Shelter

Hats, Huts, Togas, and Tents

"Your house is your larger body."

—Kahlil Gibran, *The Prophet*

Of all the artifacts that archeologists have unearthed from the campsites of prehistoric hominids, virtually no physical remains of dwellings or clothing have been found that are older than the time when anatomically modern humans appeared in Europe less than fifty thousand years ago. This lack of evidence has led many scientists to conclude that clothing and shelter appeared relatively recently in the history of the hominids.

However, the technology of fire enabled the emerging humans to live in caves as long as 1.75 million years ago, and there is plentiful evidence that other populations of emerging humans from that time period lived in places where no suitable caves are to be found. Are we therefore to conclude that numerous emerging humans lived naked and exposed for the better part of two million years with neither a shred of clothing nor a roof over their heads to protect them from the elements?

Missing from the "Stone Age"

To answer this question, we should begin by noting that all the species of hominids that existed before the invention of agriculture and metallurgy were nomadic hunter-gatherers who made most of the objects they used in daily life not from stone but rather from the plants and animals they found in their natural habitats. None of these organic materials can survive for very long once they are buried in the earth—least of all in the warm, moist earth of the tropical environments that most populations of prehistoric hominids inhabited.

The remains of dead plants and animals are the natural food of countless species of insects, mollusks, worms, fungi, and bacteria. In tropical climates, artifacts made of these materials are typically consumed by such organisms in a matter of days or weeks, and after the passage of a few years they have all completely disappeared. Even in temperate climates, such materials are unlikely to survive for more than a few hundred years at most.[1] It should therefore come as no surprise that, after the passage of hundreds of thousands of years, hardly a trace of prehistoric clothing or dwellings can be found in the remains of the most ancient archeological sites.

The hunting and gathering peoples that survived into modern times and were studied by anthropologists typically moved from place to place with the changing of the seasons and with the constantly shifting availability of plant and animal foods. For the most part, the dwellings that these nomadic people built for themselves were erected in temporary camps and abandoned after a few weeks or months. For this reason, all of the material possessions these people used in their daily lives had to be both small enough and light enough to be carried from place to place. And almost everything they made was fashioned from the perishable organic materials of plants and animals.

Spears, arrows, clubs, digging sticks, walking sticks, and the frameworks for huts and tents were made from the woody trunks and branches of trees and bushes. The walls and roofs of huts, bedding materials, and windbreaks were made from leafy vegetation, including palm fronds, grasses, and reeds. Cords, ropes, nets, bags, and hammocks were woven from the fibers found in vines, roots, and bark. Containers for food, water, tools, weapons, amulets, medicines, and pigments were made from hollowed-out gourds and seed pods as well as from the hollow stems of reeds, marsh grasses, and bamboo. Hats, robes, leggings, aprons, boots, sandals, sewing thread, cord, and tent coverings were made from the hides and sinews of animals. And drinking cups, jewelry, musical instruments, and small containers of all kinds were made from the horns, bones, feathers, and claws of birds and animals.

Like the contemporary hunter-gatherers studied by anthropologists, however, prehistoric hominids did use one type of material in their daily lives—stone—that could not decay or disappear from prehistoric archeological sites.[2] And for this reason, the survival of countless thousands of stone objects from prehistoric times has created an exaggerated impression of the importance of stone in the technologies used by prehistoric people. The numerous stone tools that were found by early archeologists in the remains of the prehistoric settlements quickly led to the general use of the phrase "the Stone Age" to describe the entire period of human history before the invention of metallurgy.

But the Stone Age was not a distinct period or age at all, since it includes the entire evolutionary history of the hominids, from their earliest appearance several million years ago to the fully modern humans of today's world. This immense period of time encompasses many of the technologies described in this book, including the domestication of fire, the invention of clothing and dwellings, the development of symbolism, the adoption of agriculture, and the beginnings of urban civilization. In fact, the Stone Age technically began to end only when the techniques of metallurgy were first developed a few thousand years ago.

Considering that the earliest stone tools date from approximately three million years ago, the Stone Age would thus comprise about 99.8 percent of all human history—and all the remaining "ages" together would amount to only one-fifth of 1 percent of the hominids' time on Earth. In fact, the advanced civilizations of the New World—including the Aztecs, Maya, and Inca, with their great urban centers, complex religions, advanced hieroglyphic writing, organized bureaucracies, and remarkable achievements in mathematics and astronomy—would probably qualify as stone-age societies and cultures, simply because these complex and sophisticated peoples neither made nor used metal tools and weapons.[3]

The concept of a "Stone Age" also created the false impression that stone tools were the most common artifacts used by prehistoric people in their daily lives. But although stone tools were used to shape, carve, and sharpen other materials, most of the artifacts used in prehistoric times were made from perishable substances that quickly disappeared from the archeological record. However, other kinds of evidence does exist, in the form of prehistoric migration patterns, the wear patterns on stone tools, and even the genetic history of the human body louse. But before considering these other forms of evidence, we should note how common it is for animals with far less brainpower than ourselves to create their own living spaces.

Living Spaces That Animals Build

Primatologists who have studied the behavior of wild populations of great apes have observed that all of these species use natural materials to construct sleeping nests for themselves, and in some cases they even cover their heads and bodies with vegetation to protect themselves from the elements. And the great apes were by no means the first or the only animals to construct their own dwellings. In fact, the creation of dwelling places by animals is an extremely ancient practice in the history of animal life on Earth.

Ants and termites—which first appeared on this planet tens of millions of years ago and have "brains" smaller than the head of a pin—build elaborate nests for their colonies of thousands of individuals by excavating tunnels and chambers in the earth and in decaying wood. Wasps and hornets build their nests from a special "paper" that they create by chewing wood fiber. And honeybees build their amazing geometrical hives from wax produced in tiny glands located under the scales that cover their abdomens.

Among the higher animals, nearly all of the ten thousand living species of birds build nests from the natural materials found in their environments—including twigs, branches, reeds, grasses, mud, and even their own feathers—which they use for sleeping, hatching their eggs, and rearing their young. And numerous species of rodents and other small animals—including mice, rats, hamsters, squirrels, rabbits, prairie dogs, gophers, and woodchucks—make nests for themselves in hollow trees or in underground burrows. Some of these burrows consist of complex networks of tunnels and chambers in which their inhabitants sleep, store food, eat, bear their offspring, and live out much of their adult lives.

Last but not least, the beaver, a rodent with a brain only slightly bigger than a walnut, routinely cuts down small trees and—by combining small logs and saplings with mud and stones—creates or enlarges ponds and wetlands to form its own special habitats of waist-deep water. Having created these artificial habitats, the beaver proceeds to construct its own dwellings in the form of elaborate lodges, complete with underwater entrances, cleverly concealed breathing-holes for ventilation, and dry, roomy chambers for eating, sleeping, giving birth, and raising their young (see Figure 4.1).

While it stands to reason that the ability to construct these animal shelters is genetically preprogrammed, their plentiful existence proves that an animal species does not require two-legged locomotion, the free use of grasping hands, or a large brain in order to build shelters from naturally occurring materials in its environment. In fact, the building of dwelling places by hominids may be largely based on an ancient genetic predisposition common both to the hominids and to the great apes and inherited from a common ancestor.

The Great Apes' Nests

In 1985, the biological anthropologist Colin P. Groves and the primatologist Jordi Sabater-Pi published a fascinating study of the nests that are constructed on a daily basis by all three living species of great apes—chimpanzees, gorillas, and orangutans—all of which are included with humans in the primate family *Hominidae*. The great apes' nests are used primarily at night for sleeping, but they are also occasionally used during the day for resting and eating. Groves and Sabater-Pi noted that all three species of great apes are nest-builders, that each of these species builds a new nest every day, that the construction of these ape nests requires considerable skill, and that the nest of each species expresses a predictable design that is repeated with great regularity by all adult individuals.[4]

The chimpanzees and gorillas of equatorial Africa and the orangutans of Southeast Asia live several thousand miles apart, and their ancestry diverged as much as fifteen million years ago, yet the nest-building habits of all three species share many common features. All of their nests are constructed by standing in one spot, bending down the branches of the leafy vegetation that surrounds them, and tramping down these leaves and branches with their hind legs, until a large, cup-shaped nest several feet in diameter has been constructed. These species may also break off other branches from nearby trees and bushes and add them to the pile, and in many cases they will finish the nest by lining it with softer vegetation to form a more comfortable bedding.

All of the great apes' nests are about the same size and shape: round or oval in form and between two and four feet in diameter. All three species construct new nests each day from plant materials near at hand and typically use them only once. Each adult member of the group constructs his or her own nest, and sleeping nests are shared only by mothers and their immature offspring and almost never by two or more adults.[5] And the nests constructed by the members of a single group of apes are typically built fairly close together within a single location, known as the "nest site." These similarities suggest that the great apes instinctively build nests for sleeping, just as birds instinctively build nests in which they lay their eggs and rear their young. But things are more complicated than that.

It turns out that only those apes that are born and raised in the wild—where they have watched their own kind building nests every night, and have slept in nests with their mothers for the first several years of their lives—appear able to build nests for themselves in captivity. Chimpanzees born and raised in captivity do not construct nests, even when they are placed in cages with wild-born chimps that construct nests every day. So while there is an instinctual predisposition for apes to build nests, it appears that in order for this predisposition to be expressed in adult behavior, it is necessary for these creatures to be exposed to certain learning experiences early in life. In short, nest-building among the great apes contains essential elements of both learning and inheritance.

Furthermore, two of the three species of great apes exhibit behaviors that seem to foreshadow the use of clothing by hominids. Chimpanzees use large leaves as "hats" to protect their heads during a tropical downpour, and orangutans often cover their bodies with leaves and branches at night—in some cases holding or balancing loose branches above their heads to form a kind of primitive roof. In fact, orangutans in captivity often cover their heads with empty sacks and may even cover their entire bodies with straw as they prepare for sleep.

Having made their case for the shared inheritance of nest-building among the great apes, Groves and Sabater-Pi went one step further. They pointed out that the ancestors of hominids and chimpanzees diverged from each other roughly seven million years ago, much more recently than the common hominid-chimpanzee ancestor diverged from the orangutans, which occurred roughly fifteen million years ago. Therefore, they argued, it is likely that the genetic predisposition to build living-spaces that is so evident among all of the great apes is also present in the DNA of the hominids, both modern and prehistoric.

"We must be perfectly clear," they wrote, "why we feel justified in searching for a common origin of human and great ape nesting/camping patterns. . . . If four species—orangutan, chimpanzee, gorilla, human—perform a certain activity whose motor components and end result are similar, parsimony suggests that their last common ancestor was itself doing something of the kind, and that its descendants have been so doing ever since. . . . Since the orangutan is less closely related to the gorilla and chimpanzee than are humans, this behaviour would have been present in the proto-human stock as well."[6] The authors then proceed to list the similarities between the living spaces constructed by the apes and those constructed by modern hunter-gatherers.

Each ape nest is used by a single individual, is round or oval in form, and averages about two or three feet in diameter. Each human dwelling is used by a single nuclear family, is round or oval in form, and averages about six to eight feet in diameter. (Dwellings with square sides were rarely constructed by hunters and gatherers. These types of dwellings did not appear until anatomically modern humans began to build permanent homes for themselves when they adopted the technology of agriculture roughly eleven thousand years ago.) Both the apes' nest sites and the humans' campsites typically contain between twenty and eighty individuals. Both are constructed within a round or oval area that averages between thirty and sixty feet in diameter. And within both the apes' nest sites and the hominids' campsites, the individual nests or dwellings are typically spaced about twelve feet apart.

The similarities between the great apes' nests and hominid's dwellings—and the fact of their common ancestry—suggests four probable conclusions. First, a genetic component of the hominid DNA may predispose all hominids to build shelters. Second, this genetic predisposition was probably inherited from prehistoric apes that were ancestral to both the hominids and the great apes. Third, if there is an essential element of learning in the apes' ability to build nests, it should not be surprising that a similar element of learning is also essential in the hominids' ability to build dwellings. And fourth, the practice of constructing dwelling-places may have always been a regular feature of hominid behavior and may have already been well-developed when the earliest hominids first evolved bipedal locomotion millions of years ago.

But here we again encounter the caution of paleontology, which tends to assume—lacking evidence to the contrary—that the absence of evidence is best viewed as evidence of absence. For in spite of the great apes' well-documented skills at nest-building—and the conclusion of primatologists that this behavior is universal among these species—many paleontologists continue to argue that hominids never constructed their own dwellings until anatomically modern humans appeared in Europe less than fifty thousand years ago.

For *Homo erectus*, with its relatively large and rapidly expanding brain, it would not have been a particularly difficult feat to build a primitive hut by weaving sticks and poles together and covering this framework with hides or with leaves and branches. Nor would it have been difficult for this resourceful creature to fashion animal skins into hats and cloaks to protect itself from the blistering sun and drenching rain of the tropics. While no material remains of such artifacts have been found in the oldest archeological sites, other types of evidence suggest that prehistoric hominids did in fact make and use dwellings and clothing long before anatomically modern humans appeared on the scene.

The Dwellings of Hominids

While few traces of hominid dwellings can be found from very ancient time periods, there are some tantalizing exceptions. One of the most interesting of these exceptions consists of certain formations of stones and other heavy objects that were found arranged into circles or ovals in some of the oldest open-air prehistoric habitation sites. Some paleontologists have interpreted these stone circles as the remains of foundations that prehistoric hominids created to anchor the walls of the dwellings they built from a framework of sticks, which they covered with hides or thatched with palm fronds, reeds, or grasses.

The oldest—and most controversial—of these stone circles is a group of large blocks of basalt (a volcanic stone) found by Louis Leakey at Olduvai Gorge in East Africa, in a deposit dated at 1.8 million years ago. The area inside this stone circle is almost entirely devoid of artifacts, while the area immediately outside the circle is littered with the remains of tools and the bones of animal prey. This suggests that the early hominids who inhabited this site tossed their trash outside of their dwelling while keeping the interior relatively free of debris—a pattern of behavior that has been frequently observed among contemporary hunters and gatherers.

More compelling evidence exists in the form of carefully documented stone circles that are associated with prehistoric sites in Europe. One of these sites is at Terra Amata, an open-air site near Nice in southern France, and the other is at Bilzingsleben, an open-air site in Central Germany. The age of both of these sites has been estimated at 380,000 years—a warm period in the global climate, when the more highly evolved populations of *Homo erectus* had established permanent residence in northern Europe.[7] In both Terra Amata and Bilzingsleben, groups of stones were found arranged in a characteristic circular formation, and in both cases the scientists who excavated these sites interpreted these stone circles as the foundations for small huts.

At Terra Amata, on the Mediterranean coast of France, the French archeologist Henry de Lumley found the remains of ashes in the center of each stone circle, indicating that the inhabitants had built fires inside their huts on a regular basis. De Lumley also identified several low walls of stones and beach pebbles that had been constructed on the northwest side of each fireplace, presumably to shelter the fires from the powerful "mistral" winds that frequently blow from the northwest in that region. De Lumley believed that the roofs of the Terra Amata huts had been supported by a kind of "tent pole" located at the center of the hut, and he theorized that each of the huts had been equipped with a hole in the roof to let out the smoke of the fire that burned inside them.

At Bilzingsleben, the German archeologists Dietrich and Ursula Mania have spent years excavating a huge open-air habitation site that has yielded the cracked and broken bones of numerous game animals, plentiful evidence of fire, and thousands of tools made of stone, bone, antlers, and ivory. The Manias found the outlines of three oval-shaped rings of stone and other heavy objects at Bilzingsleben that they believe were the foundations of dwellings. The openings of all these foundations are all oriented to the southeast, suggesting that they were built to face away from the prevailing northwest winds. Finally, in front of all three of these ovals, evidence of burned material indicates that the inhabitants of Bilzingsleben were regularly building fires at the entrances of these dwellings.

Much later, when the Neandertals began to settle in northern Europe and hunt the wooly mammoth, wooly rhinoceros, wild horses, and wild cattle during more recent ice ages, they used mammoth bones to construct walls for their fortress-like dwellings. The oldest of these sites is roughly forty-five thousand years old and located at the site of Moldova I in Ukraine, where the jaws, skulls, and large limb bones of at least fifteen mammoth skeletons were fitted together and arranged in a large circle. The Moldova I site closely resembles the remains of other mammoth bone settlements built by fully modern humans until about fifteen thousand years ago. Nothing has remained of the walls and roofs of this structure, which probably consisted of a framework of sticks and covered with the hides of the megafauna—the mammoths, rhinoceroses, bison, and wild cattle—that were hunted by its extraordinary inhabitants.

But when the scientists who excavated the sites at Terra Amata, Bilzingsleben, and Moldova concluded that these stone and mammoth bone circles were the remains of dwellings, their interpretations were promptly challenged. In spite of the advanced knowledge of woodworking demonstrated by emerging humans as early as four hundred thousand years ago, the absence of walls and roofs has led many paleontologists to propose alternative explanations for their existence. These include theories that the circles might have been formed by natural movements of the soil, by the clearing away of unwanted debris, or even that they were constructed by prehistoric people as part of some hypothetical religious ritual.

But when one considers that even the remains of dwellings dating from relatively recent historical times almost never contain evidence of the wooden frames, thatch roofing, or hide coverings that were once part of their original construction, there is no reason to expect the remains of dwellings hundreds of times older to contain such evidence. This may be another case in which absence of evidence is incorrectly interpreted as evidence of absence.[8]

There is, however, substantial evidence that the processing of animal hides goes far back into hominid prehistory. This has been confirmed by the microscopic examination of ancient stone tools, especially the large, flat flake tools with a single cutting edge called scrapers. When you look at the cutting edge of such tools under the microscope, a characteristic wear pattern can be seen, and this "microwear" indicates that many of these scrapers were used for removing the meat and fat from the underside of animal hides. In fact, microwear evidence that stone tools were used for scraping hides goes back at least 780,000 years, and this behavior continued without interruption in hunting and gathering cultures until metal tools became available in modern times.

The removal of the flesh from hides is a necessary first step in the process of turning animal skins into leather. Virtually all of the leather cured by hunters and gatherers was used for one of three purposes: first, for the roofs and coverings of tents and huts; second, for the bedding used inside of the hominid dwellings; and third, for the robes, cloaks, hats, shoes, mittens, and all the other garments that hominids must have been wearing since the time they first settled in the far northern latitudes. Evidence that raw animal skins were processed into leather is, in itself, indirect evidence that prehistoric people were constructing dwellings and fashioning garments hundreds of thousands of years ago.

No Longer Naked

There is little precedent for the wearing of clothing by any animal species except hominids. Along with the controlled use of fire, the invention of clothing by hominids ranks as one of the singular achievements that distinguishes us from all other forms of animal life. And the use of leather and other natural materials to clothe the hominid body goes back hundreds of thousands of years to the time of the emerging humans.

The strongest evidence that emerging humans made and used clothing rests upon the established fact that they long ago began to settle in geographical areas so far to the north that survival in those regions without clothing would have been essentially impossible. For although populations of *Homo erectus* had already settled in the subtropical regions of Europe and Asia nearly two million years ago, they did not expand into the northern latitudes until half a million years ago, almost 1.5 million years later.

Why would *Homo erectus* have migrated eastward across thousands of miles of grasslands to settle in the subtropical environments of Java and southern China but fail to have moved a few hundred miles north into the game-rich environments of northern Europe and Asia? And why would more than a million years have passed between the time *Homo erectus* left its African homeland and the time of its eventual expansion into the northern lands? The most likely answer is that the winters in the northern latitudes were simply too cold for a tropical animal that had become naked as part of its adaptation of living with fire. In fact, it has been calculated that no hominid would have been able to survive without clothing in climates where winter temperatures regularly fell below 53 degrees Fahrenheit.[9]

Although *Homo erectus* may have used fire to protect its kind from predators, to lengthen the time it could spend awake, and to cook its food, fire would have only been effective in protecting against the cold when the emerging humans were safe in their camps at night. During the day, when they were hunting game and gathering fruits and vegetables, setting fires out in the open would have been not only impractical but also of little use in keeping warm. Therefore, in order to pursue a nomadic hunting and gathering way of life in a cold climate, hominids would have required clothing: hats for their heads, robes for their bodies, and perhaps shoes or boots for their feet and mittens for their hands. Eventually, *Homo erectus* must have learned to fashion most or all of these things from the hides of animals.

You will recall that by four hundred thousand years ago, as the Schöningen spears have clearly shown, *Homo erectus* populations living in Germany were making sophisticated wooden objects that involved considerable skill, planning, and a complex multistage manufacturing process. There is no reason to suppose that the makers of such artifacts would not have been equally capable of fashioning hats, robes, mittens, and shoes for themselves. So the appearance of *Homo erectus* remains in northern Europe and Asia beginning about five hundred thousand years ago is more than likely the point in human history when the technology of dwellings and clothing began to take its place alongside the other key technologies of spears, digging sticks, and fire. And this was long before modern humans appeared on the scene.

Anthropologists who have studied the energy requirements of the Neandertals, who appeared after the emerging humans and who lived in northern Europe during the most recent ice ages, have concluded that although the heavy, muscular Neandertals were "cold adapted," they enjoyed only a modest advantage over fully modern humans in surviving cold weather. In fact, it has been calculated that the Neandertals could not have survived the winters of the regions they inhabited in northern Europe without covering between 50 and 90 percent of their bodies with garments of some kind.

To determine the probability that the Neandertals of northern Europe made and used clothing, the anthropologist Nathan Wales conducted an exhaustive survey of the published accounts of garments made and worn in 245 hunting and gathering societies studied by anthropologists over the past hundred years or so. Wales found that the amount of the body that was covered with clothing by hunters and gatherers corresponded closely to the winter climates in the environments these societies inhabited.

Wales noted that the Plains Cree men of Alberta and Saskatchewan—a northern Great Plains area where the winters are long and cold—typically wore leather breechcloths, hide leggings, moccasins with grass insulation, a hat partially covering the head, and a buffalo robe over the torso and one shoulder (see Figure 4.2). Wales calculated that the traditional clothing of the Cree would have covered 77.5 percent of the body, and other cold-weather cultures showed similar results. Tlingit women of the Pacific Northwest wore tailored shirts, petticoats made of cedar-bark, mountain-sheep wool blankets, and waterproof hats made from roots, covering 77 percent of the body—even though in the Northwest Coast, where winter temperatures rarely dip significantly below freezing, Tlingit women typically went without covering their lower legs and feet.

Based on his calculations of the amount of the human body that was covered by clothing in known hunting and gathering societies, Wales concluded that most of the Neandertal groups living in Europe would have had to wear clothing in order to survive. In fact, many of the archeological sites containing stone tools made and used by Neandertals are located in regions that would have required their inhabitants to cover between 70 and 90 percent of their bodies during the winter. And many of these sites were in regions in which it was likely that the Neandertals would have also covered their hands and feet during the coldest months (see Figure 4.3).

Wales theorized that Neandertal clothing had been fashioned from the skins of large animals such as wild cattle, horses, and mammoths. At the same time, he concluded that the absence of any evidence of sewing implements such as bone needles in the Neandertal sites suggests that, unlike the modern hunters and gatherers such as the Cree and the Tlingit, the Neandertals wore whole animal skins in the manner of a toga, fastened to their bodies with thongs made of leather or sinew.

Later, with the appearance in Europe of anatomically modern humans, clothing began to be tailored, as it was among contemporary hunter-gatherers. Tailored clothing must be sewed with needle and thread and fitted to the limbs and torso in ways that provide maximum protection against the cold. The remains of bone needles and other evidence from sites inhabited by Cro-Magnons and other anatomically modern humans suggest that tailored clothing originated with our own kind, *Homo sapiens sapiens.* But the natural history of a tiny parasite that infests the human body provides further evidence that the earliest, non-tailored clothing originated not during the time of the Cro-Magnons or the Neandertals tens of thousands of years ago but rather during the time of the emerging humans, hundreds of thousands of years ago.

A Tale of Three Lice

The chimpanzee and the gorilla are each infested by only a single species of louse, but the human body is home to three different types of lice. The human head louse lives in the hair of the scalp, the human body louse lives in human clothing, and the human pubic louse lives in pubic hair and is responsible for the sexually transmitted condition commonly known as "crabs" (see Figure 4.4). During the 1990s, a number of scientists conducted a series of sophisticated genetic analyses of these different types of lice in an effort to reconstruct their evolutionary history.

The human head louse, *Pediculus humanus capitis,* and the human body louse, *Pediculus humanus humanus,* are closely related subspecies. Both of them share a common ancestry with the chimpanzee louse, *Pediculus schaeffi*, from which they appear to have split off roughly six million years ago. This makes perfect sense, since it corresponds to the point in time when hominids and chimpanzees are believed to have split off from their mutual common ancestor (a prehistoric ape that has yet to be identified).

On the other hand, the human pubic louse, *Pthirus pubis,* shares its ancestry with the gorilla louse, *Pthirus gorillae,* from which it appears to have split off roughly three million years ago. This date is most curious, since the hominids split off from their mutual common ancestor with the gorilla at least seven million years ago. While it is extremely unlikely that some of our hominid ancestors were having sex with gorillas three million years ago, the ancestral pubic lice in question probably migrated from gorillas to hominids when some of the early hominids took to sleeping in abandoned gorilla nests.

That three-million-year-old date may, however, have an even more interesting significance. Since the pubic louse lives not on the head or the body but rather only in the hair of the pubic region, this may be evidence that hominids lost their body hair as long as three million years ago. If true, this might push the technology of fire even farther back into the past than the 1.75 million years we have estimated in this book.

In the quest to identify the date for the adoption of clothing, the most important event in the history of these three lice is the moment in time when the body louse split off from the head louse. The logic behind this assumption is that the body louse punctures the skin to feed on blood but does not actually live on the skin itself. Instead, the body louse lives in the clothing we wear. Therefore, it could not have evolved until hominids were wearing clothing on a regular basis.

Estimates for when the body louse split off from the head louse vary from one study to another, but all the dates range between roughly 80,000 and 170,000 years ago. You will note that this date is much more recent than the time period 500,000 years ago when populations of *Homo erectus* began expanding into the northern latitudes, but it corresponds exactly to the time period when the earliest evidence of anatomically modern humans was beginning to appear. Two possible explanations for these divergent dates have been suggested.

One possibility is that the particular strains of lice that are found today in living human populations were more modern strains that had developed in the early populations of anatomically modern humans, while the older strains of lice that were living in the heads and bodies of *Homo erectus* and the other emerging humans simply died out when the emerging humans themselves became extinct. The other possible explanation rests on the difference between the *non-tailored* clothing that was fashioned by emerging humans and Neandertals versus the *tailored* clothing made more recently by anatomically modern humans.

When hominids first began to wear clothing—which must have occurred approximately half a million years ago, when they were settling in environments too far north and too cold for survival without clothing—it is likely that their garments originally consisted of loose-fitting robes or cloaks of leather that were draped across the body and loosely fastened with thongs of leather or sinew. Such garments did not conform to the shape of the limbs and torso and would not have clung closely to the skin.

For this reason, non-tailored clothing might not have afforded a sufficiently sheltering environment to entice the head lice to abandon the scalp and begin living and reproducing on the relatively hairless hominid body. But the anatomically modern humans that appeared in northern Europe roughly fifty thousand years ago made fully tailored clothing that fitted closely to their bodies, like the clothing of the Cree and Tlingit people studied by contemporary anthropologists. We know this from the evidence of weaving and sewing that has been found in the prehistoric archeological sites of anatomically modern humans.[10]

It is therefore likely that the human body louse began to exploit its very specialized ecological niche in human clothing only when modern humans began to wear the close-fitting garments that provided a safe haven between their newly tailored clothing and their warm, bare skin. If so, this would explain why the human body louse seems to have appeared on the scene much more recently than the technology of clothing. But how the human pubic louse migrated from gorillas to hominids is a story that still remains to be told.

Protecting the "Premature" Human Infant and Its Massive Brain

The technology of dwellings and clothing gave the hominids a degree of freedom and flexibility in dealing with the environment enjoyed by no other animal species. They were no longer limited to inhabiting the caves that occurred naturally in their environments. Instead, hominids became capable of building their own artificial caves, allowing them to settle wherever food was abundant, sheltered from the wind and rain by their crude huts and protected from predators by their nightly campfires.

As we saw in the previous chapter, the ability to control fire also enabled the hominids to replace the raw food diets of their early hominid ancestors with a variety of more easily digested cooked foods. Cooking allowed the hominids to satisfy their energy requirements with smaller digestive organs and consequently to support increasingly larger brains. But this is only half of the story. As the hominid brain expanded, another problem arose—one that was completely different from the biochemical problem of supplying this increasingly "expensive" tissue with the energy it needed to function. This new problem involved the sheer mechanical necessity of finding a way for the hominid infant's rapidly expanding head to pass safely through a pelvic opening that was essentially fixed in size.

During the evolution of the hominids from the earliest forms of *Homo erectus* to the fully modern forms of *Homo sapiens*, the brain more than doubled in size, yet throughout this period the size of the human body did not increase at all. And, of course, as the rapidly evolving hominid brain increased in size, the head of the hominid infant also correspondingly increased in size. But the evolution of bipedal locomotion had required the bones of the pelvis to become shorter and thicker than the pelvis of an ape. The ability to stand, walk, and run upright made it necessary to transform the ape's flexible doughnut shaped pelvis into the hominid's solid ring of bone, roughly circular in shape, that was strong and rigid enough to support the weight of the hominid's entire upper body.

Bipedal locomotion also placed certain limitations on the width of the pelvis. If the pelvic girdle became too wide, the legs would be set too far apart, making two-legged locomotion increasingly clumsy and inefficient. Among the quadrupedal apes and monkeys, the elongated pelvic bones form a birth canal that is oval shaped. During birth, the infant primate turns its head sideways as it passes through the birth canal, allowing for a relatively quick and easy delivery. But the hominids' short, rigid pelvic bones—and the circular shape of its pelvic girdle—meant that the hominid birth canal was not capable of accommodating a larger infant's head during childbirth.

Thus, as the tremendous expansion in the size of the hominid brain took place during the past million years, it created increasing difficulties in childbirth for the emerging humans, and this problem became even more acute as the emerging humans evolved into modern humans with truly massive brains. If the birth canal remained fixed in size while the infant's head and the brain within it was growing progressively larger, how did the evolving hominids continue giving birth to babies with increasingly larger heads?

The evolutionary solution to this problem turned out to be that as the size of the brain increased, hominid babies began to be born before their brains were fully developed. In fact, if you apply the normal rules of mammalian fetal development to humans, the "normal" gestation period for humans would be at least twelve months, not nine. But at twelve months the average infant brain would simply be too large to pass through the average human birth canal, making human birth mechanically impossible.

In other mammals, including all other primates, the brain has progressed to a reasonably advanced stage of development by the time of birth, and after birth the growth of the brain slows down considerably. But the human brain at nine months is poorly developed compared with the brains of other mammals (including other primates), and as a result the brain of the human infant continues to grow rapidly during the first two years of life.

It is, in fact, not difficult to imagine how this hominid "solution" evolved. As the infant brain increased in size, only those hominid mothers whose babies who were born "prematurely" would have been likely to survive the experience, raise their offspring to maturity, and pass on their genes to the next generation. Even with this evolutionary compromise, it typically requires several hours of difficult labor before the human infant begins to emerge during birth, and only after the agonizingly slow passage of the infant's head through the birth canal does the remainder of its body emerge.

Anyone who has ever seen the birth of a cat, dog, pig, cow, or horse knows that other female mammals give birth to their babies in a matter of minutes. But giving birth to a human infant is the longest, most difficult, and most dangerous process of giving birth among any mammalian species. In fact, before the adoption of modern medical techniques at the beginning of the twentieth century, roughly one out of every hundred human births resulted in the death of the mother.[11]

What does it mean for the human infant to be born "prematurely"? Consider that the offspring of grazing mammals such as cattle, sheep, horses, giraffes, and elephants are able to stand and walk within hours of birth, while the offspring of apes and monkeys are alert to their surroundings in the first few days of life and are capable of clinging successfully to their mothers' fur within a few hours of birth. By contrast, human infants are incapable of even crawling on their hands and knees until they are several months old, and they typically cannot walk at all until they are nearly a year old.

It is no exaggeration to say that the hominid infant is born with a seriously underdeveloped brain. It is unable to grasp its mother's hair with its stubby little toes, and in any case it is born to a mother who no longer has any body hair to cling to. Hairless, defenseless, unable to move about on its own, and only dimly aware of its surroundings, the "premature" human infant is by far the most helpless and the most vulnerable of all primate offspring.

This means that as the hominid brain grew larger and the hominid newborn grew progressively more underdeveloped and vulnerable, hominid mothers would have had an increasingly acute need for clothing and shelter. Even in the tropics, where keeping a baby warm is not a pressing problem, for a hominid female to carry each of her infants in her arms throughout the first year or two of its life while spending most of the daylight hours gathering food—and probably caring for older children at the same time—would have been an impossible burden.

The Bushman mother carries her baby in a large leather cloak called a *kaross,* and a similar strategy may well have been adopted by *Homo erectus* as well. And when the emerging humans migrated out of Africa and began to settle in the more temperate regions of Eurasia, their babies would have had an increasingly difficult time surviving if they had remained naked and unprotected from the elements. But wrapped in furs or in a large cloak during the day, and securely ensconced in a warm hut at night, the increasingly premature and increasingly helpless hominid infants would have been able to survive the first vulnerable year or two of their lives. Thus it is likely that, as they evolved and their brains increased in size, the emerging humans had begun to construct dwellings and fashion clothing for themselves even before they moved north into the lands where the winters were long and cold.

In the final analysis, the technology of dwellings and clothing, as well as the technology of fire, freed the hominid brain from its natural limitations, and allowed it to grow until it was roughly triple the size of the brains possessed by other animals of comparable size. Without the technologies of fire, dwellings, and clothing, the size of the hominid brain would have been unable to expand much beyond the 650 cc size of the brain of *Homo ergaster,* the likely ancestor of *Homo erectus*—and humanity would have remained, to this day, little more than a very intelligent, meat-eating, tool-making, weapons-carrying, two-legged ape.

Hominids of the Frozen North

For hundreds of thousands of years, as the ice ages came and went, *Homo erectus* moved north during the warmer geological periods and retreated southward as the polar ice advanced during the colder geological periods. For this reason, the remains of *Homo erectus* have been found in northern Europe and Asia only during these relatively warm "interglacial" periods. In fact, for more than one million years, no evidence of continuous occupation by *Homo erectus* can be found north of the fortieth parallel (an east-west line that runs from southern Spain through the southern tip of Italy, across Greece and Turkey, and through Central Asia to the Korean Peninsula).

Nevertheless, as they learned to fashion crude garments that kept them warm and protected their naked bodies from the elements, populations of *Homo erectus* and other emerging humans gradually began to settle in the game-rich habitats of the far north. Beginning roughly half a million years ago, the remains of *Homo erectus* begin to appear in regions above the fortieth parallel, from the British Isles to northern China. Armed with fire, simple huts, and crude garments that kept them warm and protected their naked bodies from the elements, *Homo erectus* settled in habitats where the megafauna of the ice ages roamed in abundance.

Eventually, however, the *Homo erectus* populations of Europe and Asia were replaced by modern humans of our own species, *Homo sapiens,* who probably originated in Africa and spread northward through the Middle East into Europe. The first of these modern humans—which had brains as large as our own brains and successfully hunted the megafauna—were the Neandertals, that strange and amazing subspecies known to science as *Homo sapiens neanderthalensis.*

The Neandertals were thick-bodied, large-boned hominids with large and powerful muscles. The bones of the skull above their eyes had heavy brow ridges that were reminiscent of more ancient and primitive hominids. But within their long, bullet-shaped skulls, the Neandertals had brains fully as large as the brains of modern humans alive today. As hunters of large game, the Neandertals have no equal in the history of the hominids, and a biochemical analysis of Neandertal bones reveals that they lived on a diet composed almost entirely of meat. One study showed that 58 percent of the food intake of the Neandertals consisted of wild cattle—with horse, rhinoceros, reindeer, and mammoth meat making up the rest of their predominately carnivorous diet.[12]

After the arrival of the Neandertals in northern Europe, the megafauna began to dwindle in numbers, and after the appearance of the anatomically modern Cro-Magnon people roughly fifty thousand years ago, the ice age megafauna became extinct. The Eurasian climate grew significantly warmer as the most recent ice age came to an end, and the changing climate may have contributed to the extinction of some of these species. But similar extinctions of megafauna also occurred shortly after the arrival of modern humans in Australia, Tasmania, Japan, North America, and South America, where changes in climate were far less significant.

It therefore seems likely that the megafauna were hunted to extinction by modern humans. In fact, the extinction of the megafauna at the end of the last ice age marks the moment in our evolutionary history during which our species not only gained mastery over the natural environment but also began to extinguish other forms of life. This is, as we will see, a process that has accelerated, millennium by millennium, ever since.

The technology of clothing and shelter transformed humanity by providing a kind of freedom enjoyed by no other animal species. If there were no caves to provide shelter from the drenching rain of the tropics or the burning sun of the deserts, hominids could build shelters from the vegetation in their environments and the hides of the game they hunted. If the winters were cold in the frozen north, they could cover their bodies with furs and hides and huddle close to the warmth of their campfires. If the summers were hot, they could throw off their clothes and be cooled by the moisture of their plentiful sweat glands.

Armed with advanced weaponry, skilled in the control of fire, and capable of fashioning a variety of dwellings and clothing that allowed them to live in almost any habitat, anatomically modern humans with their massive brains and unparalleled intelligence soon spread throughout all the world's continents and populated almost every conceivable earthly habitat. As a result, the earth's hominid population, which for millions of years had been largely stable at less than one million individuals, began to increase dramatically. It has been estimated that by fifteen thousand years ago, when anatomically modern humans had become the only living hominid species, the earth's hominid population had already grown to several million individuals.

In the next chapter, we will see how the massive brains of the fully modern humans set in motion—through the widespread adoption of symbolic communication—a transformation in the way humans related to each other and the way human society was organized. For it was the proliferation of symbols—in language, art, design, and bodily ornamentation—that liberated our species from the genetic conformity of its animal origins. Symbols could be invented at will, easily adopted by others, and passed on to future generations through learning and tradition. Symbolic communication freed us to invent the complex cultures and powerful ethnic identities with which every human being came to identify and owe allegiance.

By replacing the slow process of biological evolution with the fast process of cultural evolution, the technology of symbolic communication allowed the anatomically modern humans to respond quickly and easily to the changing climates of the past fifty thousand years. And the unique combination of symbolic communication with the fission-fusion society enabled numerous small bands to fuse together into groups consisting of hundreds or even thousands of individuals, enabling the anatomically modern humans to form the distinct ethnic groups that have persisted to this day. By joining forces in hunting and warfare, these large tribal groups easily outcompeted the small groups of kinfolk that typified the social life of the Neandertals.

From this point forward, as will become clear in the chapters to follow, succeeding key technologies gave birth to increasingly larger forms of human society. This process began with the formation of the ethnic identities of hunting and gathering tribes and culminated with the immense industrial nation-states that now claim dominion over all humanity.

CHAPTER 5

The Technology of Symbolic Communication

Music, Art, Language, and Ethnicity

"There has never been a hunter-gatherer society . . . created by Homo sapiens that did not . . . consider itself as living in an intensely symbolic realm."

—Brian Fagan, *Cro-Magnon*

The use of pictures, designs, words, and music to communicate thoughts and ideas is surely one of the most unique of all human behaviors. While many animal species can communicate with an inherited set of vocal sounds and body language, only humans are free to invent countless thousands of visual and vocal symbols for the purpose of communication. And only humans are able to transmit these invented symbols to their offspring—and to other members of the group—entirely through the processes of teaching, learning, and imitation.

Unlike some of the other extremely ancient technologies that are unique to the hominids and that we have explored in the previous chapters, it is virtually impossible to determine the true antiquity of the use of verbal and visual symbols to communicate thoughts and ideas. This is mainly because, unlike the evolution of bipedalism or the loss of canine teeth, all of the evidence that suggests the use of language—such as symbolic drawings—is of comparatively recent origin.

There is simply no evidence that the early hominids communicated with what we would consider a true spoken language, and there is at best only scant evidence that the emerging humans were capable of this uniquely human behavior.

Although it seems reasonable to assume that modern forms of human language did not suddenly emerge full-blown out of nowhere fifty thousand years ago, the notion that more primitive forms of language were used by early hominids or emerging humans is pure speculation at this point. It was not until the appearance of anatomically modern humans that plentiful evidence of symbolic communication—in the form of drawings in Paleolithic cave sites—began to appear in the paleontological record. But is it accurate to call symbolic communication a technology?

Tools for Thought

In modern speech, we generally use the word "technology" to describe complex machinery such as spacecraft and electronic devices as well as complex processes such as computer networks and automation systems. But in this book I have used the word "technology" in its widest and most inclusive sense, meaning *the deliberate modification of any natural object or substance with forethought to achieve a specific end or serve a specific purpose.* In this larger sense, when a chimpanzee breaks off a twig for the purpose of using it to fish termites out of a nest, it is using technology, just as the early hominids were using technology when they removed a branch from a tree or bush and sharpened it to make a spear. Yet even in these examples, the concept of "technology" refers strictly to the fabrication and use of material things.

But if we define technology as "the modification of any natural substance with forethought to achieve a specific end," we must include as examples of technology the use of pigments to draw designs and pictures on the walls of caves and the use of stone tools to carve designs onto the surfaces of bones—especially when such behaviors exist as shared cultural traditions within a social group. While such technologies may not be used for any material purpose—like hunting game, making clothing, or building dwellings—they are definitely used with forethought to achieve a specific purpose: namely, the communication of human thoughts and ideas.

If we stretch the meaning of "technology" a bit further, the deliberate modification of the human voice can also be viewed as a form of technology, because the human voice is a natural phenomenon that we have deliberately modified to produce the specific sounds that are the symbolic representations of human thoughts. The same can be said for the evolution of music, dancing, and singing, primitive forms of which may have already been present, in a rudimentary form, among the early hominids and emerging humans.

In short, all forms of learned and culturally transmitted symbolism are actually tools for thought, created with forethought and deliberately used for the purpose of communicating the shared knowledge of a specific human society to any of its living members. This is true whether these tools are used to create visual symbols—such as drawings, designs, icons, or written words—or used to create auditory symbols, such as spoken words, songs, or music. In short, it was the technology of symbolism that liberated humans from the bounds of preprogrammed communication.

The Flowering of Symbolic Communication

We should note that animals as primitive as social insects have evolved ingenious methods for communicating information to each other. Ants communicate the presence of food and danger by releasing hormones from their bodies that are immediately detected by other ants. Honey bees communicate the location of flowers blooming in their environment by engaging in special waggle dances on the walls of their hives that describe these sources of nectar and pollen and their precise direction and distance from the hive. And all kinds of animals—from insects to apes—produce a tremendous variety of sounds to communicate information to other members of their species.

Most warm-blooded animals use their lungs and vocal apparatus to produce certain sounds that communicate specific messages—such as warning, territoriality, courtship, danger, and distress. And the use of vocal sounds to communicate specific messages is especially common among species that live in trees, such as birds and primates. Many species of birds—including parrots, mockingbirds, crows, and ravens—are capable of making a wide variety of different sounds, each of which is designed to communicate a specific kind of message. The same is true of most species of apes and monkeys. But most of these vocal communications are instinctual in nature and involve little or no learning.

While there is no evidence that the early hominids used any form of symbolic communication, there is some evidence that emerging humans were already beginning to experiment with drawings and designs. In at least two of the *Homo erectus* sites dating from the last half million years, paleontologists have found some simple designs on bone and shell that can be interpreted as the use of symbols. We will examine this evidence later in this chapter.

Even the Neandertals, who were modern humans with full-sized human brains, left behind only a small number of artifacts suggesting that they were just beginning to understand and experiment with objects of symbolic significance. The evidence of symbolic behavior among Neandertals consists mainly of shells with holes drilled in them and painted with natural pigments, probably to be worn as necklaces.

But when anatomically modern humans appeared in Europe between roughly fifty thousand and forty thousand years ago, the lives of prehistoric humans were suddenly rich with symbolism. Human figurines were carved out of stone or ivory and molded out of clay. Objects made for personal adornment were made from shells and animal teeth that were drilled with holes, colored with pigments, and worn as beads on bracelets and necklaces. Utilitarian objects such as spear throwers[1] were decorated with elaborate designs and carved into the likeness of animals. And the walls of numerous caves in Europe were richly decorated with a profusion of hand prints, hundreds of amazingly realistic paintings of game animals, and thousands of carved or painted designs in the form of symbols or icons, called petroglyphs.[2]

The sudden profusion of symbolism also provides evidence that for the very first time, hominids were beginning to develop distinct cultures and ethnicities. The figurines, decorations, petroglyphs, and cave paintings of this era tend to vary noticeably in style and form from one region to another and from one time period to another. These variations have made it possible to identify specific cultures in time and space. For the first time, it becomes possible to identify distinct tribal and ethnic identities in the remains of prehistoric human groups.

Finally, this was doubtless the time in prehistory when language and music—the symbolic use of sounds—became fully developed. The remains of prehistoric flutes made of bone and ivory provide concrete evidence that anatomically modern humans were making music—and probably singing and dancing as well—tens of thousands of years ago. But language, being immaterial, leaves behind no physical evidence for the prehistorian to identify, and there is no way to know precisely when hominids began to speak. Some scientists believe that early forms of language were spoken by emerging humans and possibly even by the early hominids, but this issue remains far from settled.

It is important to note that while the *capacity* for learning and creating symbols resides in the DNA of anatomically modern humans, the *form and meaning* of the symbols invented by modern humans is entirely cultural and has no basis in biology. All of the meanings that are contained in symbolic communications, whether visual or auditory, must be learned in childhood, remembered in adulthood, and passed down through the generations as the shared cultural knowledge of the social group. If we could hear prehistoric people laugh, we would know that they were happy. If we could hear them cry, we would know that they were sad. But if we could hear them speak, we would have no idea what they were saying—just as, when we look at the petroglyphs that they carved and painted on the walls of their caves, we literally have no idea what these carefully rendered designs were meant to signify.

Toward the end of this chapter, we will explore the process by which all of humanity has become divided into many distinctly different ethnic groups, a process that began when anatomically modern humans adopted symbolic communication as an integral part of their daily lives. But the point here is that because they were invented rather than inherited, these culturally determined ways of life could be re-invented, modified, and changed as the climate and environment of prehistoric times alternated—sometimes quite rapidly—as the ice ages came and went, and the earth's climate changed over and over again.

During the long period of prehistoric time when the hominids' adaptation to their environment was based largely on genetically inherited predispositions, the fundamental life-ways of the hominids changed mainly through the very slow process of biological evolution. This seems to have been the case throughout most of hominid evolutionary history, from the earliest days of the australopithecines millions of years ago to shortly before the appearance of anatomically modern humans in Europe less than fifty thousand years ago.

But the invention of symbolic communication freed humanity to base their environmental adaptations primarily on the learned behaviors of culture and tradition. And when modern humans began to live mainly by the rules of culture, entire ways of life could be changed in one or two generations through the enormously rapid process of cultural evolution, a process that we will examine in greater detail later in this chapter.

The vastly more flexible *cultural* forms of adaptation provided the Cro-Magnons and their contemporaries with a decisive advantage not only over the Neandertals but also over some of the emerging humans that succeeded in surviving into comparatively recent times. These include an isolated population of *Homo erectus* that was still living in Java as recently as thirty thousand years ago[3] as well as a *Homo floresiensis,* a dwarf human with a stature and a brain size strikingly similar to that of the early hominids, that was living on Flores Island in Indonesia as recently as twelve thousand years ago.[4] The speed and flexibility of cultural evolution is probably the reason that the anatomically modern humans prospered during the last Ice Age, while the Neandertals and all other surviving hominids, along with most of the Arctic megafauna, rapidly became extinct.[5]

Discovering Paleolithic Art

The hills of Cantabria on the northeast coast of Spain are rich with wildlife and verdant with abundant rainfall. With its freshwater streams, proximity to the ocean, upland meadows, mountain forests, and numerous caves and rock shelters, prehistoric Cantabria was home to a large population of anatomically modern humans, who left behind one of the richest collections of Paleolithic or "old Stone Age" archeological sites ever found.

The Paleolithic includes three long eras in the evolutionary history of the hominids. The oldest is the *Lower Paleolithic*, which began roughly three million years ago with the early hominids and continued throughout the time of the emerging humans. The next oldest is the *Middle Paleolithic*, which began approximately 250,000 years ago and corresponds to the era of the Neandertals. The most recent is the *Upper Paleolithic*, which began roughly 50,000 years ago and is associated with the Cro-Magnons and other anatomically modern humans. I have used the term "Upper Paleolithic" extensively in this chapter to refer to the period when anatomically modern humans lived in Europe between 50,000 and 11,000 years ago, because it helps to distinguish them from the food-producing societies of the *Neolithic* that will be described in the next chapter.[6]

Between 17,000 and 11,000 years ago, Cantabria was inhabited by the Magdalenian people[7] of Europe, who made very finely worked stone blades shaped for hafting as hand tools and projectile points. In addition to many fine examples of tools and weapons of stone, bone, and antler, the Magdalenians of Cantabria also left behind abundant evidence that they were capable of expressing their thoughts and ideas through the symbolism of artistic expression.

When the paleontologists of the late nineteenth century began to investigate the living sites of the Magdalenians in France and Spain, they discovered a kind of prehistoric remains that were, at that time, completely unknown to modern science: rich collections of drawings, paintings, symbols, and icons that these remarkable people had left on the walls and ceilings of the caves they inhabited.

One day in 1868, the Spanish hunter Modesto Peres and his dog were pursuing a fox at a place called Altamira in the hills of Cantabria, when the dog suddenly disappeared from view. Eventually, Modesto discovered that the dog had fallen through a small crevice obscured by rocks and vegetation and had landed in the interior of a large cave that no one had known existed. After rescuing the dog, Modesto returned home and reported his discovery to the owner of the land, the Spanish nobleman Marcelino Sanz de Sautuola. But the hills near Altamira were riddled with caves, and at first de Sautuola thought little of this discovery.

As the years passed, however, Marcelino de Sautuola became interested in the new science of paleontology and in the growing number of prehistoric remains being discovered in European caves, and in 1875 he began to excavate the sediments that had accumulated near the mouth of the cave at Altamira. De Sautuola soon found plentiful evidence of human habitation from the Magdalenian culture that had flourished in Europe between seventeen thousand and eleven thousand years ago, and he began to visit the cave at Altamira more frequently, eventually making it a habit of bringing his five-year-old daughter Maria with him to keep him company (see Figure 5.1). Maria, a curious child, repeatedly asked her father for permission to explore the dark interior of the cave, but de Sautuola always refused, insisting that the inhabitants of the cave would have lived only near its mouth and that nothing of interest would be found in the cave's interior.

In 1878, de Sautuola traveled to France to attend the Paris World's Fair, and while in Paris he was able to examine some of the Cro-Magnon remains that Louis Lartet had unearthed from the cave at Les Eyzies. The following summer, back in Altamira Cave, Maria once again began pleading for permission to explore the cave's interior. Finally relenting, de Sautuola gave her a candle and warned her to be careful where she walked. Maria disappeared into the darkness. A few minutes later, the sound of Maria screaming "*Toros! Toros!*" suddenly echoed from deep inside the cave. Rushing into the dark interior, de Sautuola found Maria standing in a large gallery, looking up at the ceiling. There, to his astonishment, he saw the vibrant images of extinct bison, painted with uncanny realism on the ceiling of the cave (see Figure 5.2).

De Sautuola contacted his friend Juan Vilanova y Piera, master of Geology and Paleontology at the University of Madrid, and before long Vilanova traveled to Altamira to see the paintings for himself. Awestruck by the richness and uniqueness of the art, Vilanova was convinced that they were indeed the products of Paleolithic cave dwellers. The two men prepared a scientific account of this extraordinary discovery and in 1880 published the results in several Spanish journals. Acclaimed by the Spanish public, the masterful Paleolithic art they described quickly became the talk of Europe. Later that year, at the prehistorical congress held in Lisbon, Portugal, Professor Vilanova formally presented his description of the Altamira art to the assembled paleontologists from all over Europe.

At the conclusion of Vilanova's presentation, however, an ominous silence descended upon the hall. None of the paleontologists in attendance at the prehistorical congress were prepared to believe that prehistoric cave people, whom they regarded as rude and primitive savages, could have possibly executed works of art of this caliber, and they immediately challenged the authenticity of the Altamira paintings. Led by the French anthropologist Gabriel de Mortillet, the scientific authorities ridiculed the conclusions of Vilanova and de Sautuola and denounced their publications as "works of falsehood and madness."

At that time, it was an article of faith among the intellectuals of Victorian Europe that no group of prehistoric "primitives" would have been capable of producing art that had the realism, sophistication, and mastery of technique that characterized the cave paintings of Altamira. Firmly rejecting the idea that the Altamira paintings dated from prehistoric times, the paleontologists of the prehistorical congress concluded that they must have been created by contemporary artists. Some of the conference participants even suggested that de Sautuola's "discovery" was a deliberate forgery that had been executed by a contemporary Spanish artist on de Sautuola's orders.

De Sautuola was crushed. Returning to his estate, the object of pity and ridicule, Marcelino de Sautuola suffered from depression and declining health until his death eight years later. The entrance to Altamira cave was sealed, and for years Maria de Sautuola allowed no one to enter.

But in the years that followed, examples of prehistoric art from the same Paleolithic time periods were discovered in numerous other prehistoric sites in France and Spain. And much of this prehistoric art included cave paintings done with skill and artistry comparable to those of Altamira.[8] Finally, in 1902, the French archeologist Émile Cartailhac, who had been one of de Sautuola's most vocal skeptics at the Lisbon Congress, decided to travel to Spain with the renowned French paleontologist Abbé Breuil to see for himself the paintings at Altamira.

Arriving at Altamira, the two scientists persuaded Maria de Sautuola to allow them to enter the cave and inspect the paintings. Stunned by what he saw, Cartailhac promptly reversed his position and published an official apology in the French scholarly journal *L'Anthropologie.* But neither de Sautuola nor Professor Vilanova, who had died nine years earlier, lived to see their claims vindicated or their professional honor restored. Nevertheless, the cave at Altamira has come to be called the Sistine Chapel of Paleolithic art and is now regarded as one of the largest and most important collections of prehistoric art ever found.

Troves of comparable art have since been discovered in many other cave sites throughout the world—in eastern Europe, India, Southeast Asia, and Africa, as well as in North and South America. In fact, more than three hundred caves containing works of prehistoric art have now been documented by prehistorians in France and Spain alone, most of them inhabited by Magdalenian people between eighteen thousand and eleven thousand years ago.

The most famous of these collections of prehistoric art were found in Niaux Cave (investigated by the same Émile Cartailhac in 1907), in

Pech Merle Cave (accidentally discovered by teenage boys in 1922), and Lascaux Cave (also accidentally discovered by teenage boys in 1940). The most ancient examples of prehistoric art found to date come from Chauvet Cave, which was discovered in France by cave scientists in 1994. Chauvet Cave contains numerous lifelike paintings of game animals and predators and dates from thirty thousand years ago, at least ten thousand years before the first appearance of Magdalenian culture.

While the true origins and great antiquity of Paleolithic art has now been established beyond any doubt, two important questions remain.

First, why did the Paleolithic hunters paint these scenes deep in the interior of these caves—in some cases hundreds of feet from their mouths—in places which were difficult and time-consuming to reach, where daylight could not penetrate, and where the only light available to the prehistoric artist came from the flickering flames of torches?

Second, why do so many of these paintings represent animal species—such as mammoths, rhinoceroses, and bison—that were not only relatively rare but also difficult and dangerous to hunt? Why did they rarely paint images of the deer, reindeer, and smaller game that constituted the majority of the game animals these people hunted and that constituted the mainstay of their diet?

The most common answer to the first question is that Paleolithic art was hidden deep in the interior of caves because it was intended as a form of magic, to ensure success in the hunt. The paintings may have been made in the observance of rituals so secret that they could only be performed in the most inaccessible places known to the people of those times, far away from the prying eyes and ears of the uninitiated. In fact, anthropologists who have studied the cultures of hunting and gathering people have consistently found that the most important and most sacred rituals of such people are typically carried out in great secrecy. The Cheyenne Arrow Renewal ceremony, described later in this chapter, is but one of thousands of examples of these nearly universal human customs.

The second question is more difficult to answer, but an important clue was provided in an observation by the pioneering anthropologist

Bronislaw Malinowski, who lived in the South Pacific with the seafaring people of the Trobriand Islands from 1914 to 1918. Malinowski noted that the Trobrianders always performed certain magical rituals as part of the important tasks on which their livelihoods depended, such as planting gardens, constructing dugout canoes, and fishing in the open ocean. He pointed out that all of these activities involved a significant element of uncertainty, since gardens could fail, canoes could be shipwrecked, and fishing expeditions on the high seas were not only dangerous but also often ended with the expedition returning empty-handed.

But when the Trobrianders fished in the quiet waters of their island lagoons, they used a native poison that never failed to produce an entirely predictable outcome—an abundant catch. Tellingly, no magic was performed before a fishing expedition in the lagoon. It is thus likely that the people of the Upper Paleolithic saw no need to perform magical rituals for game that was abundant and easily caught, while hunting the rare and dangerous types of game was an uncertain enterprise, involving significant risks and rewards, in which the performance of ritual magic seemed indispensable.[9]

The "Venus Figurines" of the Cro-Magnons

Among the best-known examples of prehistoric art that have survived from Paleolithic times, aside from the lifelike paintings and drawings of game animals, are the so-called Venus figurines. These voluptuous and highly symbolic statuettes of prehistoric women, only a few inches tall, are typically devoid of human faces and have legs and arms that are devoid of realism and detail. Yet they were carved with exaggerated representations of the parts of the female body associated with sex and reproduction: huge pendulous breasts, swollen bellies, massive thighs, immense buttocks, and prominent genitalia.

Typically carved from mammoth ivory or soft stone, the idea that these figurines are Venuses—that they are representations of the goddess of love—is a modern fantasy, although there is little doubt that these sculptures played an important role in prehistoric beliefs about sex and reproduction. And that the Venus figurines had a symbolic significance to the Cro-Magnon societies of the Upper Paleolithic is beyond dispute (see Figure 5.3).

Perhaps the most remarkable fact about these figurines is their incredibly wide distribution in time and space. They have been found in Paleolithic sites stretching from southwestern France in the west to Siberia in the east—a distance of over four thousand miles—and they date from time periods that begin as early as forty thousand years ago and last until as recently as ten thousand years ago. This incredible span of time and space includes almost the entire prehistoric history of anatomically modern humans in Europe. Nevertheless, each of the Venus figurines from each prehistoric time and place was executed with a particular, identifiable style. And the fact that this style varies considerably from one example to another is evidence that, tens of thousands of years ago, humanity had divided itself into distinct cultures, each associated with a particular time and place.

In addition to the exaggerated sexual symbolism of the Venus figurines, the Cro-Magnons also sculpted other highly realistic objects from the ivory they recovered from mammoth kills and from the antlers they removed from the reindeer they hunted. These include many highly decorated spear-throwers and some masterful sculptures of people and animals. One of the best known of these is the carved likeness of a woman's head, known as *La Dame à la Capuche* or "The Lady of the Hood," twenty-five thousand years old, which was found in a cave in southwestern France in 1892. Another is a horse's head of incredible realism carved from reindeer antler, which was excavated from the cave of Mas d'Azil in the French Pyrenees (see Figure 5.4).

These and other fine examples of Paleolithic art show that the Cro-Magnons employed their rapidly evolving artistic abilities to represent the animals and people that populated their universe. But the people of that era also used graphic designs of a more obscure nature to record and communicate information, thoughts, and ideas. Perhaps nothing that has survived from prehistoric times is more suggestive, and more mysterious, than the symbolic designs called petroglyphs that were painted and carved on the walls of Paleolithic caves. We will explore the subject of petroglyphs in detail later in this chapter. But first, we must grapple with the question of how humans came to use visual designs to represent thoughts and ideas in the first place.

The Symbolism of Animal Tracks

For all primates, including hominids, vision is the most important of all the senses. After tens of millions of years adapting to a life in the trees, our primate ancestors came to rely far more on the sense of sight than on the sense of smell. Animals that live on the ground, where smells linger and are plentiful, tend to have a highly developed sense of smell, and this is especially true of mammals. But in the breezy heights of the trees, smells tend to be blown away in the wind, while it is vital that the movement of a predator, or the color of ripe fruit, be seen over great distances. As a result, our tree-dwelling primate ancestors gradually lost much of their sense of smell but gained a greatly enhanced sense of sight, complete with color vision and the ability to see in three-dimensional space.

When the early hominids added the predator's strategy of hunting to the ancestral primate strategy of foraging, they possessed only the diminished sense of smell they inherited from their arboreal ancestors. So instead of tracking their prey with a highly developed sense of smell as is typical of most other mammalian predators, the hominids turned instead to their highly developed sense of vision.

Anthropologists who have studied hunting and gathering societies have consistently found the nomadic hunters to be masterful visual trackers. The most inconspicuous broken twig, the slightly bent blade of grass, or the merest trace of a footprint in the damp earth can be read by the experienced hunter as easily as you and I can read a newspaper headline. And once they have speared or poisoned their prey—even though it may gallop off out of sight—the tiniest drop of blood in a sea of grass or the most subtle change in the color of their droppings will reveal to the eye of the experienced hunter the spoor of the animal they have wounded.

Over the past million years, as the emerging humans evolved into increasingly skillful hunters, the interpretation of animal tracks developed into a high art. No longer dependent on actually seeing their prey, the emerging humans learned how to sense the presence of their prey and track their movements by recognizing the telltale signs left in the environment by the game animals that lived near them and that passed through their territories.

As they became increasingly adept at associating the visual signs of animal tracks with the identities and condition of the wild pigs, deer, antelopes, reindeer, horses, bison, and wild cattle they were pursuing, the emerging humans also developed, in addition to their tremendously sensitive and discriminating powers of visual perception, the unique ability to associate specific visual patterns with the activities of animals and people, their states of being, and the actions that those patterns represented.

The Neandertals, who were supremely gifted hunters, must have possessed a greatly advanced capability for visual tracking. But it was the Cro-Magnons and the other anatomically modern humans who took the next logical evolutionary step and *invented their own signs and symbols to represent specific animals, people, states of being, behaviors, and actions*. Once having taken this critical step, anatomically modern humans elaborated the technology of symbolic communication to a degree unprecedented in the history of the hominids. In the process, they unleashed yet another fundamental transformation in human life and society.

The Oldest Evidence of the Use of Symbols

While the cave paintings and petroglyphs of the Cro-Magnons may be the purest example of symbolic communication left behind by prehistoric people, some tantalizing evidence has survived from long before the appearance of *Homo sapiens*. This evidence suggests that *Homo erectus* and other emerging humans had begun to draw designs on bone and shell hundreds of thousands of years ago.

From Eugène Dubois' collection of artifacts from the Indonesian Island of Java in 1891, the Dutch archeologist Josephine Joordens and her colleagues recently identified freshwater shells that had been deliberately engraved with geometric designs. And a number of striking artifacts were found in 1969 in a prehistoric site in Bilzingsleben, Germany, inhabited by the emerging human *Homo heidelbergensis*.[10] (This is the same site that contained the stone circles that appear to be the foundations of extremely ancient dwellings.) The site at Bilzingsleben dates from between 380,000 and 400,000 years ago and contains evidence that *Homo heidelbergensis* may have used a primitive form of symbolism to record numbers or quantities.

One of the many artifacts from Bilzingsleben that bear deliberate markings is a percussion tool made of elephant bone, one of many such tools used by the emerging humans for putting the final touches on their classical Acheulean hand axes. The elephant bone tool is approximately sixteen inches long, and it bears twenty-one cut marks that seem to have signified some kind of counting process. The marks are divided into two groups, with seven marks in the upper group and fourteen marks in the lower group.

Dietrich and Ursula Mania, the paleontologists who excavated this site, observed that the lower third of the tool had broken off, and they proposed a reconstruction in which this missing third would have borne a series of seven marks similar to those on the upper third. The Manias further proposed that this reconstructed elephant bone with twenty-eight marks was the symbolic expression of a lunar calendar, in which each mark represented a single day in the twenty-eight-day lunar cycle. If this is correct, the first seven marks would have represented the seven-day period from the new moon to the waxing half-moon, the next fourteen marks the fourteen-day period from the waxing half-moon through the full moon to the waning half-moon, and the final seven marks the period from the waning half-moon to the following new moon.

The elephant bone percussion tool is only one of many such objects from Bilzingsleben that bear similar markings, and their abundance has inspired the independent scholar John Feliks to claim not only that *Homo erectus* possessed a spoken language but also that these emerging humans possessed highly advanced musical and mathematical knowledge.[11] However, the complete absence of any other evidence that *Homo erectus* had intellectual powers comparable to those of modern humans suggests that Feliks may have overstated his case.

While the possible use of symbolism by *Homo erectus* remains an unsettled question, there is solid evidence that Neandertals were making primitive forms of jewelry such as necklaces, hair ornaments, and possibly earrings well before the arrival of anatomically modern humans. These objects of personal adornment were not intended to keep the body warm or protect it from the elements. They were worn as symbols of status, as magical charms to ward off sickness or injury, or simply as objects of beauty. They were not used for a practical purpose; their purpose was purely symbolic.

In two Neandertal cave sites in southeastern Spain, numerous seashells were found with holes drilled in them, so they could be strung as necklaces or attached to the hair or clothing. The cave sites also contained red and yellow pigments that were used to paint the shells in bright colors. And in a cave in northern Italy used by Neandertals, remains of the wing bones of vultures, eagles, pigeons, and crows show clear evidence that the large feathers on their wings had been deliberately removed. Since there is no evidence that the Neandertals ever used any of these species for food, they were clearly plucking out those wing feathers to wear as objects of adornment, possibly as earrings or headdresses.

Finally, the Neandertals are the first prehistoric people known to have buried their dead. The Neandertal graves that have been found in Europe all have a similar pattern: the body of the deceased is placed in a small, shallow grave in a fetal position, with the knees bent upward and the head bent forward. In one Neandertal grave about seventy-five thousand years old, excavated in the 1950s in Shanidar Cave in modern

Iraq, paleontologists found the remains of pollen from flowering plants known for their medicinal qualities. It was suggested that bunches of these flowers had been deliberately placed on the body of the deceased, in a practice reminiscent of the modern European custom of placing flowers on gravesites.

The Shanidar burials have been interpreted by some scientists as evidence that the Neandertals had believed in the afterlife—a belief that is associated throughout all of human history with symbolism and the observance of religious ritual. Other paleontologists who conducted more recent analyses of this grave, however, have concluded that the pollen in the Shanidar grave may have simply blown into the grave with the wind or been carried in by small ground-living rodents.

Nevertheless, traces of red ochre—a naturally occurring pigment used by many hunting and gathering people in the course of ritual observances—have been found in many Neandertal grave sites, some as old as two hundred thousand years. This indicates that, at the very least, these prehistoric people regarded death as an important milestone in human life, and that the death of a member of the group was observed by rituals in which the use of symbolic substances such as red pigment played a significant role.

With the arrival of anatomically modern humans in Europe, however, symbolism in its purest form suddenly becomes abundant in the paleontological record. And perhaps the most fantastic collection of prehistoric symbolism in the world are the thousands of petroglyphs that were found in another Spanish cave from Cantabria, the Cueva de la Pasiega.

The Secrets of La Pasiega

In the heart of Cantabria, less than fourteen miles southeast of Altamira Cave, the cave of La Pasiega was first investigated in 1911 by the Swiss anthropologist and prehistorian Hugo Obermaier. La Pasiega was inhabited over a period of roughly six thousand years, from approximately twenty thousand to fourteen thousand years ago, first by the

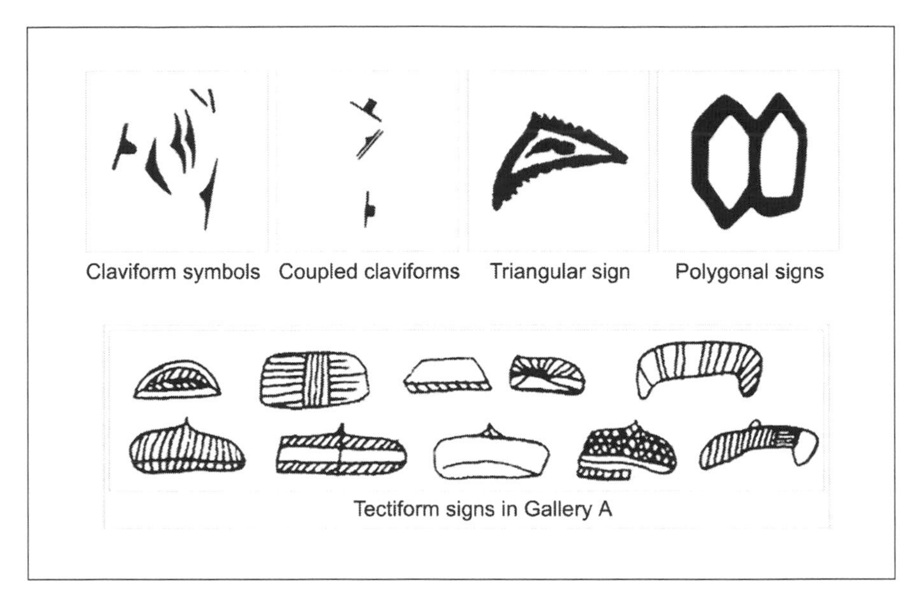

Figure 5.5. *The carefully executed designs in these petroglyphs from La Pasiega Cave leaves little doubt that they were created to record and convey specific information in symbolic form.*

Solutrean people and later by the Magdalenian people, and it contains one of the richest troves of petroglyphs ever found.

Some of the petroglyphs at La Pasiega represent animals or people, others appear to represent numbers or quantities, and still others remain entirely a mystery. La Pasiega contains thousands of petroglyphs, but aside from some images of game animals, none of the petroglyphs have ever been deciphered. Yet while we may never be able to unlock the secrets of these petroglyphs, it is clear that they are symbolic in nature. These are not meaningless lines and dots scrawled at random for no reason.

The petroglyphs found at La Pasiega are carefully executed shapes and patterns in which certain graphic elements occur repeatedly yet not in a random fashion. In this respect, they resemble the alphabets used in the true writing of civilized cultures (see Figure 5.5). The petroglyphs are also very different from the purely decorative patterns that can be found in the pottery and basketry of preindustrial societies, in which a single design often covers a large area with little or no variation.

Moreover, the petroglyphs were drawn deep in inaccessible regions of the caves, which suggests that they were not drawn casually. Finally, we know from contemporary ethnographic studies that cave paintings and petroglyphs are regarded by hunting and gathering people as forms of magic and are often associated with religious rituals.

For all of these reasons, petroglyphs must be regarded as designs that were deliberately drawn to convey specific meanings and to represent specific things and events in time and space. But what, exactly, were these things and events? Unfortunately, we have no idea.

There is no Rosetta Stone that reveals the secrets of the petroglyphs in written languages that we can translate and understand. There are no Magdalenians or Solutreans still alive to explain their meanings to us. In fact, not only do we have no idea what language these people spoke but also, even if we could hear them speak, we would still have no idea what they were saying. For the meanings that are encoded in the form of symbolic communications are, by their very natures, dependent on the living memory of those who grew up and lived in

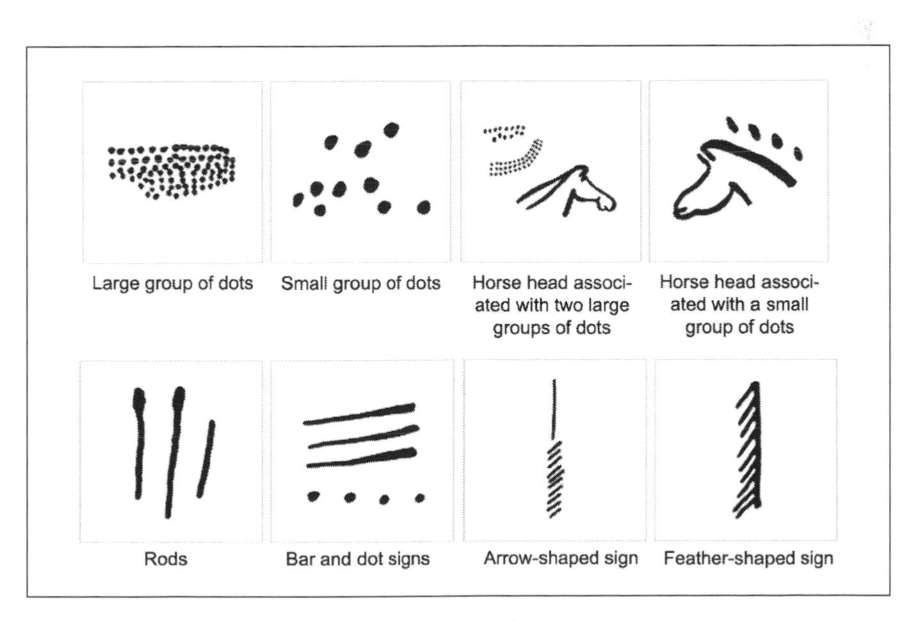

Figure 5.6. *Many of the La Pasiega petroglyphs contain dots or bars that may have signified number or quantity.*

the cultures that created them. And when all of those people disappear and their cultures pass into oblivion, we have little or nothing to go on.

We may not know exactly why the Magdalenians and other Cro-Magnon people executed the masterful cave paintings of Altamira, Lascaux, and Chauvet, but at least we know that these lifelike paintings represented real animal species that can be positively identified. Not so with the petroglyphs. There is little doubt that petroglyphs were carved and painted in order to record information and convey it to others in time and space. But what that information is, and how it was translated into symbolic form, remains a secret that the best minds of modern paleontology have been utterly unable to unlock.

If the written word—which includes both simple forms like cuneiform or the Roman alphabet and complex forms like Egyptian hieroglyphics and Mayan glyphs—is the symbolic representation of information expressed in human speech, the petroglyphs of the Paleolithic rank as the earliest and most ancient forms of human writing ever found. Furthermore, they are the best evidence imaginable that the people who created these petroglyphs were using language. In fact, some scientists believe that the ability of prehistoric people to draw meaningful images in the first place is proof that their cultures contained shared systems of meanings expressed in the form of actual spoken languages.[12] But while the possibility of hearing people speak these prehistoric languages has been lost forever, some fascinating clues about the origins of human speech have survived in the fossil remains of hominid skeletons.

The Hyoid Bone and the Origins of Language

In previous chapters, we saw how the fossil evidence of human anatomy can provide important clues to the behavior of prehistoric hominids, even when the behavior in question occurred long ago and cannot be observed directly. This raises the interesting question of whether the fossil remains of extinct hominids might offer some clues about when they first began to use spoken language. Since the tongue, larynx, and upper throat—the primary organs of speech—are composed almost entirely of soft tissues, none of these tissues have survived over the immense time periods that elapsed in the course of human evolution.

But there actually is one very small bone—located in the upper region of the human throat, just above the larynx and attached to it—that plays an important role in the movements of the muscles of the tongue and throat that shape the sounds of human speech. This bone, which is shaped like a horseshoe and is approximately one and one-half inches wide, is called the "hyoid bone."

While the hyoid bone of a modern human is dramatically different in shape from the hyoid bone of a chimpanzee, it is nearly identical to the hyoid bone of the Neandertals and very similar to that of the emerging human *Homo heidelbergensis*. It therefore seems possible that human speech began hundreds of thousands of years ago in a population of *Homo heidelbergensis* and that the Neandertals spoke an advanced form of language similar to our own modern speech. There can be little doubt that the anatomically modern humans, who left behind plentiful evidence that they routinely used visual symbolism to record and communicate information, were fully capable of human speech.

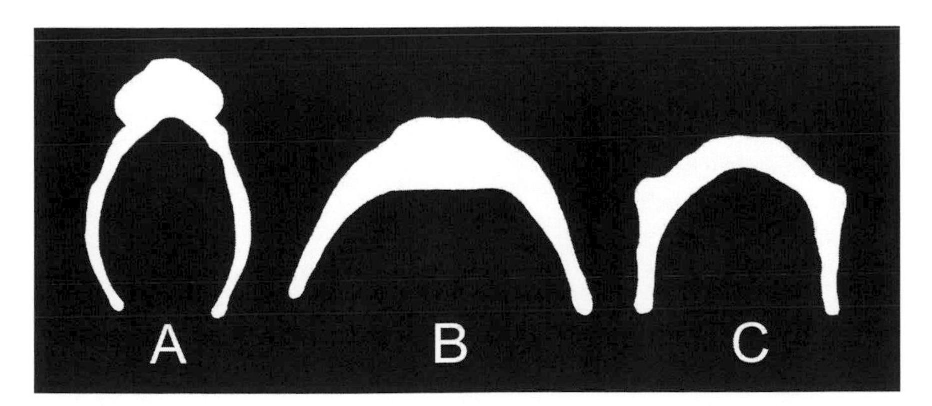

Figure 5.7. *The shape of the chimpanzee hyoid bone (A) is very different from the very similar shapes of Neandertal (B) and modern human (C) hyoid bones, suggesting that Neandertals may have had spoken languages comparable to our own.*

Some scientists have argued that language first developed when hominids began to form groups that were larger in size than the small number of individuals that are typical of the social groups of monkeys, apes, and early hominids. According to this view, language—along with an increasingly large brain—evolved so that emerging humans could bond with the larger number of individuals that were forming in the more recent human groups. While monkeys and apes bond with other members of the group through physical grooming, there is a limit to the number of individuals one can groom. Language makes it possible to expand this number dramatically.[13]

Prehistoric Music

The songs that humans sing are nothing like the songs produced by the thousands of species of birds or the few mammals, such as gibbons and whales, that sing in the course of their daily lives. The "songs" of those animals vary only slightly from one individual or population to another within each species, indicating that they are largely instinctual and preprogrammed by genetic inheritance.

The songs of humans, however, vary dramatically from one human group to another, and from one individual to another. So it seems that singing, like language, is one of those behaviors that, while characteristic of human nature, must be learned, like language. And the music that humans make by using technology to fashion musical instruments is another of those behaviors that is completely unique to our species.

Paleontologists have never been able to determine when prehistoric humans first began to make music, but humans probably developed the ability to sing first and only later invented the technology of musical instruments. If singing began with the use of language, it may have begun as long as three hundred thousand years ago. And if—as seems increasingly likely—the Neandertals had a spoken language by the time they settled in prehistoric Europe more than one hundred thousand years ago, singing is probably at least as old as that.

The oldest musical instruments found to date are the remains of several small "flutes" between five and nine inches long, that were made approximately thirty-five thousand years ago by anatomically modern humans belonging to the Aurignacian culture. The Aurignacians made most of these instruments by boring finger holes into the hollow wing bones of swans and vultures, and several of these Aurignacian flutes have been unearthed from Paleolithic cave sites located in the river valleys of southwestern Germany. When one of these bird-bone flutes was reconstructed, researchers were able to use them to produce the musical tones of B, C, D, and F.

Some of the Aurignacian flutes found at these sites were made of ivory. This required a complicated process of manufacture, in which a section of mammoth ivory tusk was split in half, each of the halves was hollowed out along its length, finger holes were carved in specific locations, and the two halves were glued together to form a perfect airtight seal. The time, care, and workmanship required to fashion these delicate instruments from bone and ivory leaves little doubt that music—and the craft of making of musical instruments—was already considered an important part of life tens of thousands of years ago.

The songs and dances of the Aurignacians, Gravettians, Solutreans, and Magdalenians probably served the same purpose in prehistoric life as they do in modern life today: as a means of emotional release, as a way of creating feelings of camaraderie and belonging among the members of a social group, and as a way of expressing one's cultural identity. It is no accident that every modern nation-state has a national anthem that its members sing together to express their solidarity as citizens of their national cultures and to reinforce their shared identity as members of these immense human groups.

Shared Symbols and Ethnic Identity

It is June on the Great Plains of western North America in the early 1800s. The heat of summer is fast approaching, and the scattered bands of Cheyenne Indians—the famed buffalo hunters of the North American Plains—are wandering slowly from the shelter of their winter camps in the wooded valleys toward the great camp circle that will soon be created in an open field on the banks of the Platte River. There, on the morning of the summer solstice, a thousand teepees, representing the entire Cheyenne tribe, will be erected in a large open field, where the Arrow Renewal Ceremony will be performed.

Throughout most of the year, the many different Cheyenne bands lived in widely scattered encampments, hunting and gathering for their food in groups composed of loosely related families. Yet, when the power of the tribe was at its height, less than four thousand Cheyenne dominated a region of prairie hundreds of miles wide that stretched from Montana in the north to Kansas in the south—an area twice the size of Texas. How did the Cheyenne successfully defend this immense territory against other Plains Indian tribes who coveted their hunting grounds and who were not only larger in size than the Cheyenne but also equally skilled at the arts of warfare? The answer seems to be that the Cheyenne had developed the ability—comparatively rare among the tribes of the North American Plains—to present a completely united front against their enemies.

The Cheyenne were formidable buffalo hunters and warriors who were skilled in the arts of lethal force. Among the Cheyenne, killing or wounding someone from an enemy tribe was considered a great achievement, a deed that a man was entitled to boast about for the rest of his life. But among men who know how to kill, there is always the danger that conflict over territory, property, or women might escalate into violence, resulting in injury or death within the tribe. Such conflicts were in fact commonplace among many Native American tribes, and it was not unusual for the members of these other tribes to become embroiled in cycles of revenge and counterattack, pitting members of the same tribe against each other.

The Cheyenne's advantage over other Plains Indian tribes was rooted in their unique culture and value system, in which tribal solidarity was considered a paramount virtue, and in which murder—defined as the killing of one Cheyenne by another—"soiled" the Sacred Arrows, their paramount symbols of tribal unity. Murder placed the entire tribe in moral jeopardy, robbing it of the protection of the divine spirits and the magical powers of the Arrows themselves. The Cheyenne believed that the body of a murderer slowly rotted away inside and became foul and putrid until the murderer died a slow, lingering death. These effects could be reversed only if the murderer atoned for his sin by sponsoring an Arrow Renewal Ceremony—an arduous task requiring months of travel on horseback to visit and inform all of the Cheyenne bands throughout the immense territory of their tribal hunting grounds.

The purpose of the Arrow Renewal Ceremony was the symbolic "cleansing"—through days of fasting, praying, and ritual observances—of the four Sacred Arrows that had been handed down through the generations from the mythical Cheyenne hero Sweet Medicine, who had received them long ago from the Great Spirit. Once every few years, all the members of the Cheyenne tribe would come together to perform the sacred Arrow Renewal Ceremony, in which the entire Cheyenne tribe would call upon the spirits of their ancestors to renew the earth, wash away the sins of the past, and restore the bonds of brotherhood among all Cheyenne—and by so doing ensure game in abundance and victory in war.

On the first day of the Arrow Renewal Ceremony—the day of the summer solstice—a thousand tepees were raised by a thousand Cheyenne families, forming the arc of a great crescent that faced the rising sun, with the door of each tepee set to catch the first rays that appeared over the horizon. A huge communal tepee, the Medicine Arrow Lodge, was raised in the center of this crescent, within which the Sacred Arrows would be renewed and purified.

On the second day, the priests took their places in the Medicine Arrow Lodge. Sacrificial gifts were placed as offerings before an altar in the center of the lodge, and after the careful performance of secret rituals, the singing of sacred songs, and the offerings of prayers, the bundle containing the Sacred Arrows was opened and examined. So sacred was this moment that all Cheyenne except those actually participating in the ceremony had to remain perfectly silent and still in their tepees, while the grounds of the great camp circle were patrolled vigilantly by the members of the Cheyenne Warrior Society. If a child began to cry, it was quickly hushed. If a dog barked, it was killed instantly with a quick blow from a warrior's club.

On the third day of the ceremony, the Sacred Arrows were cleaned and repaired, and a long willow stick was cut for each Cheyenne family. One by one, each willow stick was blessed in the smoke of incense by the tribal priests, to give each family renewed well-being. And each family would carefully keep and protect its willow stick throughout the year, as a reminder that they were Cheyenne and that they belonged to something greater than themselves, greater than their families—and greater than the local band of related families that lived and traveled together, hunting and gathering within their individual territories.

In this way, the Cheyenne harnessed the emotional power of symbolic communication to fuse dozens of small nomadic bands into a single tribal entity capable of coalescing in short notice into a fighting force of thousands. With a common language, and a common allegiance to the power of the Sacred Arrows and other tribal symbols, the many Cheyenne bands could be quickly organized into a single large group that included most of the members of the tribe. In this way, the fission-fusion society that both chimpanzees and hominids inherited from their common ancestors became a powerful social force, capable of creating the feeling of tribal solidarity among the members of many scattered nomadic bands.

The Fission-Fusion Society

During the brief periods when two distinct groups of chimpanzees temporarily fuse into one, the members of both groups mingle together with little or no overt violence or conflict. But while this fusion phase exists, the individuals of both groups display symptoms of emotional stress, which may take the form of screaming, "dominance" displays such as breaking branches, and generally rushing around with great excitement. Inevitably, however, the emotional strain of being in close contact with members of another group begins to take its toll. After mingling for a day or two, the chimpanzees sort themselves out into their original groups and depart for their home territories.

Humans, however, have taken the fission-fusion model one step further, by making the fusion cycle into an integral part of their seasonal environmental adaptation. The Inuit Eskimos spent the comparatively warm months of the brief Arctic summer in small family groups that generally numbered from two to twelve individuals, while they wandered through the tundra, gathering birds eggs and snaring rabbits. But in the winter, when they hunted larger game—such as caribou, seal, and walrus—that required the cooperation of a larger group of hunters, the Inuit would cluster in much larger groups. Settling in one location for the winter, each Inuit family or extended family would build its own igloo, creating settlements that were composed of several igloos housing fifty or more people.

The Bushmen of the Kalahari Desert displayed a similar fission-fusion pattern, gathering near a few permanent water holes in relatively large numbers during the height of the dry season and dispersing into small family groups during the rainy season, when food was more plentiful and could be found scattered throughout the desert.

When anatomically modern humans integrated symbolic communication into their daily lives, they unleashed the full power of the fission-fusion society. By adopting distinct cultural identities—and reinforcing these identities by the sharing of language, music, dance, art, design, and other forms of symbolic communication—modern humans were able to fuse the small nomadic bands in which they lived for much of the year into large groups that were able to cooperate in hunting and warfare as the need and opportunity arose.

While small groups of prehistoric nomads may have been ideally suited to the pursuit of small game and the gathering of edible plants, larger groups were better suited to the communal hunting of large animals, such as the prehistoric megafauna of mammoths, rhinoceroses, bison, and wild cattle. Although the megafauna were hunted relentlessly by the Neandertals, they continued to flourish side by side with the Neandertals for tens of thousands of years. Yet when anatomically modern humans appeared on the scene, the megafauna seem to have been hunted to extinction in the space of a few thousand years. With their tribal cultures and their ability to organize themselves into large groups when the occasion demanded, the deadly efficiency of communal hunting proved no match for the megafauna that had roamed the northern latitudes throughout the ice ages.

Seventy-five thousand years ago, the Neandertals had become well established in western Europe, and they flourished there for at least thirty-five thousand years. But a few thousand years after the arrival of anatomically modern humans, the Neandertals had completely disappeared. Unlike the anatomically modern humans, the Neandertals left behind little evidence of symbolic communication and no evidence at all of distinct cultural traditions or ethnic identities. In spite of their superior physical strength, the small bands of individual Neandertals would have been no match for an organized fighting force composed of hundreds or even thousands of assembled Châtelperronians or Aurignacians.

The Power of the Narrative

Perhaps the most significant advance in symbolic communication occurred when humans began to string together groups of words to describe events that occurred in a particular sequence over a period of time. These sequences are commonly known as "narratives."

"'Begin at the beginning,' said the King to the White Rabbit, 'and go on till you come to the end: then stop.'"[14] This deceptively simple formula describes a powerful form of communication of which no other species is capable. While the emerging humans may have had a primitive form of language, and the Neandertals almost certainly were capable of spoken language, the profound changes in human life ushered in by the anatomically modern humans may have been due to their unique ability to tell a story in narrative form.

The power of the narrative made it possible for male hunters returning to their home base with their coveted prizes of meat to recount the story of how they found, tracked, cornered, killed, and butchered their prey. The narrative made it possible for female gatherers to describe how they traveled from their home base to the exact location where choice fruits, roots, tubers, and other foods could be found. The narrative made it possible to describe the step-by-step process of making tools, weapons, containers, dwellings, clothing, and all the other myriad things that hunters and gatherers depended on. The narrative made it possible for shamans and *curanderas* to explain the causes of diseases and to teach their apprentices how to carry on the traditions of healing the sick and wounded.

Among its many important contributions to human culture, the power of the narrative allowed tribal story-tellers to recount the lives and deeds of the ancestors and the tribal gods, explain how the universe came to exist, and recite the sacred songs and rituals that gave meaning and richness to tribal life. The sum total of all these narratives constituted the "oral traditions" that anthropologists have recorded in all of the preindustrial cultures they have studied. These oral traditions functioned as the repositories of the experience and collective wisdom of every human society until the birth of civilization and the invention of the written word.

Finally, the power of the narrative gave anatomically modern humans a way of comprehending not only how events unfolded in the past but also how they would unfold in the future. In this way, the invention of the narrative gave our species the unique ability to conceive of the passage of time and to prepare for events that would not take place until days, weeks, or months in the future.

It is therefore no accident that, when the Neandertals were replaced by modern humans who possessed cultures rich in symbolic communication—and who had, through the shared symbolism of language and the oral tradition, access to vastly more information about how to live—the human population of Europe multiplied within a few thousand years until it was ten times the size that it had been during the age of the Neandertals.

The Power of Cultural Evolution

The essential mechanism of biological evolution consists of three basic processes. First, each generation inherits the characteristics of its parents through information that is encoded in the genetic material contributed by both mother and father. Second, some of this information is inevitably lost or changed through a largely random process we call "mutation." Third, most mutations have either no effect on the development of the organism or are harmful or maladaptive, making life more difficult for the individuals whose DNA has been altered.

Nevertheless, a very small number of random mutations turn out to be beneficial and help the individual organism adapt more efficiently to its environment. As a result, the more beneficial mutations become increasingly common in a breeding population and—if they are sufficiently adaptive—eventually become the new normal within the breeding population. This is the basic process that produced the ability to walk and run on two legs, use fire, cook food, construct dwellings, fashion clothing, and so on.

But biological evolution has its limitations. First of all, it is exceedingly slow. It took millions of years for our prehistoric ancestors to develop full upright posture and true bipedal locomotion, and well over a million years for the hominid brain to reach its present immense size. In the process of biological evolution, new and more beneficial genetic information can be transmitted only from a parent to its biological offspring. This means that many generations are required before a beneficial gene can spread throughout a breeding population. And before it can succeed, the children, grandchildren, and great-grandchildren of the first individual who possessed this new genetic material must consistently outcompete—and eventually outnumber—the children, grandchildren, and great-grandchildren of the other members of the group.

The DNA of a species is the complete set of instructions for constructing all the parts and pieces of the living organism, like the recipe for a cake or the blueprint for a building. If any of those instructions are changed, the cake, the building, or the organism will not be constructed according to plan. And when these changes are simply random failures in the reproduction of the instructions, they will appear as "mistakes" in the final product. It is thus necessary for a population of living organisms to discard thousands of harmful mutations before a single beneficial mutation appears that can be retained through the process of natural selection.

But cultural evolution has none of these limitations. A new kind of behavior may originate with a single individual and be rapidly transmitted to other individuals through learning and imitation. If this new behavior helps the individual to adapt more successfully to his or her environment, it can spread easily throughout an entire social group within the space of a single generation.

In addition, different groups of apes, monkeys, and humans occasionally come into contact with each other, and when these contacts are friendly, individuals from the two groups may mingle. This gives the members of one group the opportunity to observe and imitate the behaviors of individuals in the other group. In this way, new behaviors can spread not only within a single social group but also from one social group to another—and in this fashion ultimately spread throughout the population of an entire geographical region.

Finally, cultural innovations rarely originate as random events. Unlike genetic mutations, the changes in behavior that drive cultural evolution are usually purposeful and deliberate, and for that reason they are much more likely to be beneficial and adaptive than the mutations that drive biological evolution. This is true not only of the invention of potato-washing by Japanese macaques but also of the invention of cave art, the domestication of plants and animals, the development of precision machining, and the invention of the computer. It is easy to see why, when the purposeful nature of cultural innovations is combined with its very rapid means of transmission and proliferation, cultural evolution is both faster and more efficient than biological evolution.

♦

With the emergence of tribal cultures—a milestone made possible by the advent of shared symbolic communication in its many forms—humanity embarked upon a path of fusing into ever larger societies and social groups. And in each step along the way—in the agricultural village, the urban city-state, and the industrial nation-state—the size of the human group expanded exponentially. Without symbolic communication, this exponential growth would never have been possible. With symbolic communication, it was probably inevitable.

When the human social group, which for millions of years had consisted of no more than a few dozen individuals, freed itself from its primate inheritance and expanded into tribes of thousands, a process of fusion began that has culminated in the formation of vast nation-states consisting of millions of individuals and claiming dominion over the earth's entire land surface. Whether our species is capable of a final act of fusion—in which all living people achieve a shared identity as members of a single global culture and civilization—is a question that will determine the future not only of our own species but also of most forms of life on Earth. This is, in fact, the question that lies at the heart of this book.

CHAPTER 6

The Technology of Agriculture

Permanent Villages and the Accumulation of Wealth

"The transition from foraging to farming was the most profound revolution in human history."

—Graeme Barker,
The Agricultural Revolution in Prehistory

Eighteen thousand years ago, the last ice age was at its height. Massive ice sheets hundreds of feet thick covered the northern latitudes of Europe, Asia, and America. Sea levels were three hundred feet lower than they are today. Vast deserts stretched across Africa and Asia, and the rainforest was only a fraction of its present size. But great changes lay ahead. With the waning of the ice ages and the warming of the earth, humanity was about to embark on its next great metamorphosis.

Each of the four metamorphoses that had already taken place had transformed the biology of our ancestors in significant ways. The technology of spears and digging sticks transformed us from quadrupedal into bipedal animals. The technology of fire and cooking resulted in the loss of our body hair, a massive expansion in the size of our brains, and the disappearance of our tree-climbing anatomy. The technology of clothing and shelter enabled us to migrate out of the tropics and made it possible for our "premature" newborns to survive in cold climates. And the technology of symbolic communication involved significant changes in our brains, freeing us from the slow pace of biological evolution and enabling us to take advantage of the speed and flexibility of cultural evolution.

Yet throughout the millions of years it took for all of these important biological changes to take place, neither the nature of hominid society nor the relationship of the hominids to their natural environment was significantly altered. By day, adult males continued to pursue a lifestyle of predatory hunting, while adult females continued to forage for edible plants. At night, small groups of related kinfolk continued to gather in their home base for protection against predators. And every so often, the hominid group would abandon its home base and move to a new location, searching for more plentiful sources of food.

But when the Paleolithic era came to an end with the waning of the last major ice age, humanity freed itself of the need—which limits and circumscribes the lives of all other animals—to be perpetually engaged in the search for something to eat. When the technology of agriculture made it possible for humanity to produce its own food and store it for the future, our species cast off a burden that it had borne, along with all other animals, since its beginning.

Unbound from the daily search for food, our ancestors settled down in permanent settlements composed of hundreds and even thousands of people, learned to specialize in arts and crafts, and began to multiply. New and powerful technologies of transportation and communication enabled us to build cities and multiply still more, creating enormous civilizations composed of hundreds of thousands of people. The technology of precision machinery made it possible for us to create modern industrial nation-states composed of millions, and as a result we have multiplied so fast that our long-term future is now at risk. And the recent development of digital technology—which enables us to trade, visit, and communicate with all members of the human species—has made it possible, for the first time in our history, for humanity to fuse into a single global society.

None of these metamorphoses in society would have ever taken place had humans continued to pursue the hunting and gathering existence with which our species began. And yet the causes of this sudden appearance of agriculture, which made it possible for human society to embark on this road of profound social transformation, remains one of the great mysteries of the human story.

The Mystery of Agriculture

Between twelve thousand and four thousand years ago, a number of different human societies, living in widely separated locations, abandoned their former hunting and gathering lifestyle and began growing their own food. Scientists have offered many different and often competing theories to account for this remarkable coincidence, but after decades of discussion, there is little agreement about precisely why humans all over the world made the wholesale transition from hunting and gathering to agriculture at this time.

Later in this chapter, I will explain why it may have been the acquisition of language—and especially the power of the narrative—that enabled humanity to make the transition from foraging to farming at this particular moment in its history. But first, we must consider the numerous theories that have been advanced in recent years to explain the novel technology of food production. And while scientists may disagree about the causes, they do agree that agriculture first began in the Middle East roughly eleven thousand years ago, in a well-watered strip of land call the "Fertile Crescent." This is where our story begins.

The Fertile Crescent is approximately one hundred miles wide and nearly a thousand miles long. It begins in Egypt, at the eastern end of the Mediterranean Sea, and it ends in Iraq, at the northern end of the Persian Gulf. Lying directly on the path between Africa and Eurasia, and home to the earliest of the ancient civilizations, the Fertile Crescent has played an outsized role in human history since the emerging humans migrated into Europe and Asia more than 1.5 million years ago.

One of the first theories to account for the development of agriculture proposed that farming originated in the Fertile Crescent because the environment of that region was drying up, causing the dwindling

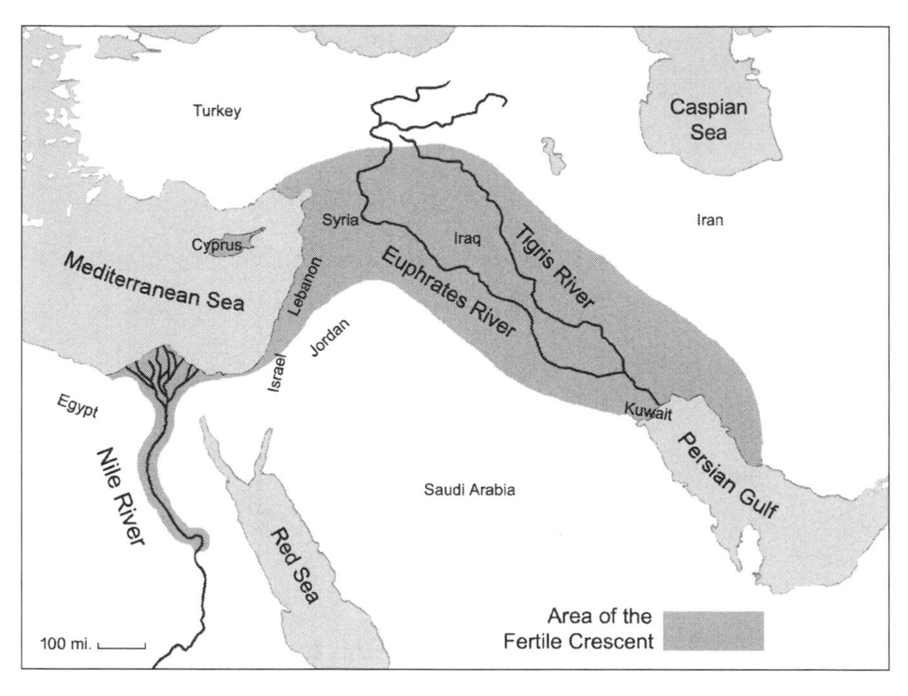

Figure 6.1. *The Fertile Crescent, which stretches from the Nile River in the west to the Persian Gulf in the east, is where the earliest evidence of agriculture appeared.*

number of people and animals to move into oases, where they were forced to live in close proximity with each other and where humans began to raise these animals for food. But a subsequent theory proposed that food-production began not in the oases but in the hilly flanks of the Fertile Crescent, because this was the area inhabited by the wild ancestors of some of the first plants and animals to be domesticated, including wheat, barley, flax, peas, lentils, cattle, goats, sheep, and pigs.

A still later theory proposed that agriculture began when it did because the Fertile Crescent was becoming overpopulated and food was becoming scarce. Before long, yet another theory proposed that humans had become dependent on certain species of plants and animals, which in turn were evolving into forms that were so useful that humans began to protect and propagate them. This was followed by the theory that pre-agricultural societies had actually created a surplus of food, allowing dominant individuals to compete for prestige and power by giving increasingly lavish feasts—and that the need to produce surplus food for such feasting stimulated the development of agriculture. Finally, a new theory proposed that after the end of the ice ages, the climate of the northern latitudes became more hospitable to plants, and this encouraged people to give up their previous nomadic ways of life, finding it simply easier to grow their own food than to hunt and gather for it.[1]

And these are only six of the best-known and most seriously discussed theories. In his exhaustive study of the origins of agriculture, the archeologist Graeme Barker listed no less than thirty-nine reasons that have been proposed over the years to explain the transition from foraging to farming, including big men, climate change, competition, desertification, diffusion, energetics, fat intake, feasting, hormones, intelligence, kitchen gardening, land ownership, marginal environments, natural selection, nutritional stress, oases, plant migration, population pressure, random genetic kicks, resource concentration, rich environments, rituals, sedentism, storage, technological innovation, water access, xenophobia, and zoological diversity.[2]

There are two major problems with these theories. The first problem is that some of the explanations actually seem to contradict each other. For example, one theory proposes that food production arose from a shortage of food while another proposes that it arose from an abundance of food. The second problem is that the events and conditions that are thought to have triggered the beginning of food production had all occurred previously in the history of the hominids. And when the same events and conditions have already occurred without producing the same results, it stands to reason that none of these conditions were sufficient, by themselves, to bring about the profound metamorphosis that took place when humans invented agriculture. Yet, within the space of a few thousand years, agriculture was invented in at least eleven different locations throughout the world.

The earliest evidence of agriculture appeared between eleven and eight thousand years ago, with the domestication of wheat, barley, flax,

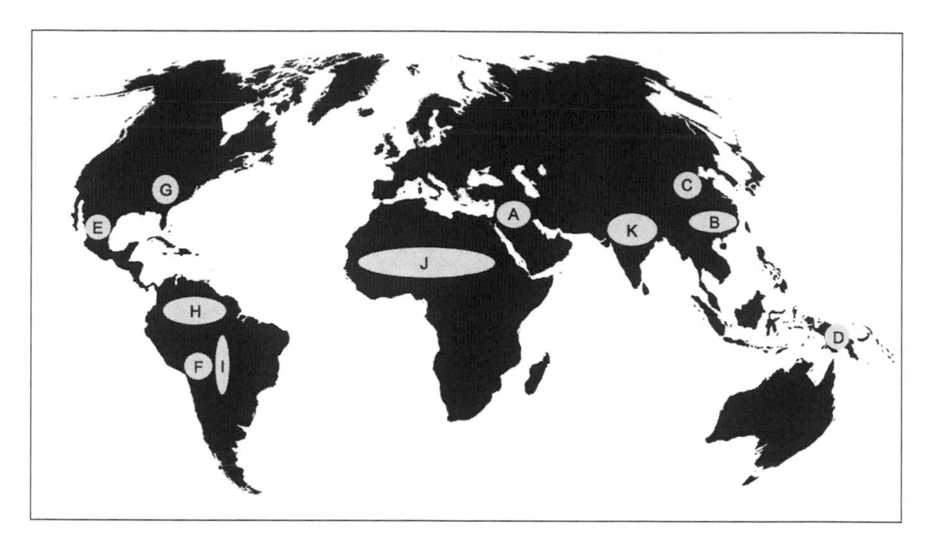

Figure 6.2. *The domestication of plants and animals began in at least eleven different locations between eleven and four thousand years ago in all the major continents except Europe. A-The Fertile Crescent; B-Central China; C-Northern China; D-New Guinea; E-The Valley of Mexico; F-The Andes Mountains; G-Eastern North America; H-Equatorial South America; I-The Western Amazon Basin; J-Sub-Saharan Africa; and K-Northern India.*

peas, lentils, cattle, goats, sheep, and pigs in the Fertile Crescent. But shortly afterward—between nine and eight thousand years ago—rice, chickens, pigs, cattle, and water buffalos were domesticated in the river basins of eastern China, and millet was being domesticated in northern China.

Maize[3] and squash were domesticated in the Valley of Mexico between eight and six thousand years ago, while yams, taro, and bananas were domesticated in the highlands of New Guinea about seven thousand years ago. At about the same time, yams, cotton, and sweet potatoes were being domesticated in the equatorial zone of present-day Ecuador and Columbia while peanuts, chili peppers, and manioc (also known as cassava and tapioca root) were being domesticated in the tropical lowlands of the Amazon Basin.

Potatoes and quinoa were domesticated in the highlands of present-day Peru and Bolivia between seven and five thousand years ago.

Sunflowers, gourds, and squash were domesticated in eastern North America between five and four thousand years ago, while at roughly the same time, millet, sorghum, yams, and coffee were being domesticated in sub-Saharan Africa and the horse was being domesticated in Central Asia.[4]

Why did such a fundamental change in the relationship between humanity and the natural environment take place all over the world between eleven and four thousand years ago? Why did agriculture appear almost simultaneously in so many different locations throughout the world? And why did agriculture begin in each region with a different mix of plants and animals from those that were domesticated in the Fertile Crescent, rather than spreading from a single point of origin in the Middle East? So the question remains: What was new or different about the humans of the Neolithic? What abilities did they have that enabled them to invent a radically different way of life, abilities that the hominids of the earlier periods did not possess?

Was a Spoken Human Language the Missing Ingredient?

"A propensity to interfere with the environment in unexpected ways," wrote the distinguished archeologist Andrew Sherratt in 1997, "seems to have been an inherent characteristic of modern humans; in this perspective, the 'origins of farming' have as much to do with the appearance of language and symbolic systems as they have to do with climatic change."[5]

Was the technology of symbolic communication the critical element that had been missing from human cultures during more ancient periods of human prehistory? Was "the appearance of language and symbolic systems," combined with the warming of the earth, the essential combination of factors that enabled humanity to finally begin producing its own food?

The evidence that has survived from the cultures of the Upper Paleolithic in Europe suggests that symbolic communication had not developed beyond a rudimentary stage even among anatomically modern humans until thirty or forty thousand years ago—after which, the evidence of music, art, design, and symbolism steadily increases in amount and variety. But throughout the entire era of the Upper Paleolithic, the earth's climate was dominated by the effects of the last ice age, and the habitats in which these prehistoric people lived were generally too cold and too dry to provide a favorable environment for growing crops.

The height of the last ice age—also known as the last glacial maximum—occurred about eighteen thousand years ago, after which the earth grew generally warmer, rainfall became more plentiful, and the conditions for growing food became more favorable. The glacial maximum previous to that occurred roughly 140,000 years ago, and the warm interglacial period that followed it lasted for about ten thousand years. But this was long before anatomically modern humans appeared in Europe—and probably well before any hominids had fully developed the true spoken languages that all humans speak today. If language was necessary to the development of agriculture, this would explain why agriculture did not begin during the interglacial period that occurred between approximately 130,000 and 120,000 years ago.

But by the time of the most recent glacial maximum, the cultures of the Upper Paleolithic had become rich with language and symbolism—a development that can be clearly seen in the plentiful remains of paintings, sculptures, and petroglyphs left behind by Upper Paleolithic people. The increasing complexity and elaboration of spoken language meant not only that these people had developed the ability to share and communicate vast quantities of complex information but also—with the development of the narrative that describes events unfolding over time—that they were able to express how events can follow each other in predictable sequences.

When humans developed fully elaborated languages, all of the detailed knowledge about the life cycles and behavior of the plants and animals in a society's environment—which had previously been limited to the knowledge that a single individual could amass in a single lifetime—became the accumulated wisdom of entire cultures. Because this complex knowledge was incorporated into the oral traditions of

Paleolithic societies, it became the collective wisdom of thousands of individuals, amassed over many generations.

The sheer amount of information needed to successfully grow crops and raise domesticated animals may have simply been beyond the capacity of any single individual to acquire and remember. It is therefore possible that the pooling of all the knowledge and experience of thousands of individuals in the form of oral traditions, combined with the warmer and wetter climates that had developed by eleven thousand years ago, was the fateful combination of factors that made possible the transition from foraging to farming.

There was, however, one more essential element that had to be added to this mix before the complete transformation to an agricultural way of life could take place: The people of those times had to give up their wandering ways and to begin living permanently in one location. Although it would make sense to suppose that people only started to live in one place when they first began to grow their own food, the archeological evidence from the Fertile Crescent shows exactly the opposite: Living in one place came first and the practice of agriculture came afterward. In fact, the farming communities of the Neolithic do not appear in the archeological record until nearly a thousand years after people began living in one place.

Living in One Place

Before the metamorphosis of agriculture, when a group of nomadic hunters and gatherers found themselves in favorable living conditions and began to grow larger than a few dozen people, the normal patterns of sharing and cooperation would have become increasingly unwieldy. People would have tended to divide into factions, and conflict among different factions would have become more open and more frequent. Eventually, these oversized nomadic groups tended to split into two or more groups and go their separate ways—an example of the "fissioning" process that occurs from time to time in the fission-fusion society typical of humans and other primates. The nomadic way of life made it easy and natural for groups to move on to new territories, and the process of fissioning acted to keep the size of the hunting and gathering group relatively small, making it easy to maintain solidarity and facilitate cooperation among the members of the group.

But agricultural societies do not have this luxury. Having become dependent on certain crops that grow every year in certain places, self-interest dictates that agricultural people stay in one place. Having invested countless hours building their large, permanent houses, they cannot simply pack up and move on to better hunting grounds. So their settlements remain fixed in place, and if food is plentiful, their social groups grow ever larger and cannot easily split apart.

As early as seventeen thousand years ago, a few isolated groups of modern humans called "Natufians"—most of whom were living in or near the Fertile Crescent—began to harvest large quantities of wild grain and store this grain for long time periods. And the archeological remains they left behind were unlike those of any other previous human groups before them.

The houses of the Natufians were solid constructions of mud brick, floored with a cement-like substance made of lime, pounded flat and smooth. They made sickles out of small, sharp microliths—extremely small stone tools hardly larger than postage stamps—which they cemented into curved wooden handles, and they used these sickles to harvest the wild emmer wheat and einkorn wheat that grew in abundance in the hills of the Fertile Crescent. They fashioned heavy mortars out of large volcanic stones and used these mortars to grind the wild grain into flour. And they built special structures to use as granaries for storing large quantities of grain, with plank floors raised above the earth to protect their precious stores from insects and rodents.

All of this evidence indicates that the Natufians had begun living in one place throughout the year—a way of life that anthropologists call "sedentary" (from the Latin word *sedentarius,* or "sitting"). By abandoning nomadism—even while they continued their hunting and gathering ways—the Natufians took the first critical step on the road to food production. But they remained foragers. The grains they harvested grew wild; there is no evidence that they either prepared the ground or planted the seeds for their cereal crops. The animals they killed and ate were also wild species; there is no evidence that they kept animals in pens or stalls and fattened them for butchering later on.

But when they settled down and began to live in permanent houses, kept their food in permanent storehouses, processed their food with devices that were too heavy to move from place to place, and established permanent villages that remained in one place for generations, the Natufians began a way of life that eventually made possible the beginnings of agriculture. And when they made the transition to producing food instead of merely hunting and gathering for it, the descendants of the Natufians had little choice but to stay in one place.

Once the agricultural way of life took hold, it was necessary to prepare the ground for planting, then to plant seeds, then to weed and (in some cases) water the growing seedlings, and eventually to harvest the mature crops. After the harvest, it was necessary to store sufficient quantities of food to last until the next growing season and to stay in one place to protect the stored food from the appetites of both animals and other humans.

By eleven thousand years ago, unmistakable evidence had begun to appear that certain societies inhabiting the Fertile Crescent were deliberately planting and growing their own crops, and the wild grains that had been harvested were gradually being replaced by domesticated varieties. The domesticated cereals had more grains on each stalk and produced bigger yields. And the heads of domesticated grain did not shatter as easily during harvesting as did the heads of wild grain. Thus these ancient farmers were able to harvest the ripe stalks without worrying that the grain would all be scattered on the ground before it could be brought back to the village for threshing.

The Pursuit of Material Wealth

There is a limit to the amount of material goods that nomads can accumulate, because people who are perpetually on the move cannot keep more material goods than they can carry with them from place to place. And since the members of a nomadic group typically cooperate in many aspects of hunting and gathering—and they have limited opportunities to store food for long periods—they tend to share most of their food with their kinfolk on a daily basis.

But sedentary people, who do not have to change their place of residence several times a year, are not bound by these limitations. Having many options for storing food for the future, they do not necessarily share it with people outside their immediate families. And because they live in one place, they can accumulate as much material wealth as they can reasonably store and protect against theft by other people. Thus, as soon as people began to live in one place, the pursuit of material wealth assumed an importance that it had never had before. And the increasing importance of wealth and status can be clearly seen in the evidence that has survived in the Fertile Crescent from Neolithic times, when the Natufians adopted the practice of harvesting and storing their wild grain in granaries.

At first, when the Natufians built permanent settlements, they constructed their granaries separately from the houses of the individual families. This suggests that the Natufians' supplies of grain were the common property of the group as a whole. After all, the custom of sharing in common the supplies of food that the group as a whole depended on would have been the natural legacy of the Natufians' long history as hunters and gatherers.

But after the passage of about a thousand years, the free-standing granaries of the Natufians disappeared, and these were replaced with storage areas located inside of the individual houses. From that time forward, Neolithic granaries were always incorporated into the houses inhabited by individual families. This indicates that the world view and value systems of these societies were changing—and that instead of regarding the bounty of nature as something to be shared among all the members of the group, the Natufians were beginning to regard food as private property to be kept by the individuals and families that had originally produced it.

In similar fashion, the territory that each group of nomadic hunters and gatherers had always claimed as its own—the territory that was analogous to the "homelands" of the ape and monkey groups described in chapter 1—began to lose its original status as the shared resource of the entire social group. Although the areas where wild game could still be hunted and wild plant foods could still be gathered probably retained their original status as communal property, the smaller areas of fertile ground that were suitable for agriculture gradually became the private property of individual families. Eventually, the best areas of arable land were passed down from parents to offspring as the inheritance of family property.

Evidence from many preindustrial societies shows that, once a society begins to live in one place—even if it continues to live by hunting and gathering and has not developed an agricultural way of life—it begins to pursue the accumulation of material wealth in ways that are remarkably similar to the traditions of agricultural people. These similarities are clearly evident in the cultures of the sedentary hunting and gathering people of the Northwest Coast.

The Nootka: Hunters and Gatherers Living in Permanent Villages

The Nootka of British Columbia were typical of the Native American tribes who inhabited the Northwest Coast of North America—all of which were hunters and gatherers and none of which ever practiced agriculture. The Nootka hunted whales, seals, otters, and porpoises in their coastal waters, as well as bear, deer, and elk in the inland forests. They fished for halibut, herring, and cod, and they gathered shellfish from the sea and the roots and berries that grew in abundance in the rain-soaked fields and forests nearby. And in addition to all this natural abundance, they harvested immense numbers of salmon that made their annual spawning migrations up the many rivers that flow into the sea. The salmon were caught in nets, traps, and weirs and then dried and smoked in quantities great enough to provide the Nootka with a larder of seafood large enough to last throughout the year, until the salmon ran again.

Because the Northwest Coast is so rich in wild foods—and because the abundant sea life was constantly being replenished by natural processes—the Nootka had no need to pursue the nomadic existence that is typical of the lifestyle of most hunting and gathering societies. The Nootka had long ago settled down in permanent villages located in favorable places along the many inlets that run down to the sea from the coastal mountains.

Since they stayed in one place, the Nootka were not limited to the simple, temporary huts of nomadic people—which were typically built in a matter of hours or days and were normally abandoned after having been lived in for a few weeks or months. Instead, the Nootka built immense "long houses" some of which were as large as forty feet wide and one hundred feet long, constructed from logs of cedar, fir, and hemlock that grew in abundance in the cool, moist forests of the Pacific Northwest. Although the Nootka long houses were dismantled and moved a few miles twice each year from the summer camps at the seashore to the winter camps in the interior and back again, they were permanent dwellings that were inhabited for years by the same group of related families.

Because rich sources of plant and animal foods were always found in the same locations, it was customary among the Nootka for the oldest male child to inherit the rights of stewardship over specific hunting and fishing grounds and over the places where plant foods grew in abundance. When another kinsman exercised his or her hereditary rights to hunt or gather food in these areas, a portion of the spoils had to be given to the individual who "owned" the rights to those resources. Since these rights were passed from the father to the eldest son, they created in effect a "class system" in which the oldest children enjoyed significant wealth, while the younger children—who were often members of the same family—lived in relative poverty.

Among the Nootka, as among the other sedentary tribes of the Northwest Coast, a complicated system of status and prestige was of paramount importance in their culture, and the chieftainships that were handed down from fathers to eldest sons carried with them not only economic advantages but also a number of symbolic rights and privileges. These included honorific titles, the right to sing specific ceremonial songs and to perform certain ceremonial dances, the right to dress in certain kinds of clothes, the right to wear certain kinds of personal ornaments, the right to sit in particular seats (carefully arranged in order of rank) that were traditionally used during ceremonial feasts, and the right to live in the most favored areas of the long house.

In these status-conscious societies, high-status individuals competed ferociously with each other by providing immense feasts, or *potlatches*, at which immense quantities of food, blankets, tools, and weapons were consumed, given away, or in some cases even destroyed by their owners in orgies of conspicuous consumption. Nor was the ownership of property limited only to material goods: The Nootka, like most of the tribes of the Northwest Coast, even practiced a form of slavery—something that we usually associate with more "advanced" civilizations. When the members of enemy tribes were captured in warfare, they became the property of the warriors who had captured them, and they remained a part of their owners' households, where they were destined to perform the most menial tasks of Nootka life.

All of this behavior would have been thoroughly distasteful to the typical society of nomadic hunters and gatherers, among whom the attainment of high status was earned by exemplary behavior and by performing acts of bravery and generosity. Hunting and gathering societies displayed little interest in accumulating material goods. They shunned ostentatious displays of material wealth, and they accorded prestige only to men who were successful hunters, warriors, and providers and only to women who were both productive gatherers and exemplary wives and mothers.

Inheriting Rank and Status

When the inheritance of wealth and property became firmly established as a traditional part of the cultures of agricultural people, it was inevitable that the offspring of the wealthiest individuals and families would begin life with more goods and privileges than people whose ancestry was less fortunate. In the societies of sedentary agriculturalists, the divisions between rich and poor that had developed among sedentary hunters and gatherers such as the Nootka became even more extreme with the passing generations.

Instead of differences in wealth being based on the difference between older and younger siblings—as was typical of the people of the Northwest Coast—differences in wealth among agricultural societies gradually became the privilege of entire families and was inherited from parents to offspring. These inherited differences in wealth and status ultimately led to the formation of permanent social classes and to the evolution of "stratified" societies, characterized by traditional class systems with inequalities in wealth, status, and privilege that endured for generations.

Clear evidence that the spread of agriculture led to the rise of social classes and to institutionalized inequalities in wealth and status has been found in the archeological remains of the earliest agricultural people of Neolithic Europe. These people, called the *Linearbandkeramik* or "linear band ceramic" people, appeared in eastern Europe about 7,500 years ago and spread rapidly during the next five hundred years until they had established their agricultural settlements in the fertile river valleys across all of Europe.

The *Linearbandkeramik* people buried all of their dead, but while about half of the men were buried with polished stone adzes, the other half were not. The adze was used by Neolithic people for important woodworking tasks, and it was a valuable item that, when included in a grave, signified that the deceased enjoyed high status in life. In 2012, a team of archeologists published the results of a study of the tooth enamel from three hundred skeletons from *Linearbandkeramik* burial grounds across Europe from France to Hungary.[6] By analyzing the relative proportions of two types, or isotopes, of strontium that are incorporated into tooth enamel during childhood, the scientists were able to determine by these "isotope signatures" whether a particular individual grew up in the fertile river valleys, where the rich loess soil provided a relatively high standard of living, or in the less fertile hills that surrounded these river valleys, where life was harder and the standard of living was more meager.

Not surprisingly, the men who were buried with stone adzes had the isotope signatures of having grown up in the rich river valleys, while the isotope signatures of those buried without stone adzes indicated that they had grown up in the poorer hill country. These differences in wealth and status became more marked and more obvious as time passed, as agricultural villages grew larger, and as agricultural families accumulated ever greater amounts of wealth and property.

Later in this chapter, we will review the evidence that the *Linearbandkeramik* people brought not only agriculture but also organized warfare into Neolithic Europe. But first, it is necessary to add two more items to the long list of changes in society that were wrought by the adoption of agriculture. The first is that children became economic assets to their families. The second is that the sexuality of women became a threat to the stability of society.

Children as Property

The children of hunters and gatherers typically had few opportunities to contribute in a meaningful way to the support of their families and kinfolk, because nearly all of the food-getting activities of hunters and gatherers required the strength, stamina, knowledge, and experience of adulthood. The children of farmers and herders, on the other hand, could easily become valuable assets to the wealth, stability, and longevity of the family unit.

The enterprise of hunting wild game was a dangerous and complicated undertaking that required discipline, skill in the use of weapons, and the strength of an adult. While the children of hunter-gatherers often played at hunting using toy weapons, they were not able to pursue and kill anything more significant than a large insect, a small reptile, or a baby bird. In fact, it took years of experience as part of many hunting parties before a young man had acquired sufficient knowledge and skill to become a productive hunter. He had to learn the habits of each of the many species of game animals. He had to learn how to track them, corner them, kill them without being wounded, butcher them, defend the kill from the appetites of other predators, and transport the edible and usable parts back to the campsite where the flesh could be cooked and eaten—and where the skins, bones, and sinews could be processed for use as hides, tools, and cords.

Similarly, the task of gathering vegetable foods required detailed knowledge of where to find the edible plants and their roots, tubers, fruits, and seeds. It required the ability to recognize their telltale signs in the immensity of the forests and in the tangled thickets of the underbrush. A woman had to know the times of year each of these foods would be available and edible, how to transport them back to the campsite, and how to process them—by grinding, mashing, soaking, roasting, and boiling—so that they could be eaten and digested in safety. Moreover, in the case of edible roots and tubers—which formed a significant part of the hunter-gatherer's food supply—it required considerable physical strength to wield a long, heavy stick and dig up the vegetable foods that could often grow at a considerable depth beneath the surface.

By contrast, the children of agricultural people could easily master many of the essential tasks involved in growing crops and tending animals. It does not require a great amount of skill, years of experience, or the strength and stamina of an adult to herd a flock of goats, feed domesticated pigs in their stalls, pull weeds out of nearby gardens, or even harvest some of the crops that farmers grow. And the children of agricultural people can easily care for their younger siblings, allowing their mothers not only greater freedom to work in the fields but also to bear children more frequently than the women of hunting and gathering societies. Thus, as foragers gradually evolved into farmers, children became increasingly capable of contributing to the support of the society as a whole—and in the process, they became increasingly valuable to their families.

The economic importance of children was further reinforced by the tendency of agricultural people to survive into their fifties and sixties much more often than their hunting and gathering counterparts. The daily activities involved in raising crops and breeding animals are generally not as dangerous as the life-threatening activities involved in hunting and killing wild animals or making long journeys through the predator-infested wilderness to gather edible plant foods and transport them back to the home base. For these reasons, as raising crops and breeding animals became more widespread, agricultural people began to regard their children as a kind of insurance policy—a means of support when they grew too old to produce their own food—and the task of providing material support to the elderly became increasingly the obligation of children toward their parents.

For all of these reasons, agricultural societies tended to cultivate strong and often exclusive bonds between parents and children. They strove to instill in their children a sense of duty toward their parents, and, with few exceptions, they placed a high value on marital fidelity—especially on the part of women—to ensure that the paternity of a man's child was never in doubt. In this way, the value of children to their parents in agricultural societies led over time to a culture of fierce possessiveness toward children and to severe restrictions on the sexual freedom of women. In fact, these cultural traits became so deeply ingrained in the cultures of our agricultural ancestors that both of them have persisted into the modern age.

The Menace of the Sexual Woman

In most hunting and gathering societies, the institution of marriage was typically a very casual affair. In fact, marriage among hunter-gatherers usually involved little more than the decision by an adult man and woman to begin publicly eating and sleeping together on a regular basis. Although the physical maturation of boys and girls at puberty was often celebrated in hunter-gatherer societies with lengthy and elaborate initiation ceremonies, it was rare for their marriages to be observed by any kind of major ceremony. And it was equally rare that the newly "married" couple required—or even requested—the consent of their parents. Instead, it was typically only necessary that the kinship relationship between the man and the woman be such that sexual relations between them was not defined as incestuous.

If—after a man and a woman had lived together for a while—they did not get along well and became estranged, took other lovers, or simply became tired of each other, divorce was achieved simply and quickly and with little ceremony. In fact, divorce usually involved little more than the simple decision by one of the partners to move out of the home they had been sharing. The striking simplicity of both marriage and divorce in hunting and gathering societies was largely due to the fact that the most important social relationships were based not on marriage but rather on membership in the clan or lineage into which every person was born. As this membership was a birthright, it was not changed or affected by marriage or divorce.

In most hunting and gathering societies, sexual experimentation began in childhood, and with the coming of sexual maturity, adolescent sexual liaisons were common. These liaisons rarely resulted in pregnancies,[7] and they were typically carried out by both sexes for several years before a preferred partner gradually emerged. Although there were differences among hunters and gatherers in the customs surrounding marriage and raising a family, it was typical of most such societies that children and adolescents, especially girls, enjoyed—at least by our standards—a shocking degree of sexual freedom.

Of the few hunting and gathering societies that survived into the twenty-first century, several have been studied in depth by cultural anthropologists. In a cross-cultural study of infancy and childhood among six surviving hunter-gatherer societies, the anthropologist Melvin J. Konner noted that, among all of these societies, sexual freedom in childhood and adolescence is socially condoned and generally practiced as the normal behavior of both sexes.[8]

Konner described the sexual experimentation among the !Kung Bushmen of South Africa as beginning in early childhood and continuing through middle childhood and adolescence. Although adults do not openly approve of sexual play and discourage it when it becomes obvious, the !Kung regard the sexuality of children and adolescents as normal, and children who openly engage in sexual behavior are lectured but not punished. In fact, the !Kung consider sexual activity to be essential for proper mental health, and they believe that sexual deprivation in adulthood is the most likely cause of mental illness.

Konner noted that among the Efe and Aka pygmies of equatorial Africa, premarital sex is practiced openly by both sexes. Among the Efe, a girl's first menstruation is an occasion for public celebration, and no serious attempt is made to restrict sexual activity in adolescence. "At a dance," a teenage Aka boy reported, "the young people flirt and get together for sex. A girl can have different boys on the same day and take turns."

Among the Hadza of Tanzania, sexual experimentation in childhood is also practiced openly, and premarital sex is a routine and expected part of adolescent life. Among the Ache of Paraguay, both boys and girls begin experimenting with sex around twelve years of age, and a girl's first menstruation is celebrated by a public ritual of initiation and purification that includes all the men who have had sex with her. Finally, among the Agta of the Philippines, premarital chastity is not expected of girls, who are able to engage in sexual activity with relative ease.

A similar pattern of premarital sexual freedom has been observed by other anthropologists among the Inuit Eskimos of the polar Arctic, where young girls were free to take lovers at will, and where married men famously shared the sexual favors of their wives and daughters with their visitors and guests.

Finally, among the Australian Aborigines, girls were often promised in marriage before they were even born, but they enjoyed considerable sexual freedom both before and after marriage.[9] And although a man in these societies was typically expected to be mature enough to provide a steady supply of food before he married, he was free until that time to pursue as many sexual liaisons as he could arrange. With the arrival of European settlers in Australia, however, the traditional patterns of betrothal in infancy and marriage before sexual maturity were largely replaced by a Western model, in which girls increasingly refused to marry the men to which they had been "assigned" and rarely married until their late teens, when they were able to choose their own partners.

But the sexual freedom enjoyed by young people, however typical it may have been throughout most of human prehistory, was only feasible in societies where a woman was free to choose her own lovers and husbands. In societies that practiced agriculture for a long time—and especially in societies in which the inheritance of land was important—the choice of a partner, and the permanence of marriage, was a very serious matter indeed.

To allow premarital sexual freedom in agricultural societies would have put the careful plans of the parental generation seriously at risk, for men and women naturally become attached to their sex partners. If those partners turn out not to be the ideal husbands or wives from the point of view of the inheritance of property, the maintenance of the farm, and the support of the parental generation, it could be very difficult for the parents to persuade their sons or daughters to marry someone else. Furthermore, if a man and woman decided later on to separate and have children by other people, the question of how their property should be inherited would become a complex and contentious issue that could easily escalate into public conflict and long-term resentment.

Thus in the oldest of the agricultural societies, where families have lived for generations in the same villages and have passed the same plots of land down to their children and grandchildren for centuries, the sexual interests of young people—and especially of women—pose a distinct threat to the harmony of village life and to the stability of the entire society. It is therefore not surprising that in the traditional agricultural societies of China, India, and Europe, premarital chastity—especially for women—was considered a paramount virtue, and it was ruthlessly enforced by the parental generation.

So thoroughly did these cultures stamp out the normal tendency of children and adolescents to engage in sexual experimentation that boys and girls were often segregated from each other from an early age. In these cultures, the virginity of a woman at the time of marriage was often considered a sacred requirement, and unmarried adolescents were never allowed to spend time together except under the watchful eye of a chaperone. In many of the traditional agricultural villages of Europe, it was even customary from medieval times until the late nineteenth century to display the sheets from the marital bed in public on the morning after the wedding night, so that all could view the bloodstains that resulted when the hymen of the virgin bride was ruptured by the groom during her first act of sexual intercourse.

But not every hunting and gathering society was permissive toward sexuality. The most common cause of serious conflict among adult males in most human societies is jealousy over the sexual favors of women, and conflict over sexual relationships with females has also been cited as the most common reason for lethal battles between male chimpanzees. While conflict between males over the sexual attention of females may produce only loud arguments and public recriminations in the more peaceful hunting and gathering societies, such conflicts among the more warlike societies could often lead to the death of a productive hunter and warrior in the prime of life. This is why, among many of the most warlike hunting and gathering cultures, female chastity was encouraged and highly valued.

The fiercely warlike Cheyenne, for example, had a sexually repressive culture when they were studied by anthropologists in the early twentieth century. The Cheyenne believed that sex robbed a man of his power in war and hunting, and some Cheyenne men were known to refrain from sex for years at a time, believing that by doing so they would eventually father powerful, warlike offspring. Not surprisingly, Cheyenne women were widely known for their "chaste demeanor," which was a sharp contrast to the often flirtatious behavior of women from other Native American tribes.

The value of this kind of sexual repression is obvious when one considers the problems caused by sex among the warlike but more sexually permissive Yanomamö of the Amazon rainforest, where sexual affairs were the cause of much male violence and where fights over women would sometimes escalate into serious warfare between Yanomamö groups.[10]

For the most part, however, sexual freedom typically became a threat to the stability of human societies when men and women became heirs to wealth and property. For this reason, nearly all agricultural societies favored permanent, lifelong marriages—typically arranged by the parents of the bride and groom—that were designed to maximize the stability and longevity of the nuclear family. Thus, over the thousands of years that people have been staying in one place and growing their own food, a general repression of sexuality—harshly enforced especially among children and females—gradually became the norm.

The Genesis of Organized Warfare

As foragers gradually became farmers, the food and material goods that were accumulating in the granaries and storehouses of agricultural villages inevitably became tempting targets for raids by enemy tribes. And the aggressive instincts of adult males, which had evolved over the millions of years that men hunted wild game, became increasingly channeled into predatory activities against other human groups.

As organized hunting by groups of males for large game became less and less important as an economic necessity, organized warfare by males for the land, material wealth, and women of other agricultural villages and rival tribes became an increasingly attractive strategy for acquiring wealth, property, and descendants. Thus, as the permanent villages of Neolithic cultures expanded throughout Europe, the appearance of increasingly massive fortifications signaled the beginnings of chronic, organized warfare.

The technologies of agriculture spread into eastern and western Europe with the *Linearbandkeramik* culture that appeared in the valleys of the Danube, Elbe, and Rhine rivers in present-day Germany beginning about 7,500 years ago from its origin in Hungary and Serbia. For the next five hundred years, the *Linearbandkeramik* people migrated rapidly up the river valleys and across Europe as far as France and Belgium in the west and as far as Ukraine in the east, displacing—and in some cases exterminating—the indigenous populations of hunter-gatherers who had been living there. They found the rich deposits of loess soils—which were particularly plentiful in the river valleys—ideal for growing their crops of wheat, barley, millet, rye, peas, lentils, and beans. In addition to hunting deer and wild boar in the open forests, the *Linearbandkeramik* people raised domesticated cattle, goats, and pigs.

Although it was once believed that these early farming people were peaceful, evidence of violent killings have recently been found in many *Linearbandkeramik* sites, including the evidence of mass executions in which entire Neolithic villages were exterminated. Archeologists have found a mass grave at Talheim in the Rhine Valley where thirty-four people—nearly half of them infants and children—were killed by being struck in the head with a blunt instrument—most likely a Neolithic stone axe.

The partial excavation of a similar grave near Vienna yielded the remains of sixty-six people who had been killed in a similar fashion—and archeologists estimated that the gravesite as a whole probably contained the remains of at least three hundred people. And at Herxheim, also in the Rhine Valley, the remains of more than three hundred people were scattered throughout a single settlement, none of them buried properly in graves, including 173 skulls of victims that were decapitated and thrown unceremoniously into ditches.

The *Linearbandkeramik* people may have dumped the corpses of the enemies they massacred into mass graves, but they buried their own dead carefully in graves. When these graves have been excavated and the skeletons examined, archeologists have found evidence of traumatic injuries to as many as one-third of the adult men, indicating that violence among these people was as great as it was among the most violent and warlike societies studied by contemporary anthropologists.

Moreover, the incidence of violent death rises dramatically as one moves from east to west, suggesting that Neolithic warfare was more common in the areas where the *Linearbandkeramik* people had more recently moved into territories inhabited by hunters and gatherers, while violence was less common in the areas that had been settled for longer periods of time. Finally, the villages of the *Linearbandkeramik* people were usually fortified with wooden palisades and often ringed with moats or ditches that sometimes completely encircled their settlements. This indicates that as these agricultural people moved up the river valleys of prehistoric Europe, they were taking over territories previously inhabited by indigenous hunters and gatherers, warring with them and in many cases exterminating them.

This is not to say that hunting and gathering nomads were always peaceful and never went to war. The Cheyenne were skilled and disciplined fighters who took great pride in their victories and boasted throughout their lives about the enemies that they killed. But the "wars" of hunters and gatherers were typically raiding parties, not unlike the raids conducted by the chimpanzees of Ngogo in Uganda described in chapter 1.

A typical raiding party by hunters and gatherers consisted of a group of males who would make a stealthy incursion into the territory of a neighboring group. If a member of the "enemy" group was encountered, he or she would be promptly killed, and the aggressors would quickly retreat to their own home territory. Even when a large raiding party attacked an enemy village or encampment in force, the aggressors rarely attempted to exterminate the entire population, appropriate all of their land, or steal all of their belongings—although it was not unusual for females of childbearing age to be captured and brought back as the spoils of war.

But when humans began to live permanently in one place and accumulate wealth, war became progressively more deadly. The Nootka of the Northwest Coast—who prided themselves on their calm demeanor and even-tempered natures—were merciless in war. When they attacked another settlement, the Nootka typically exterminated all of the inhabitants—except for the few individuals who were captured to become slaves—and made off with everything they could carry.

As a general rule, the larger and the more permanent the agricultural settlements became, they more devastating was their practice of warfare. And when the first urban civilizations arose in the valleys of the Tigris, Euphrates, Nile, Indus, Yangtze, and Yellow rivers—and the smelting of metals in the Bronze and the Iron Ages greatly increased the effectiveness of lethal weapons—the role of the lifelong, professional soldier was created. The ancient urban civilizations all organized standing armies, and warfare became a strategy of conquest, carried out on a large scale, for the purpose of dominating and controlling ever larger areas of land and ever larger numbers of people. None of this would have been possible without the transition from foraging to farming.

Seeds of Civilization

As the agricultural villages multiplied, neighboring villages began to form alliances to repel raiders and invaders from other regions. In time, individual villages began to specialize by producing certain crops or manufacturing certain types of material goods, and before long, trading places began to appear in geographically favorable locations, where people could meet on a regular basis to barter for the goods they needed and exchange their surplus products for things that were in short supply in their own village. In many cases, these markets were situated in centrally-located villages, which grew over time into market towns.

All of these events were made possible by the process of social fusion, which enabled communities to grow larger than human communities had ever been before. The proliferation of agricultural villages created densely populated regions in which everyone spoke the same language, worshipped the same gods, ate the same foods, lived in the same kinds of houses, wore the same kinds of clothes and jewelry, and observed the same customs and taboos. In the process, people came to identify not with a few dozen kinfolk but with thousands of other people who shared their common culture and ethnic identity.

♦

In time, the societies of these regions began to invent and adopt new technologies of interaction—the sailing ships, wheeled vehicles, domesticated horses, and systems of writing that made it possible for people and communities to interact over time and space. As a result, the market towns, religious centers, and communities of wealthy and powerful families gradually grew into urban centers that began to dominate the smaller and less powerful neighboring towns. Eventually, as the technologies of interaction grew in sophistication and effectiveness, these expanding urban centers knitted the settlements of entire regions into a network of allied communities, and in time these communities became the seeds of the urban civilizations that sprouted, one by one, in the cradles of civilization scattered across the globe.

In the final analysis, it was the technology of agriculture that freed our species from the endless pursuit of something to eat and made it possible for humans to settle down in one place and build enduring houses, palaces, temples, and monuments that they could bequeath to future generations. In the next chapter, we will see how, when agriculture made it possible to live in one place, human society was utterly transformed once again, and how it eventually became a force large enough and powerful enough to change the world.

CHAPTER 7

The Technologies of Interaction

Ships, Writing, the Wheel, and the Birth of Civilization

"Innovations in transportation technology are among the most powerful causes of change in human social and political life."

—David W. Anthony,
The Horse, the Wheel, and Language

In April of 2006, the American author and geographer Jared Diamond stood in line at the airport in Port Moresby, New Guinea, preparing to board a flight to the interior of the island, where he was conducting a long-term study of the people of the New Guinea Highlands. Diamond mused about the fact that although the people crowded into the Port Moresby airport were strangers to each other, there was no sign of hostility or violence. Although this is "a feature that we take for granted in the modern world, [it] would have been unimaginable in 1931," Diamond observed, "when encounters with strangers were rare, dangerous, and likely to turn violent." In 2006, New Guineans were living in a modern society, complete with officers of the law who were ready to intervene in the event of violence. But in the New Guinea of 1931 "the idea of traveling from [the villages of] Goroka to Wapenamanda, without being killed as an unknown stranger within the first 10 miles . . . would have been unthinkable."[1]

The New Guinea Highlanders were by no means unusual in their deep suspicion of strangers. For most of human history, people had lived out their lives within small nomadic bands consisting of only a few dozen individuals, and they were acquainted with at most a few hundred members of neighboring bands and fellow tribesmen. Anyone outside of this small circle—even a person who spoke one's own language and was of one's own culture—was considered not only a stranger but also a potential enemy, and was regarded as a person to be feared and avoided. Moreover, anyone who did not speak one's own language and was not of one's own culture was typically considered less than human—a reflection of the ancient division between "us" and "them" that can be seen in every human group and in every species of non-human primate.

But when hundreds of thousands of people, scattered over thousands of square miles, all became the members of a single society under the authority of a single urban civilization, humanity was liberated from the ancient fear of strangers that had been a constant menace in the simpler societies of nomads and villagers. The centralized authority of civilized societies could not tolerate the raids, abductions, murders, and revenge killings that were a regular feature of life in simpler human societies. Civilizations can only function if their citizens are able to travel, trade, and communicate with each other in confidence, free of the constant fear of violence or injury. This goal had been partly accomplished by the shared language, symbolism, and ethnic identity of tribal affiliation—but the technologies of interaction enabled civilized societies to go one step further.

The technology of agriculture made it possible for some of the people living in sedentary villages to specialize in the arts and crafts that could be traded for the food that others produced. The ancient cities were the first human settlements composed mainly of people who were free from the need to find or produce food, and the ancient civilizations were the first human societies in which large numbers of complete strangers were free to interact in an atmosphere of safety and confidence. These were entirely new developments in the history of human society.

Carpenters, Weavers, Sailors, and Scribes

As the practice of agriculture gradually spread over the Neolithic world and people settled down to live in permanent villages, the most successful farmers found themselves producing more food than they needed to consume, and surpluses began to accumulate in their granaries and storehouses. And unlike people in hunting and gathering societies—in which virtually all of the adult members of the social group devoted most of their productive time to hunting and gathering for food—some of the Neolithic villagers began to make things which they could trade for the surplus food produced by others. For the first time in human history, numerous members of the human group became liberated from the need to endlessly hunt for and gather food, and the division of labor we call "craft specialization" was born.

Some villagers began to specialize in the labor-intensive process of grinding and polishing suitable types of stone into the weapons, woodworking tools, and agricultural implements that the Neolithic way of life depended on. Others became carpenters and specialized in building the roofs, doors, windows, and furniture for the permanent dwellings that the Neolithic families lived in for years at a time. Still others became weavers and learned to spin and dye thread and weave it into cloth, while others specialized in the curing of hides and the fashioning of leather goods.

Eventually, in the later Neolithic societies, some people became potters and produced the ceramic vessels used for drinking, cooking, and storing grain. Carpenters living on the sea shores and river banks began to specialize in building and repairing boats, and they became the first shipwrights. Other carpenters learned how to build and repair wheeled carts drawn by oxen and onagers, and they became the first cartwrights and wheelwrights.

Some of the agricultural villages that originally began with a few dozen inhabitants eventually grew into towns with thousands of inhabitants. And the new technologies of transportation enabled the largest of these towns to extend their influence over the inhabitants of many smaller towns and villages in their proximity. Slowly but surely, the larger and more powerful Neolithic settlements began to dominate the smaller towns and villages that were located nearby.

In time, the steady accumulation of wealth, military power, and religious authority by communities that were located in strategic and well-defended locations led to the development of a small number of powerful, fortified urban centers. These settlements, considerably larger than any that had ever previously existed, came to exercise commercial, military, and religious authority over extensive rural populations, and the "city-state" was born. These were the first human societies in which people could be complete strangers to each other and yet could live and work side by side without hostility or distrust. We will return to this very important point later in this chapter.

With the rise of the city-state, spoken language was no longer enough to serve the interests of these new civilized societies, and true writing—that preeminent technology of communication—soon arose to enable people to communicate with each other over time and space. Writing enabled merchants to buy and sell goods from each other without personally traveling to each of the settlements they traded with. Writing enabled political and religious leaders to send information and give orders to their subordinates in far-flung communities. It enabled them to ask questions and expect answers. Finally, writing enabled the early bureaucrats to keep records of transactions, of goods purchased and stored in warehouses, and of tribute and taxes paid to them by the ordinary people over whom they ruled.

Yet even as they came under the economic influence and political control of the rising urban centers, most people continued to live and farm in the countryside, raising their own livestock and growing their own crops. This division of society between relatively impoverished and powerless farmers on the one hand and much wealthier and more powerful city dwellers on the other became a feature of every urban civilization in ancient times, and it has persisted in civilized societies throughout the rest of recorded history.

Over the course of the past five thousand years, thousands of city-states and hundreds of empires have come and gone, but the hallmarks of urban civilization have remained essentially the same until the emergence of industrial society two hundred years ago. A large but powerless agricultural population lived in the countryside and produced the food consumed by society as a whole. At the same time, a small but powerful ruling class—including an extensive bureaucracy, an exclusive priesthood, and an organized military force—lived inside the confines of an urban center that was protected by massive fortifications and filled with monuments, palaces, and temples.

Century by century, civilized societies spread from their places of origin into every continent and eventually came to dominate nearly all of humanity. But civilization itself might never have come into being had it not been for the unique combination of factors that existed in the world's great river valleys. For it was on the banks of the world's great rivers that urban civilization first took root.

The Magic of Rivers

It is no coincidence that the first urban civilizations—which arose independently within a thousand years of each other in three widely separated regions—all arose in the areas that are home to the world's largest and most fertile river valleys. The first of these areas is the Fertile Crescent, where the valleys of the Nile, Tigris, and Euphrates rivers are located. The second is the valley of the Indus River in northern India. And the third is the alluvial plains of East Central China, where the valleys of the Yellow and Yangtze rivers are located. Each of these regions eventually became its own "cradle of civilization."

There are three important reasons why urban civilization first began in these river valleys rather than in some of the other regions, such as the Highlands of New Guinea, where the transformation into a sedentary agricultural way of life also took place thousands of years ago.

The first reason is that the fertile soil and flat topography of the river valleys proved to be a richly productive agricultural environment, and the resulting increase in food production soon led to a massive increase in the populations living there. You may recall that the *Linearbandkeramik* people spread rapidly into northern Europe by colonizing the river valleys of the Danube, Rhine, and Elbe rivers, which were rich with the fine deposits of loess soil that had been left by the melting glaciers. In similar fashion, the first farmers of the early Neolithic rapidly colonized the river valleys of the Fertile Crescent, India, and China, where the domestication of many plants and animals had originally taken place.

When the hunting and gathering societies that had once inhabited these river valleys were replaced by agricultural societies, the human populations multiplied tenfold. And when many permanent settlements were established within short distances from each other, many thousands of people eventually found themselves living in areas small enough to be crossed on foot in one or two days' travel. This concentration of humanity within small geographic areas proved to be an ideal environment for the emergence of urban centers.

Second, all of these large, slow-moving rivers were flooded each year by the spring or summer rains, and the sediment carried by the annual floods renewed the farmlands along the river banks with fresh deposits of fertile soil. This ensured that the inhabitants never had to move in search of newer and more fertile farmland. The waters at the mouths of these huge rivers move very slowly as they approach the sea, allowing the fine particles of soil suspended in the river to settle to the bottom. Over time, vast quantities of sediments become deposited at the river mouths, creating immense deltas with numerous islands of rich soil, surrounded by swampy wetlands and shallow waters. The delta of the Nile River measures one hundred miles from north to south, occupies 150 miles of Mediterranean coastline, and covers an area of nearly ten thousand square miles.

The behavior of these rivers also encouraged the people who lived along their banks to make long-term plans for projects—such as draining the swamps of the deltas and building irrigation systems to water the flat river bottoms—that created new areas of fertile farmland. But such projects required the close cooperation of large numbers of people—something that was readily accomplished when many small settlements and the laborers who lived in them came under the control of a single centralized authority.

Third, the rivers provided a natural highway that linked together the hundreds of towns and villages that sprang up along its banks, setting off an unprecedented period of innovation in the technology of boat-building. Rafts, canoes, and kayaks had been used by hunting and gathering peoples long before the development of agriculture, but these early watercraft soon evolved into much larger types of boats capable of traveling long distances up and down the large rivers. The result was a great expansion of travel and trade among the many farming settlements in the river valleys, enabling the growing towns and villages to ship heavy loads of grain, hides, timber, animals, and pottery easily from one settlement to another by river transport. A few crew members sailing a single barge or river boat could transport thousands of pounds of cargo, while to move loads of this size over land required scores of humans or pack animals.

The Cradles of Civilization: Mesopotamia, Egypt, India, and China

The oldest civilized societies first appeared in the huge valley called "Mesopotamia" (literally, "between the rivers") that lies between the two great rivers of the Tigris and the Euphrates in present-day Iraq. Sumerian civilization appeared in southern Tigris-Euphrates Valley shortly after 3500 BC,[2] and Akkadian civilization appeared in the northern Tigris-Euphrates Valley several centuries later.

Both the Sumerians and the Akkadians built irrigation canals and left behind plentiful evidence of a centralized urban administration supported by a complex bureaucracy. Their early invention of writing enabled them to make records of taxes and tribute as well as to codify a written legal system, and within their fortified cities they erected elaborate temples, palaces, storehouses, and shrines. At first, Mesopotamian civilization was ruled by a priesthood, but later the rise of kings and an organized military enforced the taxation and tribute that flowed into the cities from the countryside.

At about the same time, one of the most remarkable civilizations of the ancient world arose in the valley of the Nile River. The civilization of ancient Egypt, which the ancient Greeks and Romans regarded as the oldest and wisest society in history, is well known for its massive temples, statues, and pyramids, for its extensive written texts of "hieroglyphics," and for the engineering and architectural achievements that have rarely been equaled since their time. But one of the lesser-known achievements of Egyptian civilization was their early mastery of shipbuilding and sailing. This was a natural outgrowth of Egypt's geograph-

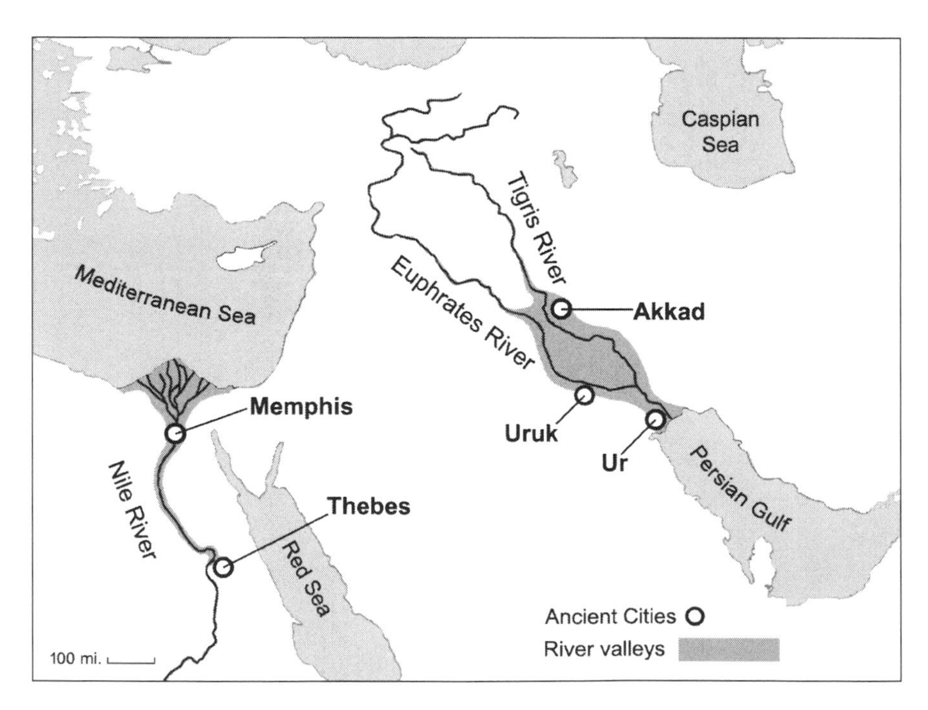

Figure 7.1. *The very earliest urban civilizations emerged in the Fertile Crescent, first in the valley of the Tigris and Euphrates rivers in Mesopotamia and later in the valley of the Nile River in Egypt.*

ical location on the banks of the Nile River, which dominated every aspect of Egyptian society and culture.

At the height of the annual floods, the Nile flows at a rate of about five miles per hour as it drains north to the Mediterranean Sea, and this current slows to slightly more than three miles per hour once the floodwaters recede. But since the prevailing winds in that part of the world blow in a southward direction, the Egyptians built ships capable of sailing southward, propelled by the north wind against the northward-flowing current. On the return trip, the crew would simply put away the sail and float northward with the current back toward the Mediterranean. Thus, the Egyptian hieroglyphic symbol for "south" was a vessel with its mast upright and its sail unfurled in the wind, while the hieroglyphic symbol for "north" was a vessel with its mast lying flat on the deck, drifting with the current of the Nile.

In time, the flat, shallow-draft river boats of both the Egyptians and the Mesopotamians were improved with the addition of high gunwales and rounded keels that allowed them to navigate the rougher waters of the open sea. These seagoing boats allowed the ancient Egyptians to sail from the mouth of the Nile to the shores of the eastern Mediterranean and from the Red Sea to destinations in Arabia, India, and Africa. And they enabled the ancient Sumerians to sail from Mesopotamia hundreds of miles south down the coasts of Africa and eastward to the coasts of India. Both the Egyptians and the Mesopotamians carried on a considerable trade, importing stone for tool-making, the semi-precious stone lapis lazuli, and—in later times—the ores of tin that were needed to make tools and weapons of bronze.

Meanwhile, more than a thousand miles east of Mesopotamia, another civilization was emerging in the valley of the Indus River in northwestern India. Two remarkable cities, Mohenjo-Daro and Harappa, arose at about 3300 BC along the banks of the Indus River. And nine hundred years later, the city of Lothal—an important center of manufacturing and trade—was established on the banks of the Sabarmati River. Evidence of trade goods from Lothal have been found as far east as Southeast Asia and as far west as the coast of East Africa.

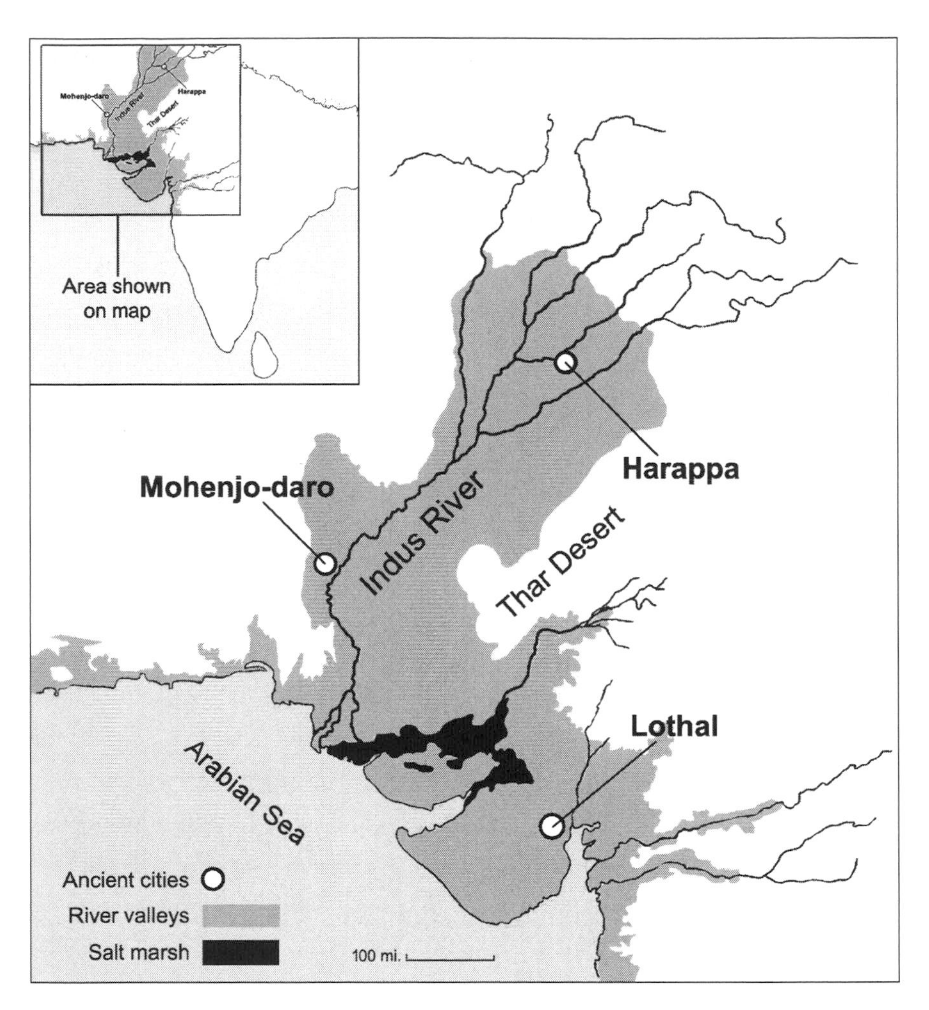

Figure 7.2. *The cities of the Indus Valley civilization that arose after 3300 BC featured advanced sanitation systems, large public granaries, and well-equipped port facilities.*

The cities of the Indus Valley civilization ruled over several million inhabitants, and they were like no other urban centers of the ancient world. They were planned communities, with streets laid out at right angles to each other—the "grid plan" that would not appear again until the rise of Roman civilization 2,500 years later. The houses were built of fired clay bricks—a marked contrast to the sun-dried mud bricks used everywhere else during this period. And an elaborate water and sanitation system included the world's earliest flush toilets, large public baths, and an underground sewage system complete with tile-lined cesspools.

Remarkably, little evidence has been found in the remains of the Indus Valley civilization of the profound differences in wealth that are typical of other ancient civilizations. Although numerous storehouses, granaries, and public baths have been excavated, no evidence has yet been found in the ruins of Harappa, Mohenjo-Daro, or Lothal of the enormous palaces, temples, and military installations so typical of the city-states of other ancient civilizations. For well over a thousand years, this society of craftsmen and traders seem to have enjoyed a degree of

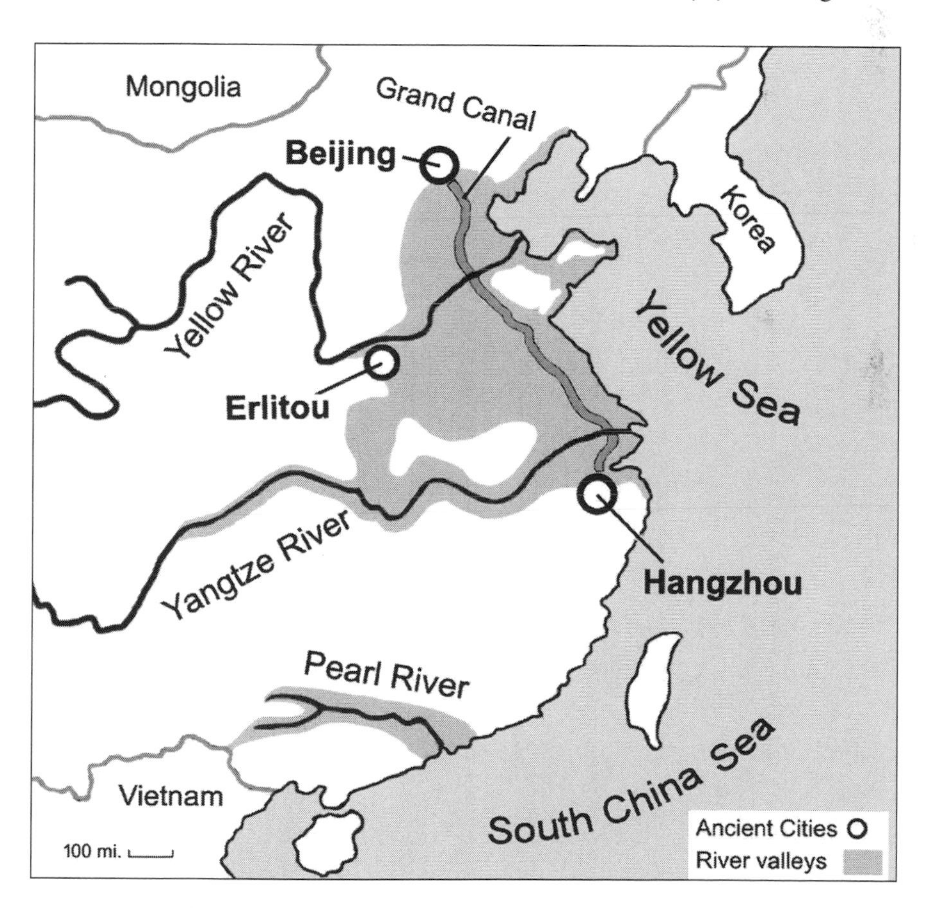

Figure 7.3. *The earliest civilizations in China arose in the valleys of the Yangtze and Yellow rivers. Later, these two rivers were linked by the Grand Canal, the most ambitious waterworks project in human history.*

equality and material comfort that was unique in ancient times. But the Indus Valley civilization declined rapidly after 1850 BC, and by 1700 BC it had all but disappeared. No other "civilization of equals" was ever to be seen again in the ancient world.

Urban civilization did not begin in China until approximately 2000 BC—about a thousand years later than it did in the river valleys of India and the Fertile Crescent—but it soon developed into the most extensive, highly organized, and enduring of all the ancient urban civilizations. The first urban center in China seems to have arisen in Erlitou, a large settlement on the banks of the Yellow River, where evidence of bronze smelting and some of the earliest remains of Chinese writing signaled the beginnings of civilized society.

In China, the fertile lowlands of the river valleys were more extensive than anywhere else in the world. A huge alluvial plain, occupying most of the area between the valleys of the Yellow and Yangtze rivers, was nearly 750 miles wide from east to west and a thousand miles long from north to south. And the valley of the Pearl River in southern China was a third important strip of rich farmland that ran inland from the South China Sea more than five hundred miles to the borders of present-day Vietnam.

The history of ancient China was marked by numerous political upheavals when political power and authority changed hands, sometimes abruptly. No less than seventeen distinct ruling dynasties and governments succeeded each other during the four thousand years that elapsed from the appearance of Erlitou culture in roughly 2000 BC until 1911, when the Qing dynasty collapsed and the modern Chinese nation was established. Throughout this time, political authority in China has been located in more than forty different capital cities at one time or another—and these do not include the multiple capitals that were established during four distinct periods of political fission, when Chinese society split into multiple warring city-states.[3]

Yet there were also long periods of stability and prosperity in China, and Chinese civilization is responsible for many remarkable innovations in technology and culture. These include the invention of writing paper, gunpowder, and firearms, the construction of the largest wooden ships of ancient times, and the construction of two of the largest public works projects in human history: the Great Wall of China and the Grand Canal.

The Great Wall of China needs little introduction. It is actually a series of walls, together measuring more than five thousand miles in length, built at different times by different Chinese emperors, beginning at approximately 700 BC and continuing intermittently to approximately the year 1600 AD. The Great Wall was constructed for the purpose of repelling the Mongols and other warlike tribes who lived in the mountains and deserts north of the Chinese heartland. It was built and rebuilt many times over the course of Chinese history, but it was decisively breached in 1644 by the armies of the Manchu warlords, who established the Qing Dynasty and ruled China until modern times.

The Grand Canal is less well known but in many ways was the more important of these monumental undertakings. It linked the Yellow and Yangtze rivers to each other, providing a means of trade and transportation throughout the entire Chinese heartland and contributing greatly to the unification of China and to its history as one of the world's oldest and largest nation-states. More than a thousand miles long, the Grand Canal is the longest man-made canal in the world. It was begun rather modestly in 486 BC to connect the Yangtze River to the Huai River, but after 600 AD, the canal was progressively enlarged and lengthened, eventually linking Beijing in the north with the important seaport and shipbuilding center of Hangzhou in the south.

At this point, we should note that while the seafaring traditions of Egypt, Mesopotamia, the Indus Valley, and China contributed immeasurably to their ultimate destinies as cradles of civilization, the evidence is clear that seafaring did not begin with the rise of civilizations. There is considerable evidence that, long before anatomically modern humans settled down in permanent villages and began to practice agriculture, *Homo erectus* and other prehistoric humans were building rafts and boats and sailing them on the open sea.

Rafts, Boats, and Sailing Ships

It is an ethnographic fact that hunters and gatherers have been fashioning dugout canoes by hollowing out logs for a very long time, and they have been building rafts even longer by lashing together logs, bamboo, or reeds with ropes made from vines or shredded bark. The Native American tribes of the Pacific Northwest—a "Stone Age" hunting and gathering people by any definition—made huge dugout canoes from the immense trunks of ancient cedar and spruce trees and sailed them on the open sea.

The true antiquity of seafaring remains a subject of considerable disagreement among prehistorians, but hominids have been constructing boats since at least the Upper Paleolithic, and the earliest boats and rafts were probably built long before that. Some scholars believe that the colonization of some of the easternmost islands of Indonesia by *Homo erectus* more than five hundred thousand years ago could never have been accomplished unless these emerging humans had been able to sail the open seas in seaworthy vessels.[4]

Whatever the final judgment about seafaring by *Homo erectus* turns out to be, it is an established fact that prehistoric people originally settled the continent of Australia at least sixty thousand years ago—a feat that required crossing fifty miles of open water that separated the Indonesian island of Timor from the Australian coast during that period of geologic history.[5] This provides conclusive evidence that people were not only building watercraft but were also sailing for days on the open sea tens of thousands of years ago.

In 1998, paleoanthropologist Robert G. Bednarik conducted an experiment in which he directed the construction of a large bamboo raft on the island of Roti off the southwestern coast of Timor. Bednarik's raft was made entirely out of native materials that had been available to Paleolithic humans. It was outfitted with sails woven from palm fiber, and it was built entirely with the types of stone tools that were commonly used sixty thousand years ago. Bednarik and a crew of five successfully sailed this craft from Timor to Australia, subsisting mainly on fish that they caught at sea using replicas of Paleolithic bone harpoons.[6]

It is nearly impossible, however, to find direct archeological evidence of seafaring dating from Paleolithic times. This is because sea levels all over the world during the ice ages were vastly lower than they are today, and the ancient shorelines of the Upper Paleolithic—along with all the evidence of human habitation they once contained—are now lying under three hundred feet of sea water.

Even if it were possible to determine exactly where prehistoric people had been living in these now-submerged shores, the action of the sea over tens of thousands of years has long since obliterated their ancient campsites, as well as the remains of any boats or rafts they might have used. But material evidence from ancient Mesopotamia proves that by seven thousand years ago, long before the earliest urban civilization began, seagoing ships were being built of reeds and sailed on the open sea.

In 2001, a team of British and Kuwaiti archeologists discovered twenty-two slabs of bitumen near the shores of the Persian Gulf, where the Euphrates River empties into the sea. Bitumen is a black, sticky substance, also called asphalt, that occurs in natural deposits throughout the Fertile Crescent and was widely used during ancient times as a waterproof caulking both for water containers and for the hulls of reed boats. The bitumen slabs have been securely dated as having originated between 5500 and 5300 BC. This was roughly two thousand years before urban civilization began in Mesopotamia.

Since the bitumen, mixed with fish oil and crushed coral, had been transported more than sixty miles from its point of origin and still bore the impressions of the rope, string, and reed bundles that were used at that time in boat building, there can be no doubt that it was being used for caulking seagoing boats. In fact, some of these bitumen slabs still have the remains of barnacles clinging to them, proving that they had spent a considerable amount of time at sea.

The Egyptians were not far behind. Small watercraft made of bundles of reed tied together with rope and used on the relatively calm waters of the Nile have survived from as early as 7000 BC, and larger reed boats later proved surprisingly seaworthy on the more treacherous waters of the open sea. In 1970, the Norwegian adventurer Thor Heyerdahl supervised the construction of a replica of an ancient Egyptian reed boat and sailed it six thousand miles across the Atlantic Ocean from the coast of Morocco in North Africa to the island of Barbados in the Caribbean Sea. Since then, there have been many other voyages undertaken in reed boats similar to those used in ancient Egypt, demonstrating that the Egyptians and Sumerians would have indeed been capable of making long sea voyages in similar reed boats.

By 3500 BC, just before the dawn of the first civilizations, the art of shipbuilding had progressed considerably. By then, both the Egyptians and the Sumerians had progressed beyond the early watercraft of reeds and were building seagoing boats out of wooden planks that were "sewn" together with straps of woven fiber. (The iron nails eventually used in boat construction would not be created for at least another three thousand years.) These plank boats were made to be disassembled, loaded on pack animals, transported long distances over dry land, and reassembled on the seashore. In fact, the Egyptians routinely transported their plank boats 125 miles across the Sinai Desert to the shores of the Red Sea, where they were sewn together, caulked with bitumen, and sailed hundreds of miles to the coasts of Africa and India on trading expeditions.

By 1500 BC, the now civilized Egyptians were building immense river barges—some as large as 230 feet long, eighty feet wide, and twenty feet deep—each capable of carrying more than a thousand *tons* of cargo. Immense barges of this type were used to transport the "Colossi of Memnon," two statues of the pharaoh Amenhotep III, each weighing approximately 320 tons, that were created near present-day Cairo and transported more than four hundred miles up the Nile to the ancient city of Thebes, where they stand to this day.

The Slow Dawn of Writing

The invention of writing is often portrayed as a unique event that occurred at a specific time and place—typically described as having begun around 3000 BC with the cuneiform writing of the ancient Mesopotamians and spread from there to other places in the ancient world. But rather than being invented at one time and in one place, all the evidence indicates that writing evolved very gradually in many places, at many times, and in many cultures.

It is beyond question that humans have been using visual symbols to convey specific meanings for more than twenty thousand years. This has been clearly demonstrated by the thousands of petroglyphs found in La Pasiega Cave as well as in hundreds of other Paleolithic sites in Europe (Les Combarelles, Lascaux, Les Eyzies, Altamira, and Val Camonica), Asia (Kapova Cave, Mgvimevi grottoes, Edakkal Caves, Ladakh, Daegokcheon Stream Petroglyphs), Africa (Akakus, Jebel Uweinat, Bidzar, Niola Doa, Blombos Cave, Wonderwerk Cave), Australia (Murujuga, Burrup Peninsula), and the Americas (Winnemucca Lake, Long Lake, Three Rivers, Cumbe Mayo).

But the petroglyphs of the Upper Paleolithic were not what we would consider full-fledged systems of writing. In our contemporary culture, the concept of "writing" implies the systematic, graphic representation of a fully developed language, including not only nouns but also verbs and ways of expressing past, present, and future tenses. And in most cases, true writing includes some indication of how the words for these various meanings were pronounced in the spoken language of the people doing the writing.

Wherever urban civilizations arose, full-fledged systems of writing were developed because they were needed. Societies composed of tens or hundreds of thousands of individuals—most of whom are strangers to each other—require technologies of communication that can transmit complex messages effectively and reliably over time and space. When these urban societies did not already possess a system of writing—or were unable to borrow a system of writing from their neighbors—they invented their own.

The Sumerian and Akkadian cultures of Mesopotamia were among the earliest people to develop a full-fledged system of writing, which began as drawings of specific actions such as walking and of concrete objects such as sheep, cows, bread, and barley. Over time, the details of these symbolic drawings became simplified, and eventually they evolved into standardized symbols that bore little resemblance to the things they had originally depicted. In the end, all of these symbols came to be drawn in a form called "cuneiform," consisting of impressions made in soft clay with a wedge-shaped reed or wooden stylus (see Figure 7.4).

Over time, the cuneiform writing of the Sumerians was adopted as a standard "alphabet" by many other ancient cultures, and cuneiform became the standard way of writing in ancient Babylonian, Assyrian, Hittite, Elamite, Hurrian, Urartian, Ugaritic, and Persian cultures. But while they may have all used the same system of writing, the people who spoke these different languages were not able to read each other's cuneiform script. Similarly, people who understand only English can read only what is written in English and cannot read Spanish, Portuguese, Dutch, Flemish, Italian, French, or German, even though all of these languages use essentially the same alphabet.

Meanwhile, systems of writing were also evolving in other ancient civilizations. In the Nile Valley, the unique system of writing that the Greeks called hieroglyphics or "sacred carvings" came into widespread use shortly after 3000 BC. Hieroglyphic writing, like cuneiform, contains a number of special symbols that represent the sounds of consonants (like many other forms of writing, Egyptian hieroglyphics had no form of notation for the sounds of vowels), but the vast majority of hieroglyphic symbols represent specific meanings. Over time, the number of hieroglyphic symbols grew from about eight hundred during the classical eras of Egyptian civilization to nearly five thousand by the time of the Roman Empire many centuries later.

While some of the best-known hieroglyphic inscriptions are those found carved into the stone walls of pyramids and tombs, most Egyptian hieroglyphics were written in "cursive" form on sheets of papyrus, made from the same marsh reeds that furnished the building materials for the early Egyptian boats (see Figure 7.5). Numerous papyrus scrolls have survived since ancient times in the dry desert climate of Egypt, providing a rich source of information about the culture and society of this extraordinary civilization.

In the Indus River valley, the emerging city-states of Harappa and Mohenjo-Daro had the same need to record transactions as the Sumerians, and evidence of a written language called the "Indus script" began to appear after 2700 BC. The Indus script has survived in the form of seals made of baked clay that contain sequences of symbols, sometimes accompanied by beautifully executed images. This form of writing consisted of more than four hundred distinct signs and symbols, reproduced in clearly defined sequences (see Figure 7.6). Unfortunately, however, cotton cloth was the standard writing material used by the people of the Indus Valley civilization, and nearly all of the records of the Indus script written on this highly perishable material have been lost to the ravages of time.

The domestication of plants and animals in China took place at roughly the same time as it did in the Fertile Crescent, and by 7000 BC, the valleys of the Yellow and Yangtze rivers were home to a large and growing population of farmers living in permanent village settlements. It is from this early period that the first evidence of visual symbols appears in China (see Figure 7.7).

The earliest evidence of Chinese writing appears in the Neolithic site of Jiahu, dating from roughly 6600 BC, and the remains of pottery basins found at Dadiwan, dating from approximately 5800 BC, were painted with symbols that are strikingly similar to some of the petroglyphs of La Pasiega Cave from Paleolithic Spain. But the oldest remains of the Chinese characters that most clearly resemble the characters of modern Chinese writing are the "oracle bones" used in the Shang Dynasty between 1500 and 1200 BC. The Chinese of that time carved inscriptions onto bones or turtle shells and then heated these objects in the fire. The future was foretold by interpreting the cracks that appeared in the oracle bones after they were heated. The "oracle bone writing" of ancient China is widely regarded as the earliest example of the written Chinese language still in use today.

Writing developed in the Americas somewhat later than it did in Asia, although both the practice of agriculture and the evolution of sedentary societies in the Americas began at least seven thousand years ago. But the absence of immense river valleys may have been the key factor that delayed the emergence of urban civilizations in those regions. Nevertheless, pre-Columbian states in the Americas did eventually arise, and their development resembles to an astonishing degree the patterns seen in the urban civilizations of the Old World.

The Maya, Aztecs, Inca, and other pre-Columbian civilizations all rose to power over their neighbors by violent means. They were all hierarchical societies divided between a lower class of peasant laborers and an upper class of hereditary aristocrats who ruled in the urban centers, while a class of professional priests performed religious rituals in elaborate temples. These societies not only kept written records but also developed advanced forms of mathematics, writing, and astronomical science, and the Maya were capable of calculating astronomical events far into the future.

By 400 BC, the Maya were using a form of hieroglyphic writing that eventually evolved in a system as complex as any that had been developed in the Old World (see Figure 7.8). Mayan writing consisted of hundreds of symbols that were combined into complex images called "glyphs"—a form of true writing used to represent numbers, objects, actions, and sounds.

Stone into Bronze

Although traditionally all of human history has long been divided into the Stone Age, the Bronze Age, and the Iron Age, the differences in human life among these three "ages" is not nearly as great as their names would suggest. This way of classifying human history originated in 1825, when the Danish antiquarian Christian Jürgensen Thomsen coined these terms as a convenient way of sorting collections of ancient artifacts into a reasonable chronological order. This was before Darwin and Wallace proposed their theories of evolution, before the discovery of the remains of Neandertals and other prehistoric hominids, and long before the development of modern archeology.

In fact, the Inca, Maya, Aztecs, and other advanced societies in the Americas were able to develop advanced civilizations while remaining largely dependent on tools and weapons made of stone. Metal-working in those societies was largely limited to the smelting and working of soft metals, such as gold, silver, and copper. These metals can be melted in ordinary wood fires that are superheated by workers blowing air through long tubes, but they lack the hardness and toughness needed for making effective tools and weapons.

Thus, where the geography was favorable, a settled, agricultural way of life was enough—even in the absence of metal tools and weapons—to stimulate the birth of civilizations, complete with state bureaucracies, organized religions, organized warfare, long-distance trade, systems of writing, and advanced knowledge of mathematics and astronomy. But the development of metallurgy in the ancient river valleys did play an important role in the development of urban civilizations, not merely by furnishing them with superior tools and weapons but especially by stimulating innovations in the technologies of transportation, which were necessary for the pursuit of long-distance trade between societies from widely separated geographical regions.

Metal working began first in the Middle East nearly ten thousand years ago with the smelting and hammering of naturally occurring nuggets of copper in the mountains of eastern Turkey and northern Syria. Copper is a soft metal that is easily worked, and through repeated heating of the nuggets of copper in a simple wood fire, they could be readily hammered into a variety of shapes. At first, the small amounts of copper that could be extracted from natural deposits were sufficient only for rings, beads, and pendants intended for personal adornment, and copper had little impact on other areas of technology or economic life.

Eventually, however, the potters of Egypt and the Indus Valley began experimenting with glazes made with mixtures of the bright green and blue minerals malachite and azurite—both of which are ores of copper—to make the enormously popular blue-green ceramic objects called faience. They soon discovered that these ores could be smelted into metallic copper by mixing them with charcoal and heating them in a fire blown to a white-hot temperature for several hours.

This discovery led to casting of molten copper in molds to make tools and weapons such as adzes, ax heads, and spear points. But copper was both rare and expensive, and it was too soft to be made into the thin, sharp tools and weapons, such as knives and arrowheads, that were important in these early technologies. Thus, for a long time, copper remained a marginal material that played only a minor role in the technology of the early agricultural societies.

But sometime around 4000 BC, the coppersmiths of the Middle East discovered that by mixing copper with small amounts of arsenic—a poisonous metallic crystal—and later with small amounts of tin—a non-toxic metal—they could make bronze, a metal that proved far superior to copper. Not only was bronze easily worked, but it also had the advantage of a lower melting point than pure copper. Yet bronze was hard enough and tough enough to be made into tools and weapons that were superior to the stone tools and weapons that humanity had been using since the dawn of prehistory. These alloys of copper and tin made metallurgy an important technology for the first time.

But the use of tin for making bronze presented a new problem. While deposits of poisonous arsenic are often found mixed with the ores of copper, the deposits of tin—a much safer alternative—were never found associated with copper. In fact, ores of tin were plentiful in only a few places in the ancient world, and none of these deposits were located in either Egypt or Mesopotamia.

The Egyptians possessed rich deposits of copper in the Sinai Desert east of the Nile Valley, but they had no deposits of tin. And neither copper nor tin occurred naturally in the valley of the Tigris and Euphrates rivers. Before long, both of these civilizations began to search for sources of tin in more distant lands, and when they found them, the heavy ores had to be transported to places where they could be smelted into ingots and where smiths could work them into useful artifacts.

Thus the need to bring copper and tin together provided a tremendous new impetus to advancing the technologies of transportation, in particular to transportation by sea. Both the Mesopotamians and the Egyptians, who had been building and sailing river boats for centuries, soon turned their attention to building ships large enough and strong enough to sail the open seas in search of sources of copper and tin.

As the use of bronze expanded, the ancient armies that were equipped with swords, spears, and arrowheads of bronze easily defeated the armies that were armed with only stone-tipped weapons. Thus, the use of bronze ensured not only that the societies with metal weapons would defeat their adversaries in warfare but that these "Bronze Age" cultures would quickly overwhelm and replace the "Stone Age" cultures with which they came into contact. The use of bronze for tools and weapons spread rapidly throughout the civilized world, becoming commonplace in the Indus Valley in 3300 BC, in Mesopotamia after 2900 BC, and in China with the emergence of the earliest urban centers in 2000 BC.

Although the "Iron Age" was destined in time to replace the "Bronze Age," bronze continued to be preferred by most ancient civilizations long after the techniques of smelting iron became commonplace and widespread, because bronze was superior to iron in many important respects.

First, bronze melted at much lower temperatures than iron. The ancient wood fires were hot enough to melt copper, tin, bronze, and gold, but—even when superheated by workers blowing air through long pipes—they did not reach temperatures high enough to melt iron.

Second, bronze can be hammered into shape when it is cold, but iron can shaped by hammering only when it is red hot. This meant that special heat-resistant tongs and anvils had to be developed for working with iron that were not required for working with bronze.

Third, the ancient process for smelting iron produced a rough, brittle, spongy mass that had to be reheated and hammered over and over again to transform it into the "wrought iron" that was suitable for making usable objects. But bronze can be worked as soon as it is smelted.

Fourth, when bronze is exposed to the weather, it combines with the oxygen in the air to form a thin but durable "patina"—a layer of oxidized metal—that seals and protects the bronze object from further decomposition. But when iron oxidizes, the result is iron oxide or "rust," a weak and powdery substance that does not protect the surface of an iron object. Thus, when something made of iron is exposed to air and water for long periods of time, the rust will penetrate deep into the object, weakening it and ultimately rendering it useless.

Why, then, did iron eventually replace bronze in all civilized societies? The answer seems to have been mainly a matter of simple economics. Iron ore is abundant throughout the inhabited world, and it can be found not only on every continent but also on most of the major inhabited islands, including Australia, Japan, New Zealand, and the Philippines. In fact, the Paleolithic societies of Europe used "red ochre"—a form of iron ore—extensively in cave paintings, in burials, and in other religious rituals.

When trade with distant lands was disrupted by warfare or the collapse of an ancient civilization—not an uncommon occurrence throughout history—ores of copper and tin would be scarce and expensive to obtain, while the more plentiful ores of iron were easier to procure. Yet, because of the difficulties of working with iron, it was not until the Middle Ages, when coal-fired furnaces came into widespread use, and when the complex process of making steel was perfected, that tools and weapons of steel—which were much harder and more durable than their iron or bronze counterparts—finally ended the practice of using bronze for making many common metal objects.

The Horse, the Wheel, and Warfare[7]

The horse was hunted as a game animal for at least four hundred thousand years, first by emerging humans, later by the Neandertals, and still later by the anatomically modern humans of the Upper Paleolithic. In fact, when the nomads of the Eurasian steppes originally domesticated the horse roughly six thousand years ago, it was not for the purpose of transportation but rather as a source of food. The reason is simple: the horse was the only animal capable of feeding itself in a snow-covered pasture.

The open grasslands of the Eurasian steppes are subject to intense cold during the winter, and for much of the time they are covered in a blanket of snow. The people of the late Neolithic had already domesticated sheep and cattle—whose meat they preferred—but both of those animals must be fed by hand when snow and ice cover the ground. Sheep can break through ice-crusted snow with their noses to reach the grass underneath, but the sharp edges of broken ice will lacerate the soft skin of a sheep's nose. Cows are even more helpless; unable to uncover the grass under a blanket of snow, cows will starve even while food is only a few inches beneath the surface. But horses, which are native to the open grasslands of the northern latitudes, will paw away the snow with their sharp hooves and graze easily in snow-covered pastures.

Once they had domesticated the horse as an alternative source of meat, the Eurasian nomads discovered that these animals could be controlled by a bit placed in their mouths and attached to a bridle made of rope or leather. In fact, by examining the wear patterns on the teeth of horses that were buried in the graves of their masters, the archeologist David W. Anthony demonstrated that horses were being controlled by simple bits and bridles as early as 4000 BC.[8]

The Eurasian nomads learned not only to ride horses but also to build wagons of wood covered with awnings of cloth, and they fashioned harnesses that enabled their horses to pull these wagons long distances over the steppes. Before long, the Eurasian herdsmen became the most mobile nomads on Earth, moving their families and all of their worldly possessions hundreds of miles every year in search of new pastures for their flocks. Five thousand years later, the pioneers of the American West crossed the untamed wilderness of the prairies using covered wagons of an almost identical design.

Although it remains to be determined whether the wheel was first invented in the steppes of Eurasia or the river valleys of Mesopotamia, there is no doubt that as soon as people began using wagons and carts, the use of the wheel spread like wildfire across the whole of Eurasia, from western Europe to China (see Figure 7.9). Evidence of wagons and carts appears suddenly throughout both Europe and Asia beginning in 3500 BC, in the form of drawings and clay models of wagons, the appearance of a written sign for "wagon," and the archeological remains of both wheels and wagon parts.

The first four-wheeled wagons and two-wheeled carts were drawn mostly by oxen, typically yoked in pairs on either side of a long draft pole that was connected to the wheeled vehicle, although the onager, a small domesticated donkey, was also used for this purpose. Wagons and carts enabled one or two people to transport heavy loads of grain, building materials, firewood, and other goods from one place to another without assistance. For the most part, these early wheeled vehicles were used for hauling goods short distances, perhaps from one village to another, from the farms and fields to the market towns, or from the countryside to the cities. They were rarely used for transporting either goods or people longer distances, since for thousands of years after the invention of the wheel, paved roads did not exist.

Between 3000 and 2000 BC, heavy battle wagons and carts with solid wheels, drawn by oxen or by teams of asses or onagers, began to be used as mobile platforms for archers and spear-throwers by the armies of ancient Mesopotamia. But these vehicles were relatively slow and difficult to maneuver. The horse-drawn chariot, however, which was also invented by the nomads of the Eurasian steppes and appeared in Mesopotamia after 2000 BC, was a different matter entirely. Unlike the slow, plodding, dim-witted ox or the undersized asses and onagers, the horse was built for speed, power, and endurance. And when a team of two or four horses was yoked to a chariot, the result was formidable indeed.

Light in weight and highly maneuverable, with open-spoked wheels and pulled by a team of fast horses, the chariot was controlled by a driver while two, three, or four warriors stood behind him, launching spears and arrows at the enemy while the chariot was driven at high speed. The sides and floor of the chariot were made of the lightest, thinnest wood or leather, its wheels were large, and the back of the chariot was open. While the chariot was nearly useless for hauling cargo, it was a perfect platform for launching attacks on enemy forces.

The ancient chariot needed no roads. It sped swiftly over open country, bouncing and rattling as its horses galloped at full speed under the lash of the charioteer, who concentrated on controlling the horses while the lancers and archers attacked the enemy. The invention of the chariot significantly changed the tactics of ancient warfare, and from the moment of its introduction, it became a fearsome and decisive weapon. Within a few centuries, the chariot had become an essential part of every ancient army from Egypt to China.

A large part of the chariot's success was due to the unique requirements of the powerful longbow, the ancient archer's weapon of choice. The longbow was as tall as a man, and it was extremely unwieldy. It required the archer to stand erect, which is why archers stood upright on the platforms of chariots. But having invented the bridle and bit—and having learned to ride on the backs of these powerful animals—the nomads of the Eurasian steppes invented a much smaller and more compact bow that was both as powerful and as accurate at the longbow.

This new "composite" bow, curved like a handlebar mustache, was made of wood, bone, and sinew glued together, and it produced as much power as a bow twice the size fashioned from a single piece of wood. The composite or sinew-backed bow enabled a rider to shoot arrows in any direction—including backward at a pursuing enemy—while mounted on a galloping horse. After 1000 BC, when archers mounted on horseback and armed with the composite bow became commonplace in ancient armies, the chariot and its spear-throwing or longbow-wielding occupants were gradually retired from battle.

It is difficult for us—as accustomed as we are to the idea of people riding on the backs of horses, mules, donkeys, camels, and elephants—to imagine how revolutionary it was for humans to think that they could so utterly dominate and control an animal many times larger and more powerful than themselves that they could even imagine riding on its back. After all, riding for hours on the back of a horse is an act that is equally "unnatural" for both the horse and the rider. By 3500 BC, several species of animals had already been domesticated, but the idea of mounting a horse and controlling its movements must have been a leap of faith on the part of the nomads who accomplished this feat sometime after 4500 BC on the Eurasian steppes north and east of the Caspian Sea.

Certainly, it was profoundly shocking to the civilized people of the Americas when the Spanish *conquistadores* appeared in full battle armor mounted on the horses that they had brought with them across the Atlantic in the early 1500s. The Native Americans had never seen horses before, and they had never witnessed the sight of human beings mounted on these great beasts. As anyone who has ever seen a horse and rider approaching at a gallop can attest, a man on a horse looks larger than life and can be an imposing and frightening figure.

Once it had been successfully bred for domestication, the horse became and remained a key element in organized warfare for more than three thousand years. In fact, it was not until the twentieth century, with the invention of the internal combustion engine and the success of tank warfare in World War I, that the long history of using horses in warfare finally came to an end.

Cities, States, and Empires

It took a long time for the technology of agriculture to develop to its full extent, a process that began with the establishment of the permanent settlements of the Natufians and was not completed until thousands of years later with the domestication of cereal grains, fruit trees, sheep, goats, cattle, and horses. By contrast, the emergence of urban civilization seems to have happened virtually overnight, with numerous cities springing up all over the ancient world in the space of just a few centuries before and after the year 3000 BC.

This rapid and dramatic transformation in human life occurred when several factors—including population density, the favorable geography of river valleys, craft specialization, and innovations in the technologies of transportation and communication—combined to produce a volatile and unstable condition known as a "positive feedback loop," which can cause major changes to take place in unusually short periods of time.

One example of a positive feedback loop is the onset of a stampede in a herd of cattle, which may begin with the panic of only a single animal, which then causes the animals next it to panic, which then causes all the animals near *them* to panic, until panic has spread throughout an entire herd within a matter of seconds.

The development of boats, wheeled vehicles, and systems of writing allowed people to interact across time and space without having to meet in person. As these technologies of interaction became increasingly effective, they facilitated the growth of large commercial and political centers. In turn, these centers stimulated new and more efficient technologies of transportation and communication, which allowed the emerging cities to grow ever larger. Before long, for the first time in human history, very large numbers of people, some in the cities and others in the countryside, began to function as members of a single integrated society.

The cities of the ancient world were beehives of activity, populated by artisans, merchants, bureaucrats, soldiers, priests, and political leaders, yet they all depended for their food on the farmers and herdsmen of the surrounding territories. At the same time, the people of the countryside depended on the cities for the benefits that civilized society made possible. These included the textiles, pottery, furniture, tools, weapons and other artifacts made in the shops of the urban craftsmen, the trade goods that were brought from distant lands, the religious authority of a dedicated priesthood, the social order created by a civil administration, and the possibility of military protection against the raids, thefts, abductions, and murders that had been commonplace among the simpler societies of previous ages.

While some of these civilized societies were called "cities," others were called "states," and still others were called "empires," the differences among them were not as great as the differences among these three names would imply. It has been estimated that roughly 85 percent of the population of all civilized societies throughout history consisted of agricultural food producers, while the remaining 15 percent consisted of the urban professionals who made civilized society possible. And every ancient city found it necessary to control the movement of goods and people between the countryside and the city, to ensure that there would always be a large and reliable supply of food for the city dwellers—and that the people of the countryside could rely on the cities for the products, services, and military protection upon which they depended.

The type of society we call the "state" came into being when the population of urban civilizations grew so large that no single family, clan, or tribe was capable of administering the affairs of society. This encouraged the formation of bureaucracies and ruling classes that were based primarily on their shared citizenship as members of a specific urban civilization rather than on their family ties, clan membership, or tribal affiliations.

Although most of the societies we consider to be "states" included multiple urban centers, there was typically a single dominant city that acted as the "capital" of the state. For example, the pharaoh Menes is widely credited with having "united" the Upper and Lower Kingdoms of ancient Egypt into a single nation, but in actuality this ancient state was formed when Menes conquered the entire Nile Valley by force. The new "united" Egypt was ruled from a single urban center, the city of Memphis, located where the Upper Nile fans out into a myriad of smaller streams and channels that forms the great Nile Delta.

The notion of "empires" was born when the armies of certain city states conquered the lands ruled by other cultures. When Menes conquered Thebes and united all of Egypt, the result was not an "empire," because Upper and Lower Egypt both shared a common culture. This culture included a common language, a common system of writing, the worship of a common religion, and similar traditions in agriculture, animal husbandry, pottery, metallurgy, water transport, and house construction. But the pharaohs were not content to rule over a single Nile Valley culture, and they eventually created true empires by expanding their rule south into Nubia, east across the Sinai Peninsula, and north along the shores of the Mediterranean Sea all the way to Syria and the southern borders of Turkey.

The history of ancient civilizations is marked by periods of peace and prosperity that alternated with wars, conquests, and subjugations, and the empires founded by military conquest arose when urban civilizations conquered other cultures and exacted tribute from them. Yet none of these empires survived for more than a few centuries. Instead, ancient history consists of a seemingly endless series of wars, conquests, consolidations, and periods of growth and stability, followed by periods of decadence, corruption, mismanagement, deterioration, and—ultimately—reversion to war. In the final analysis, the most important difference between a city, a state, and an empire was that a state usually encompassed more than one city—although a capital city was always dominant and essentially in control—while an empire typically encompassed more than one culture.

In spite of these differences, the essential nature of urban civilization was remarkably consistent. Power and wealth were centralized in the cities, while food, raw materials, and in some cases, labor flowed from the villages of the countryside to the urban centers. As one set of rulers was replaced by another set of rulers, the people of the countryside continued to grow their crops, raise their animals, and send their produce to the cities in the form of taxes and tribute. Meanwhile, the people of the city continued to manufacture things, engage in trade with distant lands, and exercise the military, political, and religious powers that knitted these civilizations together.

Dunbar's Number and the Great Fusion

Surveying the ebb and flow of the kingdoms, dynasties, city-states, and empires of civilized societies, one can easily lose sight of the fact that these were by far the largest groups of humans that had ever fused into unified social groups. With the development of agriculture, the human group had expanded from the few dozen individuals who constituted the typical nomadic band to the hundreds of individuals who lived together in the Neolithic towns and villages. But the emergence of urban civilization represented a far greater process of social fusion, the greatest in the history of humanity before the formation of the modern industrial nation-state.

Most ancient civilizations consisted of societies numbering in the hundreds of thousands of individuals. All of their citizens recognized the same leaders, considered themselves members of the same culture, and interacted as members of the same social group. This is a remarkable development indeed, considering that the human brain may be physically incapable of mentally processing more than about 150 full-fledged relationships at any given time.

In 1992 the evolutionary anthropologist Robin Dunbar studied the relationship between brain size and group size in numerous primate species. He found that the species with the largest brains were able to maintain the largest cohesive social groups while those with the smallest brains were able to maintain only the smallest cohesive social groups. His mathematical model predicted that, given the size of the human brain, human groups would retain their cohesiveness up to approximately the size of only 150 individuals. The number 150, which came to be called "Dunbar's number," turns out to be approximately the maximum number of full-fledged relationships a normal human being can maintain at any given time.[9]

Dunbar's number appears repeatedly in all kinds of human social groups. It is the average number of Christmas cards English people typically sent to their preferred list of friends and kinfolk. It is the effective upper limit of both Yanomamö and Amish villages, beyond which they typically split apart into new, smaller units. It has been the typical size of a military company for centuries. It is even the ideal size of a modern office, beyond which the formation of cliques and factions often produces disharmony and dysfunction among its occupants. Dunbar's number reflects the innate capacity of the human mind to maintain complex relationships. After all, throughout several million years of evolution, hominids needed to maintain full-fledged relationships only with the few dozen kinfolk that constituted the typical hunting and gathering group.

In addition to a few dozen full-fledged relationships, there are also those people whom you know and recognize but with whom you may not have any kind of relationship. How many different names and faces can you recognize or remember? The answer will almost certainly be less than two thousand. In the villages of agricultural societies that are smaller than two thousand people, almost everyone recognizes and is familiar with almost everyone else. But once this threshold is crossed, and a human group grows to more than two thousand individuals, there are increasing numbers of people who are strangers to each other.

This larger number of two thousand people is also an inherent mental capacity. It evolved in prehistoric times out of the need for each individual to recognize the members of other social groups with whom an individual occasionally had contact: casual acquaintances, childhood friends, occasional trading partners, people related more distantly by kinship, and the members of other hunting and gathering bands. The capacity to recognize and remember the identities of between five hundred and two thousand people was all that any member of a hunting and gathering society had ever needed.

But when the agricultural villages of the Neolithic expanded into larger towns that grew to more than two thousand inhabitants, the capacity of the human brain to know and recognize all of the members of a single community was stretched beyond its natural limits. Nevertheless, the tribal cultures that had evolved during the Upper Paleolithic with the emergence of symbolic communication enabled people who might have been strangers to feel a collective sense of belonging and solidarity. It was the formation of tribes and ethnicities that enabled the strangers of the large Neolithic towns to trust each other and interact comfortably with each other, even if they were not all personally acquainted.

The transformation of human society into urban civilizations, however, involved a great fusion of people and societies into groups so large that there was no possibility of having personal relationships with more than a tiny fraction of them. Yet the human capacity for tribal solidarity meant that there was literally no upper limit on the size that a human group could attain. And if we mark the year 3000 BC as the approximate time when all the elements of urban civilization came together to trigger this new transformation, it has taken only five thousand years for all of humanity to be swallowed up by the immense nation-states that have now taken possession of every square inch of the inhabited world.

♦

The new urban civilizations produced the study of mathematics, astronomy, philosophy, history, biology, and medicine. They greatly advanced and refined the technologies of metallurgy, masonry, architecture, carpentry, shipbuilding, and weaponry. They invented the art of writing and the practical science of engineering. They developed the modern forms of drama, poetry, music, painting, and sculpture. They built canals, roads, bridges, aqueducts, pyramids, tombs, temples, shrines, castles, and fortresses by the thousands all over the world. They built ocean-going ships that sailed the high seas and eventually circumnavigated the globe. From their cultures emerged the great universal religions of Christianity, Buddhism, Confucianism, Islam, and Hinduism. And they invented every form of state government and political system we know, from hereditary monarchies to representative democracies.

The new urban civilizations turned out to be dynamic engines of innovation, and in the course of just a few thousand years, they freed humanity from the limitations it had inherited from the hunting and gathering cultures of the past.

But when the clockmakers of Medieval Europe, in their quest to create truly accurate timepieces, created the technology of precision machinery, they unleashed a process of cultural evolution that ultimately transformed human society more completely and more profoundly than any of the technological metamorphoses that had previously taken place. We will examine this seventh metamorphosis—which continues to reshape the daily lives of all humans alive today—in the next chapter of this book.

ඏ

CHAPTER 8

The Technology of Precision Machinery

Clocks, Engines, and Industrial Society

"The enduring legacy of the pioneer clockmakers, though nothing could have been further from their minds, was the basic technology of machine tools."

—Daniel J. Boorstin, *The Discoverers*

When Father Matteo Ricci was invited to the Imperial Court of China by the Ming dynasty emperor Wan-Li in the year 1601, he and his fellow Jesuits became the first Europeans ever to enter the Forbidden City. By then, Father Ricci, a Jesuit priest from Italy, had been doing missionary work in southern China for nearly twenty years. He had survived the destruction of his mission near Canton by an angry mob, and shortly before his arrival, he had been arrested and imprisoned with his Jesuit companions while on his way to the Chinese capital at Beijing.

But Emperor Wan-Li, remembering an earlier petition in which Ricci promised to bring him a present of two clocks he had carefully carried with him all the way from Venice, ordered that Father Ricci and his companions be released from prison and their clocks delivered to the imperial palace. Wan-Li wanted to see these exotic European machines, which were said to run for days by themselves and to ring bells to announce the passing of the hours.

Wan-Li examined Father Ricci's clocks in his private residence several days before the Jesuits' arrival. The emperor was fascinated with these machines, the likes of which he had never seen. But by the time Ricci and his companions arrived at the imperial court, the larger of the two clocks had stopped working, and when the Jesuits arrived, they were warned that they must have the clock working again in no more than three days or they would have to face the consequences. Fortunately, the clock had stopped only because it needed to be rewound.

The success of Ricci's gifts—and the Jesuits' ability to predict the exact time and duration of solar eclipses with far greater accuracy than the court astronomers—guaranteed Father Ricci and his companions a place of honor in the Ming court. A special tower was built for the larger clock in one of the inner courtyards of the imperial palace, and the smaller clock was kept in the private residence of Wan-Li himself.

The emperor's clock, said Father Ricci, "struck all the Chinese dumb with astonishment." It was, he continued, "a work the like of which had never been seen, nor heard, nor even imagined in Chinese history." And the esteem that Wan-Li held for the science and workmanship of the Europeans did not go unrewarded. For the next nine years until his death in 1610, Matteo Ricci received a generous stipend from the emperor, and he occupied a privileged position in the Chinese court. Upon his death, Wan-Li broke with the Ming dynasty tradition that foreigners were not to be buried on Chinese soil, and he ordered that a Buddhist temple be built in Ricci's honor in the heart of Beijing, where his remains lie to this day.

Unknown to both Wan-Li and the Jesuits, however, an immense and fantastically elaborate clock, powered by the action of water falling from an eleven-foot waterwheel, had been built in China five hundred years earlier. This fabulous machine, housed in a tower forty feet tall, was the creation of the brilliant civil servant Su Song for Emperor Zhezong of Song dynasty China in 1094 AD. Upon Zhezong's death,

Su Song's water clock fell into disuse, and its bronze mechanisms were eventually melted down for scrap.

The numerous water clocks built by Su Song and his successors over the centuries were all single instruments, built by a few master craftsmen for the pleasure of the imperial court. Father Ricci's clocks, on the other hand, were the product of an entire industry, pursued by an army of skilled craftsmen that had been growing steadily for three hundred years and had spread throughout Europe.

In the years that followed Ricci's arrival, the clocks and pocket watches of Europe became part of a thriving trade with the Chinese imperial court, and by the 1760s, the Jesuit fathers reported that the imperial palace was "stuffed with clocks, watches, carillons, repeaters, organs, spheres, and astronomical clocks of all kinds and description—there are more than four thousand pieces from the best masters of Paris and London." By the end of the eighteenth century, the ambassador for the Dutch East India Company in Beijing reported that when traveling to Peking, "one should bring . . . especially those [clockwork] playthings that European boys use to amuse themselves. Such objects will be received here with much greater interest than scientific instruments or *objets d'art.*"[1]

Although China was far larger than any political entity the Europeans had known since the fall of Rome, and although Chinese civilization had long been superior to European civilization in numerous areas of manufacturing, transportation, government, and literature, the European clockmakers of that time were the only craftsmen in the world capable of building highly accurate timepieces small enough to sit on a table or be held in the palm of the hand. This was largely because generations of effort had driven the clockmakers of Europe to surpass the metal workers of all other civilized societies in one particular area: the manufacture of precision machinery.

The impetus for the development of mechanical clocks grew out of the unique and peculiar European obsession with keeping time, an obsession that inspired the clockmakers of medieval Europe to perfect the technology of precision machining in their quest to create truly accurate timepieces. Like ripples in a pool, the consequences of precision machining radiated into every sphere of human life, transforming everything it touched.

Precision machining made it possible to manufacture a wide variety of machines that humans had never created before: steam engines, printing presses, long-range weaponry, electric generators, telegraph wires, telescopes, microscopes—the list goes on—and the synergies created in the resulting expansion of information, science, industry, and military power gave birth to a new kind of society, based not on the limited energy of humans and animals but on the seemingly unlimited energy of fossil fuels. And it all began in the Middle Ages with the invention of the mechanical clock.

The Clockmaker's Genius

The Europeans of the Middle Ages were a pious people, and they considered it a matter of great importance that the prayers described in the Catholic Breviary be said at the correct times. For this reason, all of the many churches and monasteries of that age included a bell tower, and the monks were required to ring these bells at the prescribed hours of the day and night as a signal to the faithful when it was time to say their prayers. But the sandglasses and water clocks that the monks used during the early Middle Ages were notoriously unreliable and inaccurate. In fact, the societies of Greeks, Romans, Indians, and Chinese had used sundials, water clocks, sandglasses, candles, and incense clocks to measure time since ancient times, but each of these methods had serious limitations.

Sundials were accurate only at the specific latitude—or distance from the equator—for which they were designed, and they were completely useless throughout the night and whenever clouds hid the sun. Water clocks—except for a few that were as large as multistory buildings—depended for their accuracy on the action of water dripping slowly through a small hole drilled into the bottom of a container. But since water drips more slowly when a container is nearly empty than it does when the container is full, the water clock was seldom accurate. Moreover, every so often some impurity or piece of debris would become lodged in the drip hole, and the water clock would stop altogether.

The sandglass was hardly better. Designed to measure only short periods of time, most sandglasses measured only twenty minutes or less, since a glass large enough to measure as much as a single hour tended to be large, heavy, and dangerously fragile. Worse still, in order to measure more than one unit of time, the sandglass had to be turned upside down every time the sand ran out. Burning incense provided a surprisingly accurate way of measuring the passage of time, and incense clocks were widely used throughout Asia for centuries. But the incense clock had the unique disadvantage of consuming itself in the process of telling time. Having burned to the extent it was designed for, the incense had to be replaced with a new supply, or the clock would tell time no longer.

Thus, sometime between 1200 and 1300 AD, in response to the Church's desire for more accurate clocks, the craftsmen of medieval Europe began to build mechanical clocks out of metal. These revolutionary mechanical clocks were driven by the force of weights, hanging from chains, that turned the gears of the clockworks. The speed of the turning gears was regulated by a mechanism called an escapement that alternately locked and released each tooth on a special gear. The action of the escapement is responsible for the characteristic ticking sound made by all mechanical clocks.

The escapement made it possible for these new mechanical clocks to tell time with unprecedented accuracy, and they represented a huge advance in durability and accuracy over all the other timekeeping technologies that civilized societies had used since ancient times. But none of the first mechanical clocks had either hands or faces. Instead, they told time by the ringing of bells. (In fact, the English word "clock" comes from the German word *Glocke,* meaning "bell.") And the clock face that is familiar to us—with an hour hand and a minute hand rotating inside a circular dial bearing twelve numbers—did not come into general use until 1700, more than four hundred years after the first mechanical clocks were installed in the towers of Europe's churches and monasteries.

Moreover, none of the early clocks had pendulums. Instead, the escapement on the medieval clock was regulated by a rotating arm, called a foliot, that swung back and forth on a shaft called a verge. The speed of the clock was regulated by moving the weights that hung from the two arms of the foliot either inward—which caused the foliot to rotate more rapidly—or outward—which caused it to rotate more slowly. By our standards, the verge and foliot escapement was not very accurate, and it was not unusual for these early clocks to gain or lose several minutes each day.

But the people of the Middle Ages told time only by the hour, and they did not bother with anything as precise as a minute. For all practical purposes, the verge and foliot escapement was sufficiently accurate. It was still the only type of clock in existence when Father Ricci presented his gifts to the Emperor Wan-Li, and it was used in all mechanical clocks for at least 350 years. The basic design of the medieval clock did not become obsolete until the invention of the pendulum and the more accurate "anchor" and "deadbeat" escapements in the seventeenth century (see Figure 8.1).

As a young man of nineteen, the Italian astronomer, mathematician, and physicist Galileo Galilei had become intrigued by the motion of the swinging altar lamp in church, when he noticed that the lamp always swung at the same rate of speed, regardless of whether its arc was great or small. Galileo wrote about this phenomenon in 1602, but it was not until 1641 that the aging Galileo, now blind, drew up the plans for the first pendulum clock with the assistance of his son Vincenzio. Although Galileo's pendulum clock was never built, his concept was brought to life by the Dutch mathematician and astronomer Christiaan Huygens, who built the first pendulum clock in 1657. The pendulum clock turned out to be ten times more accurate than its predecessor, and with this greater accuracy the minute hand—which most clockmakers did not bother to include on the clock face before that time—finally came into common use.

By far the most important consequence of the European obsession with time was that the construction of accurate clocks required machines that were capable of making precisely crafted mechanical components. In order to run at a constant speed, mechanical clocks required wheels, shafts, and cylinders that were perfectly round and straight. The teeth on their gears had to be evenly spaced, and each tooth had to be exactly the same size and shape. The springs that powered the clocks of later times had to be made in a precisely uniform thickness, with a precisely uniform temper. And all of the tiny screws that held the clock parts together had to be made to fit precisely into the threaded holes for which they were intended. But the hand-made objects fashioned by metalsmiths since the dawn of metallurgy were anything but geometrically precise, no matter how fine their craftsmanship. It was not possible for a truly accurate mechanical clock to be built by a craftsman using nothing but handheld tools.

In the early days of clock making, it took months—in some cases, even years—of skilled labor to build a single clock, and for the most part only the clockmaker who actually built each clock was capable of maintaining or repairing it. For this reason, driven both by the clockmakers' need for precision machinery and the public's increasing demand for mechanical clocks, the European craftsmen began to invent the special devices called "machine tools"—which are machines designed to make parts for other machines.

The new industry of machine tools included lathes to make objects perfectly round, drills to make precise holes, milling machines to make surfaces that were perfectly flat, and a variety of saws, grinders, planers, and shapers that were guided not by hand but by special devices mounted on rails or tracks. In this way, the clockmakers of Medieval Europe were directly responsible not only for the development of the first machine tools but also for all of the many other kinds of precision parts that machine tools were capable of creating.

As the science of precision machining continued to advance, new specialties arose. While they were learning how to make clocks, the clockmakers also learned how to make scientific instruments, and they fashioned sextants and compasses for navigation, astrolabes and theodolites for surveying, and precision scales for weighing. Lens grinders created precision lenses and built telescopes for viewing the heavenly bodies, microscopes for viewing things too small to be seen by the naked eye, and eyeglasses to improve the sight of those whose eyes had grown weak.

All of these instruments and many more stimulated the great advances in science that took place as the Middle Ages came to a close and Europe embraced the new philosophies of scientific inquiry that blossomed into the Renaissance. In fact, to a great extent, the Renaissance itself came to pass as a result of the widespread dissemination of information that began in Europe at the end of the fifteenth century. And this explosion of information took place when the principles of precision machinery were applied to the ancient art of printing.

Gutenberg's Press

In the middle of the fifteenth century, the German goldsmith Johannes Gutenberg discovered that an alloy of tin, lead, and antimony in the proper proportions could be easily melted and poured into a matrix of tiny molds, one for each letter of the alphabet. This made it possible for lines of type to be assembled from individual letters, known as "movable type." By contrast, the method of woodblock printing—the most common method of printing at that time—required the printer to carve an entire page of text and illustrations from a single block of wood.

Although Gutenberg is widely credited with the invention of movable type, this kind of type, in the form of very small porcelain tiles, had actually been developed in China four hundred years earlier. But because there are at least five thousand characters commonly used in Chinese writing, the printers in China were required to maintain a huge inventory of tiles. This meant that the Chinese printer had to search slowly and laboriously through these immense inventories for each of the tiles required to complete every line of text. By contrast, the Roman alphabet used throughout western Europe contained only a few dozen distinct characters. This meant that, when Gutenberg's movable type became generally available, many lines of text could be assembled quickly and easily by the typesetter.

Johannes Gutenberg's alloy of lead, tin, and antimony proved to be so perfectly suited to casting metal type that it has remained the standard formula for typecasting to the present day. Gutenberg also developed a recipe for oil-based ink that reliably produced a clear and durable impression on paper. Last but not least, he successfully adapted the screw press to the task of printing on paper. The screw press—which operates by pressing two surfaces together by means of a large vertical screw turned by hand—had been used in a much cruder form since ancient times for processing agricultural products, including for pressing the juice from grapes and the oil from olives, and it was already in use in Europe in Gutenberg's time for printing designs on cloth.

But in order to make the screw press capable of printing texts on paper, the platen—a large, flat plate that presses the paper onto the bed of inked type—had to be machined so that it was precisely flat and parallel to the bed of type. Otherwise, some areas of the printed page would be darker than others, and some areas might be so faint as to be unreadable. The precision machining of Gutenberg's day made it possible for him to build a screw press precise enough to make clear and consistent copies of printed material (see Figure 8.2).

Gutenberg's combination of movable type, oil-based ink, and the precision-built screw press unleashed an explosive growth in printed material. Before long, the huge expansion of knowledge made possible by the flood of cheap and plentiful books and pamphlets that issued from the printing presses of Europe was a major cause of the intellectual awakening of the Renaissance. During the fifty years between 1450 and 1500, slightly more than twelve million books were printed in Europe. Three centuries later, during the fifty years between 1750 and 1800, this number had grown to over 625 million—and this was before the establishment of public schools and the spread of universal literacy.

In Medieval Europe, the few people who knew how to read and write were mainly monks working as scribes in the monasteries. But the proliferation of reading matter during the Renaissance motivated more and more people to learn how to read, and literacy—which throughout all of human history had always been the special professional skill of the scribe—eventually became a skill that every person was expected to master.

Meanwhile, the advances in precision machining that ushered in the industrial revolution made possible numerous improvements in the design of printing presses in the years after 1800. These included the invention of the steam-powered printing press, the rotary press, electrotype printing, the linotype machine, lithography, and color lithography—none of which would have been possible without the prior development of precision machinery—and all of which increased the speed of the printing process while dramatically reducing the cost of printed materials.

Before long, the phenomenon of the local newspaper became commonplace throughout the industrialized world, followed later by the phenomenon of the popular magazine. The daily newspaper first appeared in Europe in the 1600s and in the Americas in the 1700s. In the year 1800, thirteen magazines were published in America. By 1900, this number had grown to 3,500, and in that year more than eight billion copies of newspapers and magazines were published in the United States alone.[2]

Blast Furnaces and Fossil Fuels

When the town clock became a fixture of life in the late Middle Ages, no self-respecting European city or town could be long without one. By the year 1450, there were at least five hundred clocks in the bell towers of European churches and monasteries, and after another century, there were thousands. All of these clocks were exceedingly large, with great, thick gears that rotated on large shafts, pulled by the action of massive weights that hung from heavy chains. Each clock was thus not only large in size but also used an exceedingly large amount of metal: the typical town clock of those days contained a ton or more of high-quality iron and steel.

The proliferation of these large, heavy clocks placed new demands on the supply of high-quality metals, and this drove further progress in the smelting and forging of iron and steel. It was no longer sufficient to hammer red-hot chunks of crude, impure iron into rough shapes that could be shaped into the hardware, tools, and weapons of wrought iron. And the production of large quantities of iron and steel required furnaces that were considerably larger and burned much hotter than the smelting ovens that were modified versions of the pottery kilns used since ancient times. And while deposits of iron ore were relatively abundant in Europe, the furnaces that were required to smelt iron and steel were anything but simple.

Eventually, the need to produce ever greater quantities of quality iron and steel led directly to the widespread adoption of the blast furnace, a high-temperature device for smelting iron that was originally invented by the Chinese in the first century AD. The blast furnace began to appear in northern Europe in the thirteenth century and became widespread throughout Europe after 1500 (see Figure 8.3). Filled with a mixture of iron ore, charcoal, and limestone, the blast furnace was heated to very high temperatures by a fire superheated with a bellows or some other forced air supply. The ore mixture was smelted and refined by the chemical action of hot air forced into the bottom of the furnace under high pressure.

With the proliferation of blast furnaces throughout Europe, the demand for charcoal skyrocketed. But the production of charcoal required a prodigious amount of firewood. Wood not only furnished the basic raw material that was converted into charcoal by being heated in an oxygen-deprived environment, but it was also necessary to keep wood fires burning for many hours to heat the ovens that produced the finished charcoal. By the middle of the seventeenth century, the iron-works in Britain alone were consuming fifteen million feet of timber every year.[3]

In addition to the growing demand for tree bark for tanning leather, and for firewood to make charcoal, soap, and glass, the Europeans of that time were also cutting down trees to serve the needs of their expanding population. Entire forests were cut down to clear land for agriculture, to build new houses—and in some cases, entire towns—and to build the large ocean-going sailing ships the Europeans needed for their rapidly expanding sea trade. By the year 1700—to cite one of many examples—the British Royal Navy estimated that the wood of four thousand mature oaks was required to build a single ship of the line.[4]

With all of these demands for wood increasing in Europe year by year, the supply of firewood for making charcoal was soon exhausted, and by the early 1700s, the densely populated areas of Europe had become effectively deforested. The solution to this problem—the mining of coal—began in earnest in eighteenth century England, but the coal mines themselves were plagued by a number of technical problems. As we will see, one of these problems, the frequent flooding of the mines by underground sources of water, led before long to the beginning of the industrial revolution.

Coal had been used extensively in northern China and Mongolia beginning in 2000 BC, and it was known to both the ancient Greeks and Romans, who mined coal and used it for heating their houses, villas, and public baths. The Romans were well aware of the presence of coal in Britain, and they mined it extensively. But when the Romans left Britain in 410 AD, coal mining ceased, and coal was not mined again in Britain for nearly seven hundred years. Throughout the Middle Ages, firewood served perfectly well as a source of heat, and the charcoal made by firewood proved ideal for smelting iron ore. Coal, on the other hand, gave off a particularly noxious smoke, and it was considered so deleterious to human health that in 1306 the burning of coal within the city limits of London was banned by royal decree.

Throughout the ancient history of metallurgy, coal was considered entirely unsuitable for smelting iron, because unlike charcoal—which is nearly pure carbon—coal tends to be full of impurities. These include tar, sulfur, clay, quartz, chalk, salt, and numerous other minerals, all of which can easily contaminate the iron ore and interfere with the smelting process. When coal was used as the fuel for blast furnaces, it produced iron of poor quality: brittle, weak, and difficult to work with. But during the 1600s it was discovered that by roasting high-quality coal in a manner similar to the production of charcoal from wood, the coal could be freed of many of its impurities and changed into a hard, strong, porous, clean-burning fuel called coke. And in 1709, the Englishman Abraham Darby successfully smelted iron in a blast furnace that burned coke instead of charcoal.

Darby's simple innovation had far-reaching consequences. First, it took the pressure off the dwindling reserves of forest, making it possible for the first time to produce high-quality iron in many areas of Europe that had become dangerously deforested but were nevertheless rich in coal. Second, the charcoal fires traditionally used for smelting were subject to being crushed and smothered if too much iron ore was dumped in on top of them, but the greater physical strength of coke could support much heavier loads of ore. This made it possible for much larger quantities of iron ore to be loaded into the furnace at one time. As a result, blast furnaces were soon being built far larger than was previously possible, and these new furnaces produced greater yields from the same investment of time and labor. Finally, it was far cheaper to mine coal and convert it into coke than it had been to cut down entire forests and saw up the trees in order to produce charcoal from firewood.

The combined effect of these developments was to steadily lower the price of iron, which soon began to be used for making a growing variety of manufactured objects, ranging from simple pots and pans to immense iron bridges. Throughout the 1700s, coal mining underwent a tremendous expansion, and new coal mines were opened all over Europe. And the proliferation of coal mines provided more than a new source of energy for smelting iron. Although no one at the time could have foreseen it, a common problem that affected most coal mines remained unsolved until the invention of the steam engine, one of the most transformative of all precision machines, and the device that, more than any other, gave birth to the industrial revolution.

Coal Mines and Steam Engines

Most of the ancient and medieval coal mines exploited veins of coal that lay near the surface, but after centuries of mining, most of these open-pit coal mines had become exhausted. Thus, to meet the growing demand for coal, shafts were sunk far below the surface, tapping veins of coal that lay deep underground. But these deeper coal mines often had to pass through large underground aquifers, where groundwater in abundance lay beneath the surface. When the shafts and tunnels of the coal mines passed through these aquifers, great streams of water flowed into the mines, flooding the areas where the miners worked at the coal face.

Various methods were devised to pump the water out of the coal mines, chief among which was the chain pump, a series of discs or buckets attached to a chain that rotated around large pulleys, moving continuously through a large vertical pipe. The chain pump was usually pulled by a team of horses, but the power that horses could provide to pump the water up from far below the surface was limited, and before long British inventors began experimenting with the use of steam to accomplish this critical task.

In 1712, an Englishman named Thomas Newcomen built a device he called an "atmospheric engine," which used steam under pressure to power a piston and cylinder that pumped water out of the mines. Although numerous devices using steam to create mechanical energy had been made since ancient times as toys and curiosities, Newcomen's atmospheric engine was the first engine to use steam power for a practical use. By the time of Newcomen's death in 1729, more than one hundred atmospheric engines had been installed in coal and tin mines, mostly in England and Wales, but a few were also installed in other European countries. The first atmospheric engine in France was installed in 1725 to pump water from the Seine River for use by the citizens of Paris.

But the atmospheric engine had a serious design flaw. The steam inside the cylinder had to be alternately heated and cooled in order to raise and lower the piston. This not only wasted large quantities of fuel but also severely limited the speed with which the piston could move up and down inside the cylinder. In spite of its commercial success, the atmospheric engine had achieved only half of what was needed to harness the power of steam.

While it could raise and lower its piston with considerable force, the atmospheric engine was unable to translate the piston's oscillating motion into the rotary motion that made the waterwheel such a useful source of mechanical power. (In fact, one of Newcomen's engines was used to pump water from a pool below a waterwheel into a reservoir, from which it flowed back into the pool, turning the waterwheel in the process.) In an attempt to overcome this limitation, the English inventor James Pickard attached a flywheel to the atmospheric engine's piston by means of a crank, a device that had been used since ancient times, and in 1780, Pickard succeeded in patenting his crank-and-flywheel design.

Unfortunately, the atmospheric engine was ill-suited to the purpose of turning a wheel, because its primitive piston and cylinder was capable of oscillating only once every five seconds. This meant that the wheel that was connected to the piston of the atmospheric engine could attain a top speed of no more than twelve revolutions per minute, a pathetically slow speed for any rotary engine. And Pickard's patent was a serious obstacle to the Scottish instrument maker James Watt, who had already patented a much more efficient design for a steam engine.

Watt made two significant improvements to Newcomen's basic design. His first improvement was to allow the steam to be physically expelled from the cylinder into a cooling condenser with each stroke. This eliminated the wasteful requirement that the main cylinder be cooled and heated over and over again. The atmospheric engine had actually used more fuel in the process of reheating the cylinder than it used to deliver mechanical force.

Watt's second improvement was to close off the back end of the cylinder with a pressure-tight cover, allowing only the piston rod to project through an air-tight hole in the center of this cover. This allowed steam first to enter the bottom of the cylinder, pushing the piston up to the top of its stroke, and afterward to enter the top of the cylinder, pushing the piston back down again. This dual action, which gave the cylinder two power strokes in place of one, effectively doubled the power of the steam engine without increasing its size or weight.

In 1775, Watt formed a partnership with the English businessman and manufacturer Matthew Boulton to build his improved steam engine. But one last obstacle still had to be overcome. The huge pistons and cylinders of the early steam engines were not machined with sufficient precision to make a close, airtight fit. This allowed steam to escape with every stroke, significantly reducing the power and efficiency of the engine. The problem was solved by a Welsh cannon maker named John Wilkinson, who had developed a method for precision boring of cannons that he applied to the boring of the immense cylinders of Watt's early engines—the first of which had a piston more than four feet wide that traveled up and down inside a cylinder twenty-four feet long.

Watt's early steam engines were a great improvement over Newcomen's atmospheric engine, and they accomplished the same amount of work while using only one-fourth as much fuel. As a result, Watt's engine quickly became the preferred mechanism for pumping water out of the coal mines. But the simple oscillating motion of these early engines was suitable only for work that required an up-and-down motion, such as pumping water or alternately raising and dropping the hammer of a mechanical forge. The last step had yet to be taken: the oscillating motion of the piston and cylinder had to be converted to the rotary motion that was needed to power the sawmills, flour mills, and textile mills that were springing up all over eighteenth-century Europe.

It was Boulton who urged Watt to attach the piston of his improved steam engine to a flywheel by means of a crank, but Pickard had already patented that concept. Refusing to enter into a partnership with Pickard, the ingenious Watt managed to circumvent Pickard's patent by connecting the piston of his improved steam engine to a flywheel with a "sun and planet" gear that operated on a different principle from that of the simple crank. Watt's new engine proved to be a huge success, and his partnership with Boulton, which was to last for twenty-five years, made James Watt a wealthy man for life.

The impact of Watt's success at creating a steam engine that delivered rotary power at relatively high speeds cannot be overstated. It liberated European society from its age-old dependence on water wheels—which could only be placed in favorable locations where a reliable source of falling water could be harnessed—as well as on windmills, which were dependent on the unpredictable coming and going of favorable winds. For the first time, a source of rotary power could be located wherever it was most advantageous.

Textile mills could be built wherever the location was most favorable. Sawmills and flour mills could be built along the banks of quiet, navigable rivers, ready to receive loads of grain and wood brought by barges and riverboats. And the numerous deposits of both coal and iron ore that existed throughout the British Isles made Britain a favored place for the development of industries based on iron, steel, coal, and steam.

Of all the consequences that flowed from the creation of an efficient steam engine with rotary power, none was greater than its enormous impact on the means of transportation by land and by sea. When the steam engine was attached to a huge wheel fitted with wooden paddles, the steamship was born. When the steam engine was mounted on a large and heavy wagon made entirely of steel and placed on steel rails, the railroad was born. And of all the many offspring of the rotary steam engine, the steamship and the railroad had to be counted among the most transformative. Together, they revolutionized long-distance travel and trade, and they made possible the greatest transformation in the nature of human society since the birth of urban civilization thousands of years ago.

Engines of Mobility

For thousands of years, the sailing ship had been at the mercy of the winds at sea, which could blow fitfully and unpredictably from many different directions. In the early 1800s, before the first ocean-going steamship was built, a typical trip from Europe to America by sailing ship lasted an average of *two months* at sea. By contrast, the Great Western, the first paddle-wheel steamer built expressly for crossing the Atlantic, made its maiden voyage from England to New York in April of 1838 in sixteen days, and by 1875 a flotilla of iron-hulled steamships—with advanced steam engines and screw propellers instead of paddle wheels—were regularly crossing the Atlantic in seven days. Other trips by sea that had previously taken months, to South America, Africa, India, and the Orient, were also made in a fraction of the former time. As a result, intercontinental travel and trade increased tremendously in the century between 1800 and 1900.

The first regular railroad service between two cities was established in 1830 in England between Manchester and Liverpool, and the success of the railroads in the ensuing decades set off a frenzy of building tracks, trains, and locomotives throughout Europe and North America that consumed ever larger quantities of iron and steel. Trips that had once taken days were accomplished in a matter of hours, and heavy cargos that had previously been transported slowly by ships, barges, and riverboats were being loaded onto freight cars and shipped over land in a fraction of the time that travel by boat had previously required.

When the machinists and inventors of the late eighteenth century perfected the art of creating the precisely fitting piston and cylinder—and linked the piston with a precisely machined crankshaft—the back-and-forth motion of the piston was converted into continuously rotating motion, and the "reciprocating engine" was born. The steam engine was the quintessential reciprocating engine of the nineteenth century, and it quickly became the primary source of power for the mills, ships, and railroads of Europe. But the reciprocating engine was destined for an even more exotic and transformative future.

The steam engine was a precisely machined reciprocating engine in which the combustion of fuel took place *outside* of the piston and cylinder—burning wood, coal, or coke to heat water in a boiler and turn it into steam. Once they had perfected the steam engine, the European inventors began tinkering with the idea of having the combustion of fuel take place *inside* the engine. Before long, engineers from several different countries had completed designs for the radically new idea of an *internal* combustion engine. Most of these engines operated by having a small amount of fuel ignited directly inside the cylinder. The idea was that the force of the expanding gases heated by the combustion of fuel would drive the piston forward and turn the crank on a flywheel.

As early as 1807, the French brothers Nicéphore and Claude Niépce successfully sailed a boat up the Saône River that was powered by an internal combustion engine running on a mixture of moss spores, coal dust, and resin.[5] For this achievement, their engine was granted a patent by Napoleon Bonaparte. Other experiments followed, using hydrogen gas and spirits of turpentine as fuel. This was followed by increasingly successful attempts throughout the nineteenth century by Swiss, American, British, French, German, and Belgian inventors to use coal gas—a byproduct of the process that turned coal into coke, and the principal gas used for lighting during the Gaslight Era—to power an internal combustion engine.

Ultimately, the German inventor Nikolaus Otto succeeded in designing a gas-powered engine sufficiently powerful and reliable for commercial production. In 1864, Otto opened a factory to produce internal combustion engines, and by 1875, he had sold over six hundred gas-powered engines. But because coal gas was supplied primarily by metal pipes buried in the ground, the gas-powered engine was suitable only for stationary installations and was not a practical source of power for motor vehicles.

In 1884, however, the British engineer Edward Butler designed and built the first modern gasoline engine, complete with an ignition coil, spark plug, and carburetor. Butler installed his engine in a three-wheeled vehicle he called the Velocycle, which had a top speed of ten miles per hour. But due to Britain's infamous Red Flag Act of 1865, which established a speed limit for self-propelled vehicles at two miles per hour in residential areas and four miles per hour in the countryside—and that further required a man to walk in front of the vehicle waving a red flag—Butler was unable to find customers for his Velocycle. Butler destroyed his invention in 1896, sold the metal for scrap, abandoned his quest to build a motor vehicle, and devoted the rest of his life to designing internal combustion engines for boats.

Ultimately, it was the German engineers who succeeded in solving the technical problems that had stood in the way of producing a practical, affordable automobile. Gottlieb Daimler and Wilhelm Maybach, both former employees of Otto, built a small gasoline engine that they mounted on a two-wheeled vehicle in 1885, creating the world's first motorcycle. Karl Benz began producing the first commercial automobile in 1886, and by 1890 Daimler and Maybach had also begun to produce automobiles for the commercial market.

Finally, Rudolph Diesel successfully developed the first working version of the Carnot engine, using the thermodynamic principles established by the French military engineer and physicist Nicolas Carnot. These principles had been published as early as 1824, when Carnot was only twenty-six years old, and they described a process in which the combustion of fuel in the cylinder was ignited not by an electrical spark but rather by the heat that was generated when the piston compressed the gases inside the cylinder. Although Rudolf Diesel's life was cut short when he died mysteriously at sea in 1913, his concept was immortalized in the diesel engine that has since replaced the steam engine as the primary source of power in the largest and heaviest of modern vehicles: ships, locomotives, heavy trucks, and earth-moving equipment.

By the late nineteenth century, the petroleum industry, which had begun very modestly in the 1850s in Europe with the manufacture of paraffin wax from crude oil, had matured into a major industrial force, and by 1880, more than twenty million barrels of crude oil per year were being pumped from oil wells in the US alone. The combination of the gasoline distilled from petroleum with the precision-machined internal combustion engine made possible the creation of the modern gasoline engine: a lightweight, compact, powerful, dependable, and economical source of power. And it was the modern gasoline engine, more than any other device, that enabled the early twentieth-century inventors to create both the automobile and the airplane—two technologies of long-distance transportation that reduced the distances separating human societies to a fraction of their former size.

From Fire Lances to Firearms

Warfare is neither a new nor a recent phenomenon in human history. Even some groups of chimpanzees, as we saw in chapter 1, have been observed acquiring territory by force, using a process that is strikingly similar to human warfare. And the archeological evidence described in chapter 6 suggests that organized warfare—including the formation of armies and the wholesale slaughter of civilian populations—has been a common occurrence in human life since long before the evolution of urban civilization. But when the science of chemistry and the techniques of precision machinery were applied to the design and manufacture of weapons and their delivery systems, the violence and destruction of warfare soon reached a level unprecedented in human history.

The invention of firearms began in China in the ninth century AD with the discovery that saltpeter—a naturally occurring nitrate of potassium—could be easily ignited and would burn with a characteristic blue flame. Before long, the Chinese had discovered that by mixing powdered saltpeter with smaller amounts of powdered sulfur and charcoal, it was possible to produce an explosive mixture that they called black powder—the forerunner of modern gunpowder. The beauty of black powder was that, in the process of burning, the saltpeter actually released more oxygen than it consumed, and this extra oxygen allowed the sulfur and charcoal to burn even when the mixture was ignited within a closed container, such as inside the barrel of a gun.

The widespread belief that the Chinese invented gunpowder but used it only to fashion entertaining fireworks is entirely false. By the tenth century, the Chinese were already filling hollow bamboo tubes with black powder to shoot poison-tipped or burning projectiles called fire lances, and by the early thirteenth century, they were using gunpowder to build bombs, flame throwers, rockets, guns, and cannons—all of which were used extensively in warfare. At this time, the Europeans had no firearms, and European weaponry consisted mainly of daggers, swords, spears, lances, maces, arrows, and crossbows, plus a variety of ramrods and catapults designed to batter down the walls of military fortresses and fortified towns.

Firearms appeared in Europe after the middle of the thirteenth century, but for several centuries all firearms had some significant disadvantages that seriously limited their use. Hand-held guns were heavy, and they required several minutes to load with gunpowder and projectiles. They had to be held in one hand while they were lit with a burning wick by the other hand, they were notoriously inaccurate, and they sometimes exploded in the hands of their owners.

As late as 1600 AD, the muzzle-loading "matchlock" musket was fired by a burning wick that was thrust against a small amount of priming powder when the trigger was pulled. A Dutch soldier's manual published in 1607 listed no less than twenty-eight individual steps that a musketeer was required to perform before he was ready to fire a single shot from a matchlock musket.[6] In fact, the matchlock musket took so long to load and fire that it was not until the introduction of the flintlock musket after 1650—in which the gunpowder was ignited by a spark produced by a spring-loaded flint striking a steel plate—that the use of firearms in warfare finally began to replace the traditional handheld weapons.

Cannons were more deadly than the catapults that had been used since ancient times to hurl huge stones at fortified targets, but the early cannons were brutally large and heavy. A single cannon typically weighed thousands of pounds, required a large team of horses to move it into firing range, and could take as long as several hours before it could be cleaned, packed with gunpowder, loaded with cannonballs, and was ready to fire a single shot. Cannons were not only highly inaccurate, but also, like handheld guns, they occasionally exploded and killed their operators instead of firing their projectiles.

The problem with all of the early firearms was in their method of manufacture. The barrels of the early guns and cannons were created by casting molten bronze or iron in molds, but the process of casting molten metal does not produce a smooth or precisely straight barrel, and small irregularities inside the barrel would affect the trajectory of the projectile. In addition, the early firearms lacked the "rifling" that is incorporated into all modern firearms. Rifling consists of spiral grooves that cause the projectile to begin spinning as it travels down the length of the barrel, which greatly straightens the trajectory of the bullet, immensely increasing the range and accuracy of the weapon.

But the technology of precision machining made it possible to bore the barrel of a firearm into a precisely smooth, straight tube, and it was this development, more than any other, that gave the firearm its deadly effectiveness and that finally rendered obsolete the swords, spears, and arrows of earlier times. This is why, when James Watt needed to build a precisely fitting piston and cylinder for his new steam engine, he turned for a solution to John Wilkinson, a cannon maker.

When the Chinese invention of gunpowder and firearms was married to the European advances in precision machining, the firearm emerged as the most deadly weapon in human history. The standard issue M-16 rifle used by the US Army is capable of firing ten rounds of ammunition *per second* with an effective range of six football fields laid end to end. The US Army's M-61 machine gun can fire one hundred rounds of ammunition per second, and a modern howitzer can shoot an explosive shell six inches in diameter and destroy a target eighteen miles away. And these are only three of the tens of thousands of different firearms manufactured and used at the present time, from tiny pistols less than four inches long to immense naval guns more than sixty feet long with a range of twenty-two miles that can fire an explosive shell sixteen inches in diameter weighing more than two thousand pounds.

For millions of years, the early hominids, emerging humans, and modern humans had all faced the problem of defending themselves and their offspring from the threat posed by the large predators that inhabited their world and their natural environments. The lethal weapons that prehistoric humans learned to fashion may have enabled them to kill and injure both their prey and their natural enemies, but armed only with the spears, javelins, harpoons, and arrows of the past, preindustrial people had no guarantee that they would survive a confrontation with the lions, tigers, bears, wolves, buffalos, elephants, rhinoceroses, walruses, wild horses, and other dangerous beasts that they encountered while hunting or moving their camps from one place to another. Before the invention of firearms, countless hunters, gatherers, farmers, and herders had been killed by the attacks of frightened, enraged, or wounded beasts.

But the development of precision-machined firearms changed all of that. For the first time in human history, the contemporary hunter came to be armed with a lethal weapon that could wound or kill other animals from a considerable distance. The effectiveness of modern firearms—combined with the dramatic reduction in the extent of wilderness areas and the steady movement of the human population from the countryside to the cities—has effectively ended the threat to human life once posed by the presence of wild animals in our environment. In the modern world, the threat to human life and safety now comes almost exclusively from other human beings.

All Things Electrical

In 1820, the Danish physicist Hans Christian Ørsted was delivering a lecture when he noticed that the needle of a compass reacted when an electric current from a battery was switched on and off. Ørsted became the first person to publish the discovery that when electricity flows through a wire, it produces a magnetic field. Although other scientists experimented with Ørsted's discovery of the relationship between electricity and magnetism, it was not until the 1830s that Michael Faraday, the son of an English blacksmith, proved that Ørsted's discovery worked in reverse: that when a magnetic field moves across an electrical conductor such as a wire, it creates an electric current in the conductor. This phenomenon, known as electromagnetic induction, made it possible for Faraday to create working models of the two essential devices of the electric age: the electric generator and the electric motor.

Before Faraday's pioneering work, the only reliable sources of electricity were primitive forms of the wet cell battery—large, heavy, and fragile objects filled with dangerous quantities of sulfuric acid. Yet it is also important to remember that Faraday would have been unable to create either the electric motor or the electric generator if it had not been for the techniques of precision machining. These techniques made it possible to build mechanisms with smooth, durable bearings that enabled their rotors to spin at high speeds with a minimum of vibration, which were essential for the proper functioning of both motors and generators.

Once the principles of electromagnetic induction were firmly established and electric generators had been successfully designed and built, any source of rotary power could be used to generate a steady electric current. The first public power station was built in southern England in 1881, powered by a water wheel. By the end of the nineteenth century, electric generators, driven by steam and water power, were being built throughout Europe and North America, and by the early twentieth century, electricity began to spread like wildfire throughout the inhabited world.

The arrival of affordable electric power quickly led to the development of a host of new technologies that have completely revolutionized human life. Electric elevators—vastly more powerful and efficient than their hydraulic and steam-powered predecessors—made it possible to build true high-rise buildings, and the skyscraper was born. The telegraph—and later, the telephone—made it possible for people to communicate instantly across distances of thousands of miles for the first time in human history. Subway trains carried millions of people to work every day through tunnels in the ground, while electric trolleys trundled through the streets of the world's major cities.

Electric lighting transformed the night from a dim, shadowy world of flickering lamps and candles into a world of electric bulbs that lit up the night as it had never been lit before, freeing humanity from its ancient dependence on the rising and setting of the sun. Refrigeration completely transformed the food industry, allowing meat, poultry, and seafood to be transported thousands of miles and to be frozen for weeks and months without damage. Air conditioning transformed the lives of everyone living in climates that grow hot, freeing them to work and play in invigorating coolness. And the arrival of radio, movies, and television—none of which would have been possible without electricity—ushered in the age of mass communications.

Meanwhile, the machine tools of the industrial world went into high gear, freed from the bulky and dangerous belts, pulleys, and shafts that had long been used to transfer power from steam engines to factory assembly lines. As late as 1900, steam engines still provided more than 80 percent of the mechanical power used in manufacturing, while electric motors provided less than 10 percent. Forty years later, these proportions had become exactly reversed.[7] By that time, modern machine tools were all driven by electric motors, and in the process they became smaller, lighter, more portable, and more effective than the previous generation of machine tools that had been driven by steam engines—or the generation before that, which were powered by the action of human hands and feet turning cranks and treadmills. In less than two hundred years, the widespread use of fossil fuels, reciprocating engines, and electricity—all of which were made possible by the technology of precision machinery—had thoroughly transformed every aspect of human life.

Only during those rare moments when a severe storm or a construction accident causes the lights to go out for a few hours do those of us who live in the developed world get so much as a partial, temporary glimpse of the central role played by electricity in our daily lives. Heating and air conditioning systems grind to a halt. Refrigerators cease to function. Stoves will not light. Elevators stop working. Television is nonexistent. Computers go dark. Mobile devices cannot be recharged. And the workplaces and transportation systems of the modern world—the offices, factories, warehouses, shops, hospitals, schools, airports, trains, subways, and elevators that make urban life possible in contemporary society—cease to function and must be abandoned until power is restored. And when a power failure occurs at night, we are plunged into the unfamiliar and inconvenient darkness of a world lit only by candles and lanterns.

And these are just the technologies that depend on the electricity produced in power stations and delivered by power lines to buildings and other permanent structures. If electricity itself were suddenly to disappear, then cars, trucks, buses, and trains would neither start nor run, airplanes would not fly, the forces of law and order would be unable to communicate, and with neither telephones, mobile devices, nor radios operational, only people who happened to be in the same place at the same time would be able to speak to each other at all.

Given our extreme dependence on electricity in contemporary life, it is astonishing to think that electricity itself was not generally available until late in the nineteenth century. And of all the many ways that electricity has revolutionized human life, perhaps the most significant—in terms of its cultural and social consequences—has been the proliferation of technology in the house and home and the effect of domestic technology on the position of women in modern society.

The Machinery of House and Home

For fifteen months, beginning in the fall of 1967, I lived with my family on the Greek island of Ios, where I studied the traditional way of life of Greek island villagers for my doctorate in cultural anthropology. I chose Ios Island because at the time it had no harbor facilities capable of serving cruise ships, and neither electricity nor motor vehicles had yet come to the island. As a result, the islanders were living largely as most other European villagers had lived before the coming of the industrial revolution.

The people of Ios were still growing their own grain and grinding it into flour in the village windmill. The bread dough, which they prepared at home, was baked every morning by the village baker in a huge stone oven heated by wood fires that were lit every evening and burned through the night. The islanders grew their own olives, which they pressed for oil, and their own grapes, which they pressed for wine. All of the pork, lamb, goat, chicken, and eggs consumed on the island, as well as all the seafood and vegetables, were raised by local farmers and caught by local fishermen. Beef was not available. The women of Ios still spun thread from cotton and wool, and they knitted much of their clothing by hand.

On Sunday mornings, weather permitting, a small steamer would arrive from Athens, bringing those who had left the island the week before—to visit relatives or conduct business—back to Ios. The island's only automobile, an aging station wagon, was used on those mornings to carry the old and infirm passengers up the long, winding dirt road from the harbor to the village. Unwilling to pay for the ride, most people loaded their luggage on a pack animal and endured the long, steep walk up an ancient stone stairway. For the rest of the week, the station wagon sat idle. Mules and the occasional donkey were the only means of transport.

In the traditional village society of Ios, people worked from early morning to sundown. After dark, they ate supper by the light of candles and kerosene lanterns. The more affluent villagers used gas hotplates for cooking; the less affluent used charcoal braziers. Cold water was carried into the houses in cans and pails that were filled at public water taps on the street. Hot water was heated on the stovetop. Clothes were washed outdoors by hand with a scrubbing board in a metal pan. Floors were cleaned with brooms and mops. Washing machines, refrigerators, vacuum cleaners, gas ranges, air conditioners, and dishwashers were unknown. Canned food was sold in a few small general stores, but it was purchased only for special occasions. Frozen food did not exist.

In this traditional and largely preindustrial society, the division of labor between men and women was strict and all-encompassing. Men raised the crops of grain and tended to the orchards and the vineyards. Men took care of the animals, built the houses, and repaired them when necessary. Men dealt with politics and government officialdom. Men went abroad to earn money as employees on ships, in factories, and in restaurants.

Women raised the children and took care of the house and home. Women tended the vegetable gardens, purchased the food and other household necessities, and prepared all the meals from unprocessed raw ingredients. Women washed the dishes, swept and mopped the floors, carried the rugs outside and beat the dust out of them, washed the clothes by hand and mended them if they were worn, made the beds, washed the linens, spun thread from cotton and yarn from wool, knitted socks and sweaters, whitewashed the houses, nursed and washed the babies, fed and clothed the children, cared for the sick, infirm, and elderly members of the family, and did their best to keep their homes cool in the summer and warm and dry in the winter.

Without a partner of the opposite sex, a man or woman was a second-class citizen in island society. By himself, a man had no one to bear his children, prepare his food, mend and wash his clothes, or keep his house clean and orderly. By herself, a woman had no hope of employment and no reliable source of money, food, or protection. Living alone was out of the question. Unmarried men and women lived with their aging parents or with the families of relatives. They lacked not only a home of their own but also the social status, self-respect, and personal autonomy that came with it. And their greatest misfortune, in the value system of island culture, was they had no children to bring joy and laughter into their lives, to provide them with a vision for the future, or to support and care for them in their old age.

Coming as I did from a developed, urban society in which domestic technology was commonplace, I did not, at first, comprehend the irreplaceable position that the women of this Greek island occupied in everyday life. But a man without a wife in that traditional society was only half a man. Even if he were to defy convention and attempt to do "women's work" by himself, he would be hard pressed to find enough time to do so. The village women of the island spent the entire day performing the myriad tasks that were required to maintain a respectable home. And in addition to running a respectable household, the younger women also had the burden of caring for their children: washing them, feeding them, clothing them, and keeping them out of danger. There was literally no way a man could live a normal life without a woman to take care of his domestic needs, and the few adult men who were not married were the objects of pity—and occasionally of ridicule.

The mechanization of the household in the twentieth century has changed all of that. Washing machines, dryers, vacuum cleaners, and dishwashers make it possible to do all the washing and cleaning that a household requires in far less time than was required in preindustrial times. Frozen food, prepared food, take-out food, refrigerators, gas ranges, microwave ovens, and a gallery of electrical appliances from simple blenders to complicated food processors have all reduced the preparation of food to far simpler tasks that can be executed in a fraction of the time that was once required. And all of these technologies depend on the availability of electric power.

With the proliferation of employment opportunities for women—and the growing acceptance of men performing routine household chores—living alone has become a realistic option for both sexes. In fact, the single-person household has now become the chosen way of life for roughly 30 percent of the households in Europe and the United States (in Sweden, the number is 47 percent). And the rest of the world is not far behind. Single-person households are even more common in Japan than they are in the United States. But not every nation has embraced this way of life. In India, where family life has remained strong and vibrant, only 3 percent of households consist of only one person.

No one would argue that it would be better to return to the conditions of the past, when men and women were so utterly dependent on each other that they were incapable of leading a respectable life without a marriage partner, or when women spent the better part of their adult lives cooking, cleaning, and raising children. But—paradoxically—the very labor-saving devices that modern people have acquired to make women's work lighter and easier have had the effect of depriving women of the indispensable position they formerly enjoyed in their relationships with men and in society in general. Meanwhile, the economic importance of men—who traditionally were the only members of the family who were free to pursue gainful employment—was, if anything, magnified by the advent of the so-called employment society.

As a direct result of the mechanization of the house and home—and the complete reversal in the economic significance of children from economic assets to economic liabilities—the former importance of women as bearers of children and keepers of the house and home was seriously diminished. During the early years of the twentieth century, the mechanization of the home did not proceed much beyond the advent of mechanical cleaning technologies such as the washing machine and vacuum cleaner. And throughout the years of the Great Depression and the Second World War, massive unemployment and compulsory military service meant that a large number of men were unable to adequately support their families. This was the time when women first began to work outside the home in large numbers, and many women found the experience both empowering and liberating.

But when the war ended, the men returned to their jobs. And when the women left their wartime jobs and went back to keeping house and raising children, the status of women and the prestige of women's work plummeted to historic lows. This is why, beginning in the middle of the twentieth century—when the mechanization of the household was in full swing, but before it was considered acceptable for married women with children to work for wages at full-time jobs—the women of the modern world began to experience the existential crisis in their status and self-respect that has continued to the present day.

In the hunting and gathering societies of the past, women not only gathered the firewood and cooked the food but also brought home most of the vegetable foods that the group as a whole could not live without—and that enabled adult men to spend their days hunting for the equally vital supplies of meat. In agricultural societies, women not only worked in the gardens and did all the cooking, cleaning, and washing, but also bore and raised the children that brought wealth, status, and the promise of future security to the family as a whole.

But in modern industrial society, the only job that is truly indispensable is earning money through gainful employment, and because this began as "men's work" during the first decades of industrialization, men have traditionally dominated this sphere and still enjoy a considerable advantage over women as employees. Meanwhile, the importance of women as the bearers of children has lost much of its former significance. The people of the developed world have increasingly come to rely not on their children but rather on their investments, pensions, social security payments, and health insurance policies to provide them with economic security in their old age.

In short, cultural inertia—the tendency for a culture to retain values and attitudes that are traditional but have become obsolete due to other changes in society—has dealt the women of the modern world a cruel hand. On the one hand, women are still expected to perform most of the work involved in cooking, cleaning, maintaining an orderly household, and raising children. On the other hand, these tasks have lost much of their former importance, both to their husbands and to society at large.

Trapped between a traditional role that is no longer indispensable and that no longer commands the respect it once did and a nontraditional role in which they are newcomers and often unwelcome competitors, modern women struggle to find the social acceptance and personal self-respect that the male sex never lost and that most men continue to take for granted. As we will see in the next chapter, however, the advent of digital technology has challenged the traditional role of adult men more than any previous transformation, and it has diminished their status in society in ways that the inventors of the first computers could never have foreseen.

Love, Sex, and Marriage in Industrial Society

As the inheritance of farmland has become largely irrelevant to the life of the average modern citizen, it is no longer important for the parental generation to arrange the marriages of their children and heirs. Modern people rarely marry to enhance the economic relationship between their families, and since the ancient custom of arranged marriage is effectively obsolete, the expectations of married life have changed profoundly. As a result of these changes, the novel (but actually very ancient) idea of "marrying for love" has become the primary reason for the people of modern society to marry.

As the people of the modern world have moved from the farms to the cities, children are no longer in a position to contribute meaningfully to the family fortunes. And as financial support in old age has gradually become an obligation of government throughout the industrialized world, children have lost their importance even as a source of security for the parents in their old age. Since the children of modern urban society have become a net economic burden to their families, it is not surprising that families have become smaller with each generation of urbanization, and that many married couples have chosen to have no children at all.

Rather than regarding marriage as the means to an economic end, young people in the developed world have instead come to regard marriage as a pathway to companionship and sexual satisfaction. And as modern people have increasingly chosen their marriage partners on the basis of physical attraction and emotional compatibility, the traditional values of premarital chastity and virginity at marriage have rapidly lost their appeal.

The independence and personal freedom that accompanied the arrival of the automobile and the anonymity of urban life provided young people with unprecedented opportunities for privacy, and the traditional practice of providing a chaperone for every interpersonal contact between young men and women has been largely abandoned. In fact, young people in modern society are now generally expected to engage in some form of sexual interaction when they pair off as part of their normal social lives. And it has become increasingly accepted throughout the developed world for adults of all ages to live together in complete sexual freedom without benefit of the legal status of marriage. By 2013, 75 percent of American women had lived with a partner without being married by the time they were thirty years old.

Modern society's tolerance of this general increase in sexual freedom has been powerfully augmented by the development of effective contraceptives, which has liberated both sexes from risking an unwanted pregnancy when they have sex. The potent combination of the new ideals of marrying for love, the new standards of social and sexual freedom for unmarried couples, and the development of truly reliable contraceptives triggered a transformation in sexual values in modern society, as the ancient human sexual instincts, themselves the product of millions of years of evolution, have reasserted themselves.

In fact, the traditional institution of marriage that we inherited from ancient societies was never designed to provide intimacy, companionship, mutual attraction, or sexual satisfaction. Traditional marriage evolved in agricultural societies as a way to create lifelong partnerships, establish mutually beneficial economic relationships between families, and maximize the stability of land ownership in agricultural society. These goals were achieved by a set of customs that made both men and women socially, economically, and psychologically dependent on each other. And it was these customs—not lasting affection or mutual attraction—that ensured the permanence of marriage.

Nevertheless, ever since the idea of marrying for love became the norm in modern society, men and women expect the intimacy, companionship, mutual attraction, and sexual satisfaction that typically accompanies new partnerships to continue indefinitely, and there is much bitterness and disappointment when, as so often happens, these benefits tend to disappear as the years go by.

In the hunting and gathering societies that were universal in human life until ten thousand years ago, men and women formed relationships with each other quite casually, and for the most part when these relationships no longer gave them satisfaction, they ended them just as casually. In most hunting and gathering societies, as we noted in chapter 6, it was normal for adolescents to be sexually promiscuous, and in most cases promiscuity gave way to monogamy only after several years of casual sex. Divorce in these societies typically involved little more than one of the partners moving out of their shared home—which was, in any case, usually nothing more than a temporary dwelling that would be taken down or abandoned when the group moved on in search of new food supplies.

In hunting and gathering societies, children typically "belonged" not to their parents but to the kinship group or lineage of either the mother or the father, depending on whether the society was matrilineal (in which a person belonged to the kinship group of the mother and inherited his or her identity through the female line) or patrilineal (in which a person belonged to the kinship group of the father and inherited his or her identity through the male line). And mothers and fathers rarely or never belonged to the same kinship group, because hunters and gatherers typically regarded any sexual relationship between members of the same kinship group to be a form of incest, and such relationships were strongly prohibited. Thus, the integrity of the kinship group was not affected by the sexual relationships of its members, nor was it threatened if one of those sexual relationships was terminated.

In the final analysis, the technology of precision machinery—and the urbanization of human society that that technology made possible—has rendered the so-called traditional family, with its assumption of lifelong permanence, essentially obsolete. The arranged marriage has become a thing of the past. The economic benefits of having children have been replaced by significant economic costs. The economic and social interdependence between men and women has weakened considerably. And not only virginity in adolescence but also the traditional prohibitions against premarital and extramarital sex have lost most or all of their former social stigma.

Yet the human family survives and will continue to survive. Men and women will continue to form sexual partnerships, and they will continue to have children. Those who are not inclined to procreate will contribute neither their inherited biological predispositions nor their learned cultural preferences to future generations. While the "traditional family" has become obsolete, the human family has not. Our innate human natures, the product of millions of years of human evolution, ensure that the human family will survive as long as our species endures.

Strangers to the Natural Universe

With the full blossoming of industrial society, most humans have ceased to acquire or produce food for themselves. For generations, they have been moving away from the farms that their families owned and lived on for generations, and they have settled in the cities to work for wages as the employees of strangers. For the first time in human history, a human society has emerged in which only a small fraction of the population is engaged in producing food, while a majority of the population pursues work that has nothing to do with either finding or producing food.

In 1870, between 70 and 80 percent of the population of the United States still lived and worked on farms. By 2010, this proportion had dwindled to less than 1 percent of the population. Yet as a result of the mechanization of agriculture made possible by the internal combustion engine, the 1 percent of agricultural workers remaining in the United States produce more food than is consumed by the remaining 99 percent of the population.

Before the twentieth century, most people woke up to the sound of roosters and wild birds, and at night their fields and pastures were illuminated, if at all, by the light of the moon. When they stepped out of their houses in the morning, their nostrils were filled with the smells of animals, manure, hay, grain, and the sweetness of pasture land. Throughout their lives, they were surrounded by dogs, cats, horses, cows, pigs, sheep, goats, and a host of other birds and animals.

City dwellers in the twenty-first century experience very little of these things. For the most part, they wake up to the sound of the alarm clock, and at night their world is illuminated by streetlights. When they step outside in the morning, their nostrils are filled with the smells of cars, trucks, buses, and trains. Throughout their daily lives, they are surrounded by machines: motor vehicles, computers, mobile phones, television sets, radios, stoves, refrigerators, washing machines, vacuum cleaners, and a myriad of electronic devices. Unless they own a pet animal or bird, human beings are the only species they come in contact with on a regular basis.[8]

As agriculture has been transformed by the proliferation of agricultural machinery and the consolidation of millions of small family farms into a small number of immense factory farms,[9] even the life of the typical farm worker has lost its former naturalness and variety. Whether sitting for hours at a time in the air-conditioned cab of a giant farm machine, high above immense fields of corn, wheat, soybeans, or alfalfa, or stooping for hours at a time among endless rows of tomatoes or strawberries, the typical farm worker increasingly spends all day at the kinds of monotonous, repetitive tasks once limited to the urban factory worker on an assembly line.

Thus, as it becomes increasingly urbanized, humanity is losing the intimate relationship with the natural environment that characterized human life throughout virtually the entire history of our species. And while this process of urbanization has proceeded furthest in the most developed nations, the rest of the world is not far behind. As recently as 1950, roughly 70 percent of the world's people still lived in rural communities, while only 30 percent lived in cities. At present, the numbers of rural and urban dwellers are roughly equal. By the year 2050, these proportions will be exactly reversed: roughly 70 percent of the world's people will live in cities, while only 30 percent will live in the countryside.

At the present rate, the overwhelming majority of humans will be urbanites by the end of this century, and they will have only limited contact with the natural environments within which all hominids evolved. We can only guess at the effect this unprecedented loss of contact between humanity and nature will have on the worldview of future generations.

A World of Employees

Perhaps the most profound change in human life that has occurred with the transformation into an industrial society is the general disappearance of the ancient human activity of finding or producing food and its nearly total replacement by employment for monetary gain. Before 1800, only a tiny fraction of the earth's human population worked at jobs for wages. Aside from hereditary rulers, most people were farmers, and the rest were tradespeople, craftsmen, learned professionals, bureaucrats, soldiers, and clergymen. But with the proliferation of factories and offices in the nineteenth and twentieth centuries, working for wages rapidly became the most common form of work in the new employment society.

We may take it for granted that people need jobs, and that the health of our society rises and falls on the availability of well-paid employment, but the fact remains that a society composed almost entirely of people who work under the direction of other people—who in turn provide them with a regular supply of money—is an entirely new phenomenon in human history. And the values and traditions that modern cultures have developed to deal with the social and psychological problems created by the employment society remain immature and embryonic.

The typical wage earner spends the major part of every working day performing tasks under the scrutiny of an employer. Often, these tasks have little or no relevance to the wage earner's personal interests or relationships with others. If he or she fails to perform the assigned tasks according to the standards established by the employer, the result will be criticism, threats of dismissal, or—if the failure is significant—termination of employment. Since every wage earner depends on continued employment to provide all the goods and services he or she requires for material support, comfort, and security, the threat of dismissal is a frightening prospect, and termination from employment can be one of the most traumatic experiences of modern life.

It is therefore not surprising that the people of modern society are plagued by levels of anxiety, depression, and self-doubt far beyond anything that was observed by the anthropologists who studied the preindustrial cultures of either hunters and gatherers or agriculturalists. It is surely one of the great paradoxes of human history that the evolution of modern life, with all of its safety, security, comfort, and diversions, has been plagued by an epidemic of eating disorders, heart disease, insomnia, drug addiction, neurosis, psychosis, and pathological discontent that is unprecedented in the history of the human species.

♦

Even as the effects of mechanization, urbanization, the employment society, and the demise of traditional values are now spreading rapidly throughout all human populations, a new metamorphosis is brewing. And the changes that it will bring to human life may dwarf even the massive changes that have already taken place from the development of precision machinery.

This latest transformation, triggered by the invention of digital technology, has the potential to replace all forms of human work—both physical and mental—with machines that never complain, never disobey, need no food, water, sleep, or rest, and that demand no wages. The earth's population has grown so large that it is no longer possible for humanity to go back to hunting, gathering, or growing its own food. What kind of world, then, will we go forward into? These are the questions that we will struggle to answer in the next chapter of this book.

☙

CHAPTER 9

The Technology of Digital Information

The World Wide Web of Human Interaction

"[M]an's most human characteristic is not his ability to learn, which he shares with many other species, but his ability to teach and store what others have developed and taught him."

—Margaret Mead, *Culture and Commitment*

In 1822, the British mathematician, engineer, and inventor Charles Babbage designed a machine that would be capable of performing the complex logarithmic and trigonometric calculations that were used by navigators, engineers, and scientists—and doing so without the possibility of error. Babbage called his invention a difference engine, and it was to be built entirely of precision machinery, with thousands of wheels, shafts, and gears but with no electrical parts. The reason is simple: it would be years before Michael Faraday and others would succeed in building the first electric generators. Until then, an affordable and reliable supply of electric power was not available, and none of the relays, vacuum tubes, and other electrical components needed to make an electronic calculator had yet been invented.

The difference engine was a monumental undertaking. It was to stand eight feet tall, to be composed of twenty-five thousand metal parts, and to weigh approximately fifteen tons. In 1823, Babbage hired the distinguished British engineer Joseph Clement to build his difference engine, but after several years of effort, work on the machine had been only partly completed. Babbage and Clement quarreled over the cost of the work, and eventually, eight years after the project had begun, they parted company. By then, the British government had spent £17,000 on Babbage's difference engine—equivalent, at that time, to the cost of twenty-two new steam locomotives.

Undaunted, Babbage set to work designing a far more sophisticated computing machine he called the analytical engine, which is generally recognized as the world's first true computer. Unlike the difference engine, which was capable only of mathematical calculations, the analytical engine incorporated most of the basic functions that are considered essential in modern computers today. It could be programmed using punched cards; it could store information in a form of memory; and it was capable of branching logic. Had it been built, the analytical engine would have been too large to be operated by human power. Babbage therefore planned to use a steam engine to turn the analytical engine's thousands of wheels and gears.

Unfortunately, neither the difference engine nor the analytical engine was ever constructed in Babbage's lifetime. But using a set of plans that Babbage produced in 1849 for an improved version of the difference engine, his brainchild was finally constructed in 1989 by the London Science Museum (see Figure 9.1). Babbage's difference engine worked perfectly, ending more than a century of speculation about whether his outlandish machines were actually capable of functioning as intended. And in 2010, a campaign was launched by the British software engineer John Graham-Cumming to raise £100,000 from the public and build a working model of the analytical engine in time for the 150th anniversary of Babbage's death in October of 2021. A computer model must be created first, and as of this writing, actual construction had not yet begun.[1]

Although Babbage's mechanical computers were never able to perform useful work, his concept of using machines to perform mathematical computations so complex that humans were essentially unable to execute them without errors was, in itself, a milestone of momentous importance. Babbage and his contemporaries freed humanity from the limitations of the human brain as a calculating device and launched a new age, based on a radically new technology for processing information.

Machines That Think

In the year 1890, the US Census Bureau faced a serious logistical problem. Due to the rapid growth in the US population, it was taking so long to tabulate the data collected in the census that by the time the results could be published, the census was hopelessly out of date. The results of the census of 1880, tabulated by hand, were not ready for publication until 1888, and the Census Bureau estimated that if it continued to tabulate the data by hand, the results of the larger and more complicated 1890 census would not become available for thirteen years. This would delay the publication of the 1890 census until 1903—three years after the following census, to be conducted in 1900, would have already been completed. The Census Bureau needed a machine that could perform huge amounts of mathematical calculations faster, more reliably, and more accurately than could be accomplished by any number of human brains.

In a bold attempt to solve the Census Bureau's problem, the American engineer Herman Hollerith built a tabulating calculator that used a keypunch machine to create decks of punched cards that his machine could read automatically. Hollerith's machine created a sensation when it succeeded in tabulating the results of the 1890 census in a single year. Hollerith went on to found the Tabulating Machine Company, leasing his machines to countries across Europe to tabulate their censuses. In 1911, the Tabulating Machine Company merged with three other corporations, and in 1924 its name was changed to the International Business Machine Company, better known as "IBM."

Hollerith's machines were all mechanical devices, with mechanical memories. And although the German engineer Konrad Zuse built a fully programmable computer as early as 1939, its calculations were performed by electromechanical relays, and its memory system was still mechanical in nature. It was not until 1946, with the public unveiling of the Electronic Numerical Integrator And Computer or ENIAC—which had been built at the University of Pennsylvania in secret during World War II for the US Army—that the age of the all-electric computer finally dawned.

Instead of mechanical relays, ENIAC executed its computations with thousands of vacuum tubes, similar to those that powered the radios, television sets, and other electronic devices during the middle of the twentieth century. ENIAC contained over seventeen thousand vacuum tubes, weighed thirty tons, and was three feet wide, eight feet high, and one hundred feet long. It was one thousand times faster than the best electro-mechanical computers of its day.

Although ENIAC remained in use until 1955, its vacuum-tube technology presented some daunting problems. The vacuum tube was constructed on the same basic principles as the electric lightbulb: a glass tube, with the air removed, contained a metal filament that was heated to an incandescent glow when an electric current passed through it. But thousands of vacuum tubes, with their thousands of glowing filaments, generated an enormous amount of heat, and when these tubes were switched on and off each day, the alternating states of heating and cooling created stresses that eventually caused some of these delicate mechanisms to fail or burn out. In fact, on each day of operation, several of ENIAC's vacuum tubes would typically fail, leaving the great thinking machine unable to operate until all of its faulty tubes had been located and replaced—a process that could take as long as several hours.

As the years passed, a more advanced vacuum-tube computer, the Universal Automatic Computer or UNIVAC was designed and produced for the commercial market by the Remington Rand Corporation and went on the market in 1951. UNIVAC weighed eight tons, was the size of a small garage, and cost one million dollars apiece. Although its computing power was miniscule by today's standards, a strange event involving UNIVAC marked a turning point in the public acceptance of these fascinating new machines.

In the summer of 1952, with the US presidential election between the Democrat Adlai Stevenson and the Republican Dwight Eisenhower fast approaching, the Remington Rand Corporation approached CBS News with the novel idea of using UNIVAC on live television to predict the outcome of the election as the results from the polling places were being received. Although CBS News anchor Walter Cronkite and News Chief Sig Mickelson were skeptical, they decided there would at least be some entertainment value in having an "electronic brain" on hand to analyze the election returns as they came in. As Election Day approached, most of the polls were predicting everything from a Stevenson landslide to a tight race, but the general consensus was that Stevenson would win the election.

When the polls closed and the election returns began coming in, they were fed into UNIVAC by teletype (see Figure 9.2). To everyone's astonishment, UNIVAC predicted at 8:30 p.m. on Election Day that Eisenhower would win in a landslide and that he would garner 438 electoral votes to Stevenson's 93. The bottom line was that UNIVAC was giving Eisenhower 100-1 odds of being elected President. This prediction seemed so improbable, and so incompatible with conventional wisdom, that CBS News decided not to report it. Then, a few minutes later, UNIVAC suddenly appeared to change its mind. At 9:00 p.m. Eastern Standard Time, CBS Correspondent Charles Collingwood announced that the "mechanical brain" was giving Eisenhower only 8-7 odds of winning.

But UNIVAC's apparent change of mind was due to a major error that had been made in entering the data. When the error was corrected, UNIVAC went back to the 100-1 odds of before. Yet still afraid to report this seemingly outlandish prediction, CBS News remained silent. But as the night wore on, the immensity of the Eisenhower landslide slowly became apparent. The final tally was 442 electoral votes for Eisenhower versus 89 for Stevenson. UNIVAC had predicted the results to within less than 1 percent of the final result. Late that night, an embarrassed Charles Collingwood announced on live television that although CBS News had been unwilling to report it, UNIVAC had actually predicted the outcome of the election hours ago. From that day forward, the computer's place in election coverage—and in countless other areas of modern life—was transformed from a novelty into an absolute necessity.

The Incredibly Shrunken Computer

For all of their remarkable abilities, the generation of vacuum-tube computers such as ENIAC and UNIVAC posed some extremely challenging problems. Their size was immense. Their computing speeds were relatively slow. Their appetite for electric power was enormous. The amount of heat generated by their thousands of vacuum tubes was staggering. And their thousands of large and fragile vacuum tubes had a propensity for failing frequently and needed constant replacement. In fact, the first generation of vacuum-tube computers, consisting of several different models built by several different companies, typically were capable of continuous operation for only a few hours before a failure in one of their many thousands of electrical components would require them to be shut down for repairs.[2]

But the proliferation of computer technology after mid-century accelerated enormously when, in 1947, John Bardeen, Walter Brattain, and William Shockley invented the transistor, a small wafer of germanium, silicon, or gallium arsenide. The transistor performed the same functions as a vacuum tube, but it was both vastly smaller and used only a fraction as much power as a vacuum tube. For this invention, Bardeen, Brattain, and Shockley were awarded the Nobel Prize in Physics in 1956.

It did not take long for the transistor to revolutionize modern electronics, and it soon replaced the vacuum tube in all of the electronic devices that had previously used them, including computers, recording equipment, audio amplifiers, radios, television sets, radar installations, aviation systems, and a myriad of other electronic devices. In 1954, the Texas Instruments Corporation began commercial production of the transistor radio, and before long the old power-hungry portable radio, which had typically drained all the power from its batteries in a matter of hours, was replaced by the small, lightweight transistor radio, which would run for days or weeks before its batteries needed replacing.

By the early 1950s, other scientists and engineers were hard at work developing an even more amazing device, called the integrated circuit. This combination of transistors, resistors, capacitors, and other electronic components were connected to each other within the near-microscopic structure of a single wafer made of a semiconductor of germanium or silicon. In September of 1958, the Texas Instruments engineer Jack Kilby produced the first operational integrated circuit, using germanium wafers, and a few months later, Robert Noyes of Fairchild Semiconductor produced an improved version of the integrated circuit using silicon wafers.

If the transistor succeeded in replacing the vacuum tube with something infinitely smaller, lighter, more durable, and requiring only a fraction of the power, the integrated circuit succeeded in replacing entire circuit boards—each containing multiple electronic components connected together by hand-soldered wires—with a single microchip that could be mass-produced by the thousands. The integrated circuit was literally a circuit board on a chip.

As time passed, the complexity and processing power of the integrated circuit or microchip increased steadily, as scientists and engineers developed ways of making these microchips ever smaller, cramming more electronics on each one of them, and manufacturing them at a lower cost. In 1965, Gordon Moore, cofounder of the Intel Corporation, published a now-famous article in the magazine *Electronics* entitled "Cramming More Components Onto Integrated Circuits." In this article, Moore predicted that ten years in the future (i.e., by 1975) as many as sixty-five thousand electronic components would be fabricated on a single microchip. "I believe," he added confidently, "that such a large circuit can be built on a single wafer."

Years later, when Moore's prediction had proven unnervingly prophetic, the phenomenon he described was dubbed "Moore's Law." It states that the number of components contained in a single microchip—and thus its processing speed and power—will double approximately every two years. Yet Moore's Law turns out to have been, if anything, slightly conservative, as Moore himself had guessed.[3] Since 1965, the processing power of the microchip has doubled almost every eighteen months, and in 2014, Intel announced the 15-Core Xeon E7 v2, an integrated circuit with more than 4.3 *billion* transistors on a single microchip roughly the size of a soda cracker. The list price of this monster microchip was $7,700.

The invention of the integrated circuit, and the steady progress in both the affordability and the miniaturization of computer systems since the 1960s, has led to a proliferation of computer systems in every aspect of daily life that could scarcely have been anticipated in the vacuum tube days of ENIAC and UNIVAC. In order for a vacuum-tube computer such as ENIAC to equal the power of Intel's soda-cracker-sized Xeon microprocessor, it would have to be almost 247,000 times as large. This means that, at three feet wide and eight feet high, an ENIAC computer of equivalent power would have to be more than 4,600 miles long, would cost 1.48 *trillion* dollars, and would weigh as much as sixty-six Nimitz-class supercarriers—the largest aircraft carriers in the US Navy and the largest warships ever built.[4] Of course, an ENIAC as powerful as an Intel 15-Core Xeon E7 v2 would also have to contain 4.31 billion vacuum tubes, and since the probability was that at least seven of those vacuum tubes would fail during the first second of operation, it is clear that such a monster would never remain operational long enough to complete a single computing assignment.[5]

The incredible miniaturization of computer circuitry and reduction in the cost of computational power has led to a proliferation of small, lightweight computers that have been integrated into a wide variety of vehicles, machines, and devices. All modern motor vehicles, including automobiles, trucks, buses, and locomotives, are now built with on-board computers that control their operations, monitor their performance, and diagnose their problems. And modern jet aircraft and rocket-propelled spacecraft—as well as pilotless aircraft or drones—are so dependent on computerized guidance and control systems that without their on-board computers, they would literally be unable to fly.

Multi-engine passenger jets rely on computer systems—both in the cockpit and in ground-based air traffic control systems—not only to travel to their destinations but also for guidance during routine landings on airport runways. In fact, airline pilots have become so dependent on these systems that in July of 2013 a high-profile accident occurred at San Francisco International Airport when an Asiana Airlines Boeing 777 fell short of the runway, in clear weather, because the computerized landing system at the runway had been shut down and the pilot, flying by hand, was unable to land the aircraft properly.[6]

Although computer systems have become vastly smaller and more affordable over the past seventy-five years, Moore's Law seems destined to eventually run its course. Semiconductor industry experts predicted that the rate of doubling of integrated circuit components would begin slowing to once every three years after 2013, and after 2020, they believe it will slow even more.[7] The reason Moore's Law cannot hold true indefinitely is that—as physicists and computer scientists have pointed out in recent years—the size of the components within an integrated circuit is fast approaching the limit beyond which they cannot continue to shrink without losing their ability to function.

As the integrated circuit becomes ever more microscopic, its components are beginning to approach the size of atoms and molecules. For example, some of the connecting "wires" in integrated circuits are now less than fifty nanometers in width (a nanometer is equal to one billionth of a meter). This is only twenty-five times the width of a DNA molecule. When an electric current passes through a wire as thin as this, there is a tendency for the current to "leak" out of the wire and affect neighboring components.

Moreover, when certain substances are reduced to sizes below fifty nanometers, the laws of physics become subject to quantum effects, and many familiar elements begin to display properties that they would never possess in more normal sizes. At these submicroscopic dimensions, copper becomes as transparent as glass, aluminum can burn like paper, and gold can dissolve like a sugar cube in hot water. Needless to say, as the physical properties of components becomes transformed at these submicroscopic dimensions, the integrated circuits that are built of them will no longer operate as intended.

Yet we must remember how many technologies once considered impossible have eventually seen the light of day, and the ultimate history of computer miniaturization is yet to be written. A new domain of physics and engineering called nanotechnology is now devoted entirely to the fabrication of devices with molecular-sized components. Carbon nanotubes, gold and zinc oxide nanorods, DNA nanostructures, and revolutionary techniques of molecular self-assembly are already being used to fabricate exotic submicroscopic components capable of performing some of the basic computer functions.

From our present vantage point, we have no way of telling whether nanotechnology will achieve the kind of real-word breakthrough that will allow Moore's Law to continue in effect much beyond the year 2025. Nevertheless, with integrated circuits that contain billions of transistors already on the market, further miniaturization will not be needed in order for computers to continue their seemingly inexorable invasion into every area of technology and, with it, into every area of human life.

In the early decades of information technology, no one imagined that the computer was destined to become a personal device that ordinary people would use constantly in the course of their daily lives. After all, the computer was invented specifically to perform mathematical computations that by their nature were simply too massive and too time-consuming for human brains to execute with any hope of precision, reliability, or efficiency. Yet the seemingly inexorable advances in computing power that Moore's Law had predicted not only made computer systems small enough to sit on a desktop or fit in the palm of the hand but also made them generally affordable.

One of the most important consequences of the increasing affordability of computing power is that all of the forms of storing information can now be preserved in computer-friendly formats. The very large memory systems that are required to store the space-hungry files of digitized text, photographs, audio recordings, and videos have become so inexpensive that everything written, photographed, or recorded is now routinely stored in digital form. And this development has created a new horizon in the history of human life.

The Horizon of Digital History

In the domain of human communications, nothing will determine the way our age will be perceived by future generations more than the digital systems that now exist for recording, storing, and retrieving the cultural records of our time. A complete record of the behaviors, customs, and value systems of our time—preserved in exquisite fidelity in the form of digitized text, photography, audio, film, and video—is now accumulating, day by day, in quantities that were simply unimaginable to the people of former generations.

The essential difference between history and prehistory is that written records have survived from historical periods, while only archeological artifacts—and no written records—have survived from prehistoric times. Defined in this way, history begins at different times in different places: in Egypt and the Fertile Crescent with the invention of cuneiform writing and hieroglyphics around 3200 BC, in China with the development of the Longshan script around 3000 BC, in India with the invention of the Indus Valley script around 2700 BC, and in the Yucatan Peninsula of Mexico, with the development of Mayan hieroglyphics, around 400 AD.

Unfortunately, this common definition of history presents some thorny problems. For example, do the histories of China and India begin with the development of their earliest systems of writing, even though these forms of writing have never been deciphered? Or should it begin with the earliest forms of writing from these regions that have been successfully deciphered, even though they date from many centuries after writing was first widely used?

And what about cultures that were culturally and technologically advanced but did not leave behind written records? For example, the Celtic cultures of ancient Europe were an advanced agricultural people with a well-developed class system and a sophisticated knowledge of metalworking in both bronze and iron. By the first century AD, Celtic culture had spread throughout the whole of Europe from Hungary to the British Isles. Yet while the Greeks and Romans were familiar with the Celts and wrote extensively about them, the Celts themselves did not use writing. But it would be misleading indeed to call Celtic culture prehistoric.

Future historians will arrive at the study of the twenty-first century with a sigh of relief. The year 2000 marks the approximate beginning of the horizon of digital history, after which the recorded information of all human societies will be available to future generations with unparalleled completeness and fidelity. By contrast, most of recorded history before the year 2000 is fragmentary and incomplete.

From the invention of writing in the early civilizations of the Nile Valley and the Fertile Crescent until the development of photography in the late nineteenth century, human history can be studied only from the incomplete and imperfect remains of hand-drawn art, written manuscripts, printed books, and periodicals. And from the late nineteenth century until the late twentieth century, all that has survived from the phonograph records, movies, and television programs of those times are the blurred, grainy, and often deteriorated physical archives that were originally stored on recording masters, film, and magnetic tape. However, most of these twentieth-century archives have now been digitized and converted into computer files, ensuring that they will at least not continue deteriorating further.

Until the invention of digital technologies, the methods that were used for making copies of recorded information all suffered from the same failing: each time a copy was made, some of the information was lost. This meant that a copy of something could never be as good as the original, and a copy of a copy was worse yet, and so on. Each generation of copies was of poorer quality than the generation before it, so the more times something was copied, the worse the quality of the copy became. This was true of the microfilm and photocopies that were made of material written or printed on paper, of the photographic copies that were made of photographs and films, and of the duplicate copies made of recording masters, magnetic audiotapes, and videotapes. And since all of these forms of storage media were made of either paper or plastic, they typically became brittle, discolored, and ultimately useless with the passage of time.

But when techniques were developed for converting all forms of media into digital formats, the problems of making and storing copies of text and images virtually disappeared. Whenever a line of text, an image, a music track, or a video is stored on a computer or mobile device, it is converted into a digital format—meaning that it is converted into a mass of ones and zeroes that have been arranged in a specific order to represent the text or image in question. When the computer file is copied, the specific pattern of ones and zeroes is reproduced exactly as it was in the original, with no mistakes and therefore no loss of fidelity. Therefore, no matter how many times it is copied, the computer file will always consist of exactly the same pattern of ones and zeroes as before—and the text, music, photos, or videos that these ones and zeroes represent will also be exactly the same.

As long as the medium on which these files are typically stored remains physically intact and readable—whether it is a computer chip, a hard disk, a compact disc, or a storage medium yet to be invented—the copies will never deteriorate or be lost. And even if the storage medium itself begins to age after the passage of many years, the files can be given a new lease on life by simply copying them again onto a fresh storage medium. Even in the unlikely event that no further improvement will be made in computer storage technologies, every written word, drawing, music track, photograph, or video that exists today in the form of a computer file can potentially be stored for hundreds of thousands of years into the future.

If 3200 BC is remembered as the time when humanity first crossed the horizon of written history, 2000 AD will be remembered as the time when humanity first crossed the horizon of digital history. Every book, journal, newspaper, magazine, pamphlet, blog, or website that is published today—as well as every movie, record album, and television program that is aired or broadcast—has been immortalized somewhere on a computer storage device in the form of ones and zeroes. And yet this stupendous trove of information, as massive as it is, will be dwarfed by the infinitely larger mass of emails, texts, tweets, photos, and videos that are being created and transmitted every day by millions of individuals all over the world—and that for the most part are being stored and preserved by the Internet service providers that are transmitting them.

The principal problem future historians will face is how to search through the stupendous volume of material that will have survived. The cultural record of the present day that is now being recorded and stored in digital form will bestow upon our time a singular distinction in the annals of human history. We have become, and will always remain, the first generation whose life and times will have been preserved in exquisite detail and uncompromising fidelity. All of our most soaring achievements, as well as our most banal frivolities, will live on in a mass of ones and zeroes long after we have gone, and they will be available for the enlightenment and amusement of future generations as long as human civilization itself endures.

Robots, Automation, and the Future of Meaningful Work

If the cultural transformations brought about by precision machinery ultimately reduced the status of women in the mid-twentieth century by allowing much of the work of house and home to be done by machine, the transformations of digital technology have also reduced the status of men in the twenty-first century by allowing most of the mental and physical work traditionally done by men to be performed by computers, robots, and automation systems.

In the 1950s, when the phenomenon of automation was in its infancy, the question foremost on the minds of social scientists was:

"How will people in future society adjust to the vast increase in leisure time that the new technologies of automation are destined to create?" It was assumed—rather naively, it now appears—that if machines were able to take over most of the tasks then performed by wage earners, the work week of all workers would simply shrink in proportion to the amount of their work that was performed by machines. And the problem that this would cause was always framed in terms of what the average worker would do with all the free time that resulted from this liberation from the daily grind.[8]

But the expansion of free time turns out not to be much of a problem. Although the average American worker's free time has increased by several hours per week since the 1950s, most of this extra free time is spent watching television. In addition, millions of men and women retire from their jobs every year, and many retirees live on for ten, twenty, or thirty years before they die of old age. And most retired people find meaningful and rewarding things to do with their time. Rather than posing a problem that has to be solved, the expansion of free time has come to be seen as an opportunity that can be exploited. But the same cannot be said of the men and women in the prime of their lives who have been deprived of their jobs, their salaries, and much of their self-esteem when the work they once did is taken over by computers, robots, and automation systems.

It seems not to have occurred to the social commentators of the 1950s that when machines began doing the work previously performed by live humans, the result would not be a gradual shrinkage of the work week but rather a trend toward growing unemployment and underemployment. No one seems to have understood that when computers became capable of performing mental work and robots became capable of performing physical work, most of the people who had been doing this work would simply be relieved of their jobs—and that they would be replaced by a much smaller number of highly trained technicians required to program, operate, and maintain these new computer-driven systems.

Furthermore, the new generation of knowledge workers has come for the most part not from the ranks of manual laborers but are instead college graduates who have been educated in such subjects as computer science, mathematics, project management, and process control. Yet this is the world we live in today, and increasingly this appears to be the fate of the millions of people who have been employed in business and manufacturing. Therefore, the question we actually must answer now is this: "How will all the people who are losing their jobs from the proliferation of automation systems find meaningful, well-paid employment elsewhere and regain their former self-respect and status in society?"

Although this problem has been steadily worsening since the appearance of affordable and reliable computer systems in the mid-twentieth century, this fundamental question seems to have no satisfactory answer. Automation systems are increasingly performing all of the routine tasks that were once performed by live humans. Robots are gradually taking over the physical tasks involved in manufacturing. And the computer systems that have been developed to fly pilotless aircraft will soon be driving the vehicles of the modern world to and from their destinations.

This raises the question of how the wealth of modern society will be distributed to those members of society who are unable to find gainful employment. It is not reasonable to assume that the shrinking population of workers who still have adequate incomes will be taxed at increasingly higher rates and that governments will redistribute this money to the members of society who cannot find jobs. For one thing, there will come a point at which the haves will no longer be numerous enough to support the have-nots. For another, the absence of meaningful work destroys the self-esteem of all adults—and self-esteem cannot be restored by handouts.

This problem may ultimately become even more acute for men than for women, because most women have the option of bearing and raising children, and this role is universally accorded social approval and is recognized as vital to the future survival of every human society. Nevertheless, the status and self-esteem of women in contemporary society does not rest only on their reproductive roles. Were they to be deprived of their own independent sources of money and economic security, the status of women would soon decline again to the abysmal levels of mid-twentieth-century society.

The grand socialist experiment of the twentieth century, in which Russia, eastern Europe, and China attempted to put all the economic activities of society under the control of state governments, failed to produce either general prosperity or greater personal satisfaction in life for two reasons. First, it interfered with the ancient and natural human motivation to work and create for personal gain. Second, a system based on the concept of "from each according to his ability, to each according to his need" can only work in a close-knit human group such as the family, where people depend for their self-esteem on the love and approval of others—and in which they perceive their own well-being in terms of the well-being of their closest kin and life partners. However noble they may be, such concepts do not work at the level of the nation-state, where most people are strangers to each other.

Meanwhile, the age of information technology has already produced some notable advances in human culture. Foremost among these is an unprecedented ability to communicate with others, to share information more freely and widely, and to pool our collective wisdom for the ultimate benefit of humanity. These abilities begin with the individual, who has now been given the power to speak his or her mind in a way that allows anyone else on Earth to listen.

Social Media and the Digital Soapbox

Of all the human capabilities that information technology has spawned thus far, perhaps the least anticipated has been the sudden proliferation of the communication system called social media, in which each individual establishes a personal network of other people and communicates with the other people in this network on a regular basis. And while social media have been hailed as an entirely new form of human communication, the principles of human interaction that they embody are as old as prehistory.

The simpler societies that were studied by anthropologists throughout the twentieth century—the hunting and gathering societies and the slash-and-burn agriculturists—had a characteristic way of communicating their thoughts and feelings to the other members of their group. Since the dwellings of these people were typically small and made of permeable materials such as hides or palm thatch—and since they were typically located only a few feet apart—the members of these societies enjoyed little or no personal privacy. When they were sitting in their own tents or huts, anything that they said in a normal speaking voice could be heard by their neighbors, and anything said in a loud voice was easily heard by all the people living in their small, close-knit encampments.

When people living in such communities had a complaint about a spouse, neighbor, relative, or rival, they might park themselves in the middle of a public space—for example, in the clearing that is often located in the center of a group of huts or tents—and air their grievances in a loud, accusatory tone. The speeches they made in this way were directed not to any one individual but rather to the community as a whole, and members of the community typically listened to them with great interest. Often, one or another member of the group would shout out some comment of approval or disapproval from inside the confines of a dwelling. Many times, these comments from the largely unseen audience were meant to be humorous, and it was not unusual for these comments to provoke laughter from other members of the group.

These impassioned speeches not only provided the speaker with a way to relieve his or her frustrations but also made it possible for individual complaints to be aired in the arena of public discourse. The comments of the listeners—which were also typically heard by other members of the community—functioned as a sounding board of public opinion, and they provided an opportunity for the group as a whole to weigh in on the pros and cons of the speaker's complaint. In this way, the values and attitudes of the group itself were aired, reinforced, refined, and sharpened by these common acts of speechmaking.

When sedentary villages arose during the agricultural revolution, however, the physical nature of these more permanent dwellings turned out to be fairly inhospitable to this ancient human practice. In many agricultural communities, such as in the dispersed rural communities of northern Europe and the Far East, individual families tended to live on the fields and gardens they owned, and their houses were typically dozens or hundreds of yards apart—too far away from each other for one person's lament to be heard by a neighboring family. Moreover, these types of settlements often lacked the kind of common area that would be easily accessed by all the members of the community. People who lived in these communities were rarely in close contact with each other except during the occasional communal ritual such as weddings, funerals, or religious ceremonies.

In other types of agricultural communities, typical of the rural villages of Mediterranean Greece, Italy, and North Africa, families lived in dense communities consisting of fortified villages. Farmers spent the nights in their village houses and went out of the villages to their gardens, pastures, vineyards, and orchards during the day to work. The houses in these communities tended to be permanent dwellings built of mud brick or stone, with thick, solid walls, and with small windows and doors. This kind of architecture is not conducive to carrying the sound of an individual farther than the nearest neighbor—and then only if the individual's voice is raised nearly to a shout. In these kinds of communities, the grievances and opinions of husbands and wives, debtors and creditors, allies and enemies, lovers and rivals were rarely aired in public but instead tended to be whispered in private. They were often the subject of gossip but were rarely the subject of open, public discussion.

With the urbanization of humanity that arrived with the industrial revolution, neighbors were not necessarily even members of one's own community, and they rarely included the kinfolk that once formed the communities of hunters and gatherers. Thus the cultures of modern people—who are typically descendants of agricultural villagers and townspeople—had long ago lost the hunters' and gatherers' tradition of openly airing their grievances, complaints, or opinions. But the invention of social media has now provided a mechanism by which modern people have discovered the pleasures and perils of airing their opinions to a network of peers. This is probably the reason for the unprecedented numbers of people who have joined social media networks in recent years.

Modern social media not only enable all of the people included in each person's network to offer comments and responses of their own, but also make each of these comments and responses available to everyone else in each person's network. Thus, the digital soapbox that has been created by the development of social media networks has reestablished the ancient human practice—found among all hunting and gathering nomads but severely diminished by millennia of agricultural and industrial civilization—to share the most intimate questions, observations, and opinions in their daily lives with their peers on a daily basis. And the universality of this practice among hunters, gatherers, and primitive villagers all over the world suggests that it is a both fundamental human need and a natural human practice of sharing information, reinforcing values, and testing public opinion about all the pleasures and vicissitudes of daily life.

Of all the many changes in human life that have been wrought by the development of information technology, perhaps none is as important—and ultimately transformative—as the unprecedented expansion in the ability to communicate with others. This ability goes far beyond the public airing of private thoughts, feelings, and attitudes made possible by social media. In fact, it encompasses every kind of human interaction: the sharing of knowledge among people with similar interests, the transmission of private messages from one individual to another, the trading of goods and services, and the ability to travel far from one's own homeland into the homelands of other societies and cultures.

The World Wide Web of Human Interaction

Our rapidly evolving universe of personal computers, mobile devices, the Internet, communications satellites, and worldwide commercial aviation has already produced unprecedented advances in the technologies of interaction. Information technology has created a world in which people are increasingly able to write to, talk to, visit with, and trade with anyone else on Earth, in real time, twenty-four hours a day.

It was digital technology that enabled spacecraft to put communications satellites in low-earth orbit. These satellites have not only reduced the cost of international telephone calls to an affordable level but have also made it possible for all of humanity to watch athletic events, natural disasters, and outbreaks of warfare in real time on electronic devices. And for the first time in human history, a global transportation network has been established that enables any person on Earth to trade with or visit any other person on Earth in a matter of hours or days. This freedom of interaction is without precedent in all of human history.

The development of the steamship and the railroad in the early 1800s led to a huge increase in long-distance travel and was partly responsible for knitting together all of the many thousands of towns, villages, and

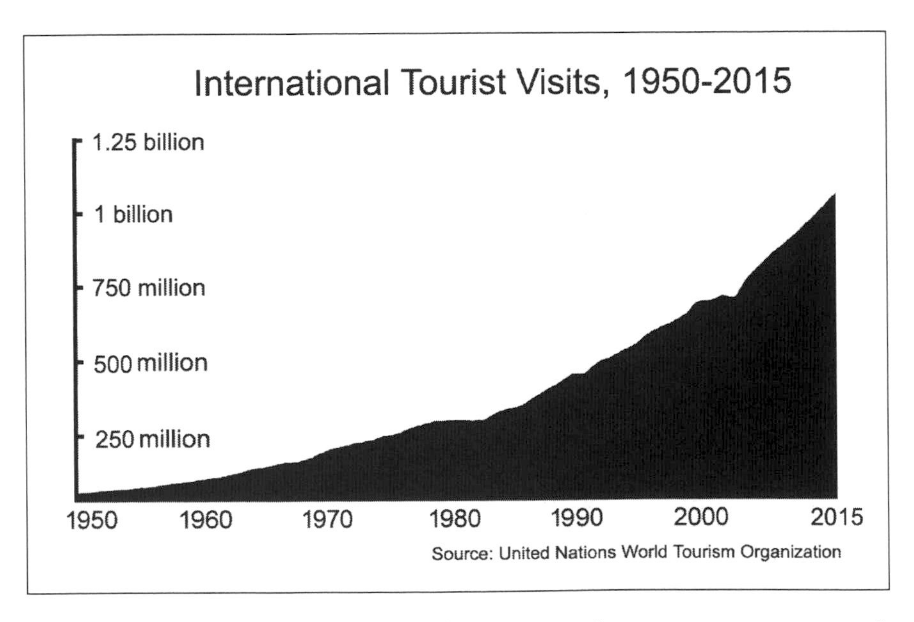

Figure 9.3. *Since the development of computerized transportation systems, the number of international tourist visits has skyrocketed, growing from approximately twenty-five million visits per year in 1950 to more than one billion visits in 2015.*

city-states that had existed since ancient times into the small number of nation-states that exist today. More recently, the number of people who have traveled to other nations and other cultures, both as tourists and for business purposes, has increased exponentially—largely as a result of the advances that have taken place in the technologies of interaction since 1950. In that year, roughly twenty-five million people visited other countries as tourists. By the year 2011, this number had grown to roughly one *billion*. This is an increase of 4,000 percent in the space of a single human lifetime (see Figure 9.3). And at its present rate of increase, according to the United Nations World Tourism Organization, the number of international tourist visits is likely to reach 1.6 billion by the year 2020.[9]

The explosion in the growth of international tourism has been accompanied by a corresponding increase in the number of people who have been migrating from one continent to another and from one nation-state to another. By the year 2013, approximately 232 million people living on Earth were migrants who had been born in one nation and had moved their permanent residence to another nation.[10] This is more than twice as many people as inhabited the earth when the emergence of urban civilizations had already begun.

This world wide web of human interaction will produce a transformation in human life and society as profound as any of the metamorphoses of the past. In fact, if we look at the changes in society and culture that were set in motion by the technology of precision machinery, it is clear that the ultimate effects of a key technology can hardly be predicted by looking at its most immediate consequences.

Who would have guessed that the development of precision machinery by medieval clockmakers—driven by the simple desire to produce more accurate clocks—would lead before long to the invention of the modern printing press, an explosion of knowledge, and a Renaissance in the arts and sciences throughout the Western world? Or that when the forests of Europe were being consumed by the clockmakers' rising demand for iron and steel, the civilized world would begin mining coal, which would lead before long to the invention of the steam engine with all of its multiple effects on the nature of manufacturing, travel, and trade? Or that the success of the steam engine would spawn an industrial revolution that would, within the course of a single century, produce the internal combustion engine, the automobile, the airplane, and the widespread use of electric power?

Who would have guessed that these industrial technologies would ultimately result in the abandonment of food production and the wholesale movement by the great majority of the earth's people from the countryside to the cities? Who would have predicted that precision machinery would ultimately produce the media of mass communications, the rise of the employment society, the mechanization of house and home, the deforestation of Earth's land areas, the destabilization of global climates, the mass extinction of plants and animals, and all of the fundamental changes that have taken place in the customs and traditions that govern the institutions of marriage, the structure of the family, and even the norms of human sexual behavior?

Now that digital technology is freeing our species from the limitations of time and space to which we have previously been bound, the changes that lie in the distant future are largely unimaginable. In the near future, however, we are certain to witness at least one major transformation of human society: the birth of a global civilization.

The Ultimate Act of Fusion

The phenomenon of social fusion among humans first began during the prehistoric period of the Upper Paleolithic, set in motion by the technology of symbolic communication. The evolution of language, art, and visual symbolism made it possible for the nomadic bands of that era—each composed of only a few dozen kinfolk—to share a common cultural identity with other nomadic bands and, by doing so, fuse into tribes composed of thousands of individuals.

A fusion of tribes on a larger scale occurred late in Neolithic times, when the agricultural villagers of that period developed river boats, wheeled vehicles, and systems of writing—and fused into urban civilizations that embraced many tribes and created communities numbering in the hundreds of thousands of inhabitants. Finally, when the industrial revolution created more powerful and more efficient technologies of travel, trade, and communication, the many thousands of city-states that had survived since the dawn of civilization fused in less than two centuries into less than two hundred nation-states, most of which now contain populations that number in the millions.

The reasons for these fusions are not a mystery. As people find it easier to interact across geographical space, they can more easily visualize other people as belonging to their own familiar worlds. The more easily they communicate with each other, the more likely they are to seek and find a common language. The more they travel to each other's lands and settle in each other's territories, the more likely they are to adopt each other's customs and traditions. Over time, they become more likely to eat the same foods, wear the same clothes, and even observe the same holidays. They become more likely to intermarry, and to produce children that grow up exposed to more than one set of customs, values, and attitudes.

All of these things tend to reduce the cultural differences between groups that were formerly distinct and separate. Eventually, people who had previously belonged to distinctly different cultures begin to identify with a new and larger culture, especially if this newer culture offers a set of values, institutions, and traditions capable of reconciling the differences among the cultures of past eras. Thus, among the many consequences that flow from the significant advances wrought by information technology, one result can be predicted with confidence: human groups will fuse into larger groups than have ever previously existed.

Nevertheless, there are always powerful forces in human society that resist change. Even as many areas of human life are in the process of fusing, many others are undergoing the opposite process, and are fissioning into the component parts of what was formerly a unified whole. Former empires break up into their constituent parts, and these become independent nations. The dissolution of the Spanish Empire, the British Empire, the Austro-Hungarian Empire, the Ottoman Empire, and the Soviet Union all furnish vivid examples of the process of political fission in modern times. And nation-states can also lose their political unity and break up into their constituent parts as occurred in the 1990s, when the former Republic of Yugoslavia fractured into the separate nations of Serbia, Croatia, Slovenia, Macedonia, Montenegro, and Bosnia-Herzegovina.

Finally, while the languages of dying cultures are being lost, at least some "dead" languages have been revived and have become living languages again. The Hebrew that for centuries survived only as a liturgical language has become a modern language that millions of Israeli citizens speak on a daily basis. And we should not forget that even as the political world has been consolidated into less than two hundred sovereign nation-states, more than seven thousand distinct languages are still actively spoken in the world today. Each of those languages represents a distinct ethnic group, with distinct customs and traditions and a distinct identity that still survives. Yet language experts estimate that more than one-third of existing languages are currently either in trouble or dying.[11] The process of cultural fusion is now moving faster than at any previous time in human history.

In the final analysis, as the history of past transformations has so clearly demonstrated, each significant advance in the technologies of interaction has eventually spawned a larger and more inclusive form of human society. And the advances in the technologies of interaction that have flowed from the development of digital technologies are the most momentous in all of human history. In all likelihood, this fundamental process will eventually continue to its final end point: an ultimate act of fusion, as humanity slowly unites into the larger entity of a new universal human culture—in short, the birth of a global civilization.

♦

When a global society and culture begins to take shape, no further fusion of human beings will be either possible or necessary. Nor will this ultimate act of fusion mean that any of the human groups that have already come into being will cease to exist. History shows us that all of the traditional human groups that emerged from former metamorphoses—the families, clans, tribes, neighborhoods, villages, towns, cities, regions, states, and nations—have kept their identities, maintained their integrity, and continued to govern their appropriate spheres of human life, even as larger and more inclusive entities have emerged.

In our next and final chapter, we will examine the numerous forces—ultimately unleashed by the very technologies that have made us human—that have come to threaten the long-term viability of our planet and its fragile web of life forms. We will also consider the likelihood that humanity will use the power of digital technology to create new global entities with the ability to manage the earth's resources—not for the benefit of any one nation but rather for the well-being of all humanity and all other forms of life on Earth.

ઉ৪ঌ

CHAPTER 10

Our World at the Brink

Is Humanity Drifting Toward a Planetary Catastrophe?

"The world is a dangerous place to live; not because of those who do evil, but because of those who look on and do nothing."

—Albert Einstein, *Conversations with Casals*

On July 21, 1969, the American astronauts Neil Armstrong and Buzz Aldrin became the first human beings—indeed, the first terrestrial organisms—to set foot on the moon. They were followed over the next three and one-half years by ten others. And of all the unique sights and experiences that the moon landings provided, perhaps the most powerful, in its effect on the astronauts themselves, was the sight of the earth from the depths of space.

Frank Borman, the commander of the Apollo 8 mission, recalled the experience of seeing the earth, in all its multicolored glory, floating in space nearly a quarter of a million miles away. "I happened to glance out of one of the still-clear windows," he wrote, "just at the moment the earth appeared over the lunar horizon. It was the most beautiful, heart-catching sight of my life, one that sent a torrent of nostalgia, of sheer home-sickness, surging through me."[1] And James Lovell, the pilot of the command module, once remarked, "The most impressive sight I saw was not the moon, not the far side that we never see, or the craters. It was Earth . . ."[2]

Yet the future of our unique, irreplaceable planet is now seriously threatened as never before by the very technologies that made us human. The human scourges of war, pollution, deforestation, species extinction, and climate change—all of which have flowed from our technological prowess—have put the living world at risk. But before we review each of these threats in detail, we must consider a fundamental question of our age. Can we escape the ills we have created for ourselves by leaving the earth behind and starting over? Can we build a better life for humanity on the virgin soil of another planet?

Can We Colonize Other Planets?

There is an idea that has become popular in recent years, in which it is imagined that future generations of humans will escape the earth's problems by using advanced technologies to colonize other planets. But even if we ignore the sheer logistical challenge of launching hundreds of thousands of tons of supplies and equipment into space—and we consider only the environmental conditions that we know to exist on other planets—the goal of colonizing other heavenly bodies seems not only unrealistic but also, for all practical purposes, literally unattainable.

Our moon is a silent, airless world of lifeless rock and dust. A single day on the moon lasts for twenty-eight Earth days, and temperatures on the surface of the moon during these lunar "days" are hot enough to boil water, while surface temperatures during the lunar "nights" can plunge to nearly 300 degrees below zero Fahrenheit.

The planet Mercury is an airless ball of iron and rock rotating so slowly that a single day on Mercury lasts almost as long as two months on Earth. For this reason, surface temperatures on Mercury rise to 650 degrees Fahrenheit during Mercury's "day" and drop to 274 degrees below zero Fahrenheit during Mercury's "night."

The planet Venus is smothered in rolling clouds of sulfuric acid, and its atmosphere is so dense that atmospheric pressure on the planet's surface is a crushing 1,350 pounds per square inch. This is ninety-two times greater than the 14.7 pounds per square inch on Earth at sea level and is, in fact, equivalent to the pressure that a diver would feel by descending half a mile into the sea. Due to the "runaway greenhouse effect" of its carbon dioxide atmosphere, surface temperatures on Venus remain uniformly above 800 degrees Fahrenheit. This is hot enough to melt most soft metals, including lead and zinc.

The planet Mars is a frozen wasteland of rocks and dust that are red from their high concentration of iron oxide. Surface temperatures on Mars average 80 degrees below zero Fahrenheit. The Martian atmosphere is one hundred times thinner than the atmosphere of Earth and 95 percent of this atmosphere consists of carbon dioxide, with only a trace of oxygen. In addition to its inhospitable temperatures and unbreathable atmosphere, Mars is regularly pounded by gigantic dust storms that can last for months at a time and often grow large enough to envelop the entire planet.

Jupiter, Saturn, Neptune, and Uranus—the "gas giants" of our solar system—are composed of cores of ice and rock larger than Earth that are covered with thick atmospheres of hydrogen and helium and buried under immense oceans of liquefied hydrogen and helium thousands of miles deep. None of these planets have any real "surfaces" in the normal sense of the word—only mushy regions where gases become compressed into liquids and where liquids become compressed into solids, all of which are hidden in total and perpetual darkness.

Considering the hostility of their environments to all known forms of life, none of the other planets in our solar system are reasonable candidates for human colonization. Surely it would be far more practical—and pleasant—to colonize the vast, uninhabited continent of Antarctica, where temperatures average only between 20 and 50 degrees below zero, the air is eminently breathable and rich with oxygen, and the skies are blue. But lacking the fictional romance of the Red Planet, there has been no general stampede to colonize the earth's fifth-largest continent.

But what about the "Earth-like planets" that astronomers have been discovering in other, nearby solar systems? Could one of them provide a second home for the surplus hominid populations that will—given the current rate of human population increase—soon be overrunning the earth?

While astronomers generally agree that the universe contains many other Earth-like planets, these worlds are all so distant that we know very little about their climates or surface characteristics. The Earth-like planet nearest to our solar system is believed to orbit the star Tau Ceti, located twelve light-years from Earth, but the planet believed most likely to have an Earth-like climate is called "Gliese 832 c," located at a distance of sixteen light-years from Earth.[3]

Gliese 832 c has been estimated to be five times the size of the earth. Thus, a person who weighed 160 pounds on Earth would weigh 800 pounds on the surface of Gliese 832 c. This would prevent a normal human being from either standing or walking. But let us suppose for the sake of argument that the gravity problem could be solved somehow—perhaps by strapping on metal braces to help support the body under these crushing loads. Even so, the daunting problem of how a group of humans would survive the long journey from our solar system to these "neighboring" solar systems would still have to be solved.

The typical space rocket escapes the earth's gravitational pull by achieving a launch speed of approximately 18,000 miles per hour. NASA's "New Horizons" mission—designed to explore the region outside of our own solar system—achieved a launch velocity of 36,000 miles per hour, which, combined with the speed of the earth's orbit around the sun, boosted the spacecraft to a velocity of 100,000 miles per hour. Although this was fast enough to escape the sun's gravity, New Horizons had slowed to 31,000 miles per hour by the time it actually left the solar system.

In 2018, a planned mission by NASA called the "Solar Probe Plus" is projected to use the "slingshot effect" of the sun's gravity to reach a staggering 450,000 miles per hour while orbiting the sun. This is fast enough to travel from the earth to the moon in thirty minutes. But even if a spacecraft capable of carrying live humans plus all their cargo and equipment could somehow—even while fighting the sun's gravity—attain a speed of 450,000 miles per hour on its way out of the solar system, it would have to travel through space for nearly *twenty-four thousand years* before arriving in the vicinity of Gliese 832 c.[4] It is difficult to imagine how a handful of human beings could survive inside the confines of a spaceship for roughly five times longer than the entire history of human civilization. It is even more difficult to imagine what the tiny population of such a spaceship would look like after more than seven hundred generations of inbreeding.

Let us suppose that some future civilization, having developed a technology unknown to us, will defy the known laws of physics and succeed in constructing a functioning space vehicle capable of travelling at nearly the speed of light. Would such a technology then open up the universe to interstellar colonization by hominids?

Even in the case of this highly improbable scenario, such interstellar colonists would still be required to survive for many years in space without life support from their home planet before they arrived at the vicinity of the nearest Earth-like planet. These colonists would therefore need a technological infrastructure capable of providing them with a reliable supply of food, warmth, and breathable air while they traveled for years through the blackness of space. And thus far, even the best efforts of modern technology have utterly failed to provide human beings with a means of surviving indefinitely once all physical contact with the biosphere—the sum total of all life forms and ecological systems that cover the surface of the earth—has been severed.[5]

Life Without the Biosphere

Any attempt to colonize other worlds would require the creation of an artificial ecosystem capable of supporting human life without being connected to the earth's biosphere, and an attempt to do exactly that was made in 1991—not on another planet but in the relatively benign terrestrial environment of the southwestern United States. At a cost of $200 million, a three-acre enclosure called Biosphere 2[6] was constructed in the Sonoran Desert near Tucson, Arizona, to serve as the model for a self-sustaining environment that could be replicated in an extraterrestrial colony (see Figure 10.1). It was to be inhabited by a group of eight people who called themselves the Biospherians.

The four men and four women who volunteered for this mission intended to live inside this enclosure for two years, sustaining themselves without any supply of air, food, or water from the outside. Biosphere 2 was stocked with soil, water, plants, and animals, and it included a small sea, a savanna environment, a mangrove swamp, a rainforest, a desert, and a farm. The idea was that these various environments and their atmospheres would interact to form a totally independent life support system within which humans could live indefinitely.

In September of 1991, the Biospherians passed through the airlocks of Biosphere 2 and began their two-year mission. But in spite of the availability of massive technological and financial support from outside the enclosure, the Biosphere 2 experiment demonstrated how quickly an ecosystem can collapse when its connection with the natural biosphere has been severed.

Throughout the entire first year of the mission, the farm that had been established inside Biosphere 2 failed to provide sufficient food for the crew. During the first twelve months, the Biospherians experienced continual hunger, were obsessed about the scarcity of food, and lost a significant amount of weight. By the end of the first year, the eight Biospherians had split into two opposing factions that were barely on speaking terms with each other.

In spite of a profusion of green plants, oxygen levels inside the enclosure steadily declined, ultimately falling to the level normally found at an elevation of 17,500 feet. Meanwhile, carbon dioxide levels skyrocketed, fluctuating wildly from one day to the next. Fearing for the health of the crew, project administrators were forced to pump oxygen into the enclosure repeatedly, beginning seventeen months into the experiment.

Over time, the atmosphere inside Biosphere 2 also became permeated with nitrous oxide, ultimately reaching levels that threatened the crew with permanent brain damage. In addition, the stillness of the air inside the enclosure caused the trunks and branches of the trees—normally strengthened by the action of the wind—to grow weak and brittle, and they became prone to what scientists later reported as "catastrophic dangerous collapses." At the same time, morning glory vines grew wildly, smothering the other plants and trees and requiring constant weeding.

All the species of pollinating insects that had been brought into the Biosphere died out, preventing most of the agricultural plants from reproducing and ensuring that they would not survive beyond their normal life spans. Most of the other insects also died, ultimately leaving Biosphere 2 completely overrun by vast swarms of cockroaches and "longhorn crazy ants" running wildly in all directions.[7]

Areas that had been intended as deserts turned into chaparral and grasslands, and the water system became so loaded with chemical nutrients that it was necessary to circulate all of the water over thick mats of algae that had to be periodically harvested, dried, and stored inside the enclosure. Finally, of the twenty-five species of birds, mammals, fish, and reptiles originally introduced into Biosphere 2, all the animals except for six species had died by the time the experiment ended twenty-four months later.

In a sobering report on the lessons of Biosphere 2 published in 1996, biologist Joel E. Cohen and ecologist G. David Tilman concluded, "At present there is no demonstrated alternative to maintaining the viability of Earth. No one knows yet how to engineer systems that provide humans with the life-supporting services that natural ecosystems produce for free."[8]

There is only one world we know of that can sustain human life, and it is our blue-green planet that so thrilled and delighted the astronauts during their visits to the moon. We have no other home; we can breathe no other air; no other planet can feed us. The earth is quite literally the only life support system available to the human species. We have no alternative to keeping the earth's biosphere healthy and alive so that we ourselves can remain healthy and alive. All else is science fiction and fantasy.

War Machines and Doomsday Machines

When modern firearms were combined with vehicles and airplanes powered by internal combustion engines, the modern war machine achieved a capacity to kill and destroy that the military leaders of ancient and medieval armies could not have imagined in their wildest dreams. More than one hundred million people were killed in warfare during the twentieth century alone—at least four times as many people as were living on Earth when the first urban civilizations arose. Yet as fearsome as the modern war machine has become, its destructive power pales in comparison with that of nuclear weapons, the doomsday machine of contemporary civilization.

Although we rarely seem to think about it any more, the risk of thermonuclear annihilation is the most immediate threat to intelligent life in the history of our species. Hundreds of intercontinental ballistic missiles, nuclear submarines, and strategic bombers, armed with thousands of thermonuclear weapons, are now targeted at the world's major population centers with the express purpose of destroying them. Should these weapons actually be used for their intended purpose, they would certainly destroy human civilization, probably exterminate the human species, and possibly annihilate all of the other intelligent life forms with which we share this planet.

Nine nation-states—the United States, Russia, Britain, France, China, India, Pakistan, Israel, and North Korea—currently possess all of the nuclear weapons in the world today.[9] Although the total number of thermonuclear weapons possessed by these nine nuclear nations is a closely guarded secret, estimates have been compiled regularly by the Federation of American Scientists based in Washington, DC, and published periodically in the *Bulletin of Atomic Scientists*.[10] These scientists estimate that, as of 2013, the United States and Russia jointly possessed at least sixteen thousand nuclear warheads, while the other seven nuclear nations collectively possessed at least one thousand more.

Taken together, these nuclear weapons represent a total explosive power of at least 160 billion tons of TNT—the equivalent of approximately twenty-three tons of TNT for every human being on Earth. But since only about 25 percent of these weapons are actually deployed at any given time, the explosive force that could be used immediately actually amounts to about six tons of TNT for every person alive today. Six tons of TNT is equivalent to approximately fifteen thousand sticks of dynamite, which is more than enough to kill you and your family members, destroy your home, and annihilate your workplace, your schools, the businesses you patronize, and all the places you frequent for food, entertainment, contemplation, and worship. In short, the nuclear forces currently deployed are more than sufficient to destroy everything that humanity has built since the dawn of civilization.

History shows us that people and societies typically do not ordinarily take action to remove a threat to their well-being until some event occurs that vividly demonstrates the power of that threat. It therefore seems possible that the infrastructure of thermonuclear war will remain in place until some unexpected and tragic event galvanizes public opinion and brings the world to its senses, such as a thermonuclear weapon being detonated in a major metropolitan area, killing millions of people. Such an event would doubtless produce an urgent and powerful demand for nuclear disarmament. But this would be a heavy price indeed to pay for an action that should have been taken long before as a matter of common sense.

The construction of the nuclear doomsday machine—and its continued maintenance and development since the mid-twentieth century—is surely one of the most astounding acts of collective insanity in the history of the human species. Never in all of the millions of years that hominids inhabited this planet has there existed another technology even remotely capable of an equivalent act of collective self-annihilation. In what is surely the greatest irony in human history, humanity has embraced the doomsday machine as a "necessary" means for achieving national "security"—and modern society has lived with this grotesque reality for so long that it has developed a bizarre complacency about the threat of nuclear annihilation. Nevertheless, the doomsday machine is real. It is active, operational, and ready to destroy all of humanity—and it could be activated in a matter of minutes by any of the world's nuclear powers.

And yet—the likelihood that any of us will die in warfare has actually declined significantly since Paleolithic times. The hundred million people that died in warfare during the twentieth century represented less than1 percent of the people who lived on Earth during that hundred-year period. By contrast, ethnographic studies have shown that in the most warlike of hunting and gathering societies, up to half of all adult males die as the result of violent combat with other humans. And while the scale of killing in warfare can be horrendous in modern societies, modern warfare tends to be intermittent, in contrast to the more or less continuous warfare that was observed among many preindustrial societies. Modern nation-states typically experience decades of peace in between their bloody wars, and some nations—Switzerland, for example—have managed to avoid warfare with other nations for generations.

Before the unspeakable tragedy of a nuclear accident occurs, the world's nuclear arsenal—including the ballistic missiles designed to deliver them—can and should be repurposed. Rather than aiming them at each other and creating an ever-present risk of self-annihilation, we should aim our nuclear missiles at a potential threat to all human life: one of the many large asteroids that are floating in space. If one of these asteroids, miles in diameter, were to collide with the earth at some future time—as has already happened numerous times in the geologic past—it would raise enough dust into the atmosphere to block most of the sunlight from reaching the earth's surface.

Although astronomers have assured us that a collision with a large asteroid is a very rare event that is unlikely to occur in the foreseeable future, what has happened before can happen again. And such a collision would have disastrous effects on the biosphere. It could annihilate most of the earth's green plants and seriously deplete oxygen levels in the atmosphere. Vast clouds of dust would linger in the atmosphere for years, blocking the sun and triggering an ice age that might be more severe than any in recent geologic history.

But a series of well-timed explosions on the surface of an Earth-bound asteroid could either deflect it from its course and cause it to pass harmlessly by or break it up into millions of smaller pieces, most of which would burn up in the atmosphere before reaching the ground. Surely this would be a more sensible mission for the nuclear arsenal than its present assignment of bringing doomsday to human civilization.

Pollution and Plastic

Serious atmospheric pollution by humans on a global scale began at least two thousand years ago, when the Romans began smelting large amounts of copper. Further evidence of atmospheric pollution by copper also dates from the Middle Ages, when smelting of copper was widespread in both Europe and China. It is estimated that roughly two thousand tons of copper were released into the atmosphere every year during both of these periods.

Pollution from human activity, however, accelerated dramatically with advent of the industrial revolution. Immense quantities of sulfur dioxide, nitrogen dioxide, carbon monoxide, methane, butadiene, benzene, toluene, xylene, and other petroleum solvents, detergents, chloroform, insecticides, herbicides, ammonia, acids, nitrates, phosphates, heavy metals, and pharmaceuticals have all flowed into the earth's atmosphere, soils, waterways, and oceans. And much of this pollution ends up in the sea.

In addition to growing amounts of toxic chemicals, the oceans today contain an estimated 100 million tons of plastic, virtually all of it deposited since the middle of the twentieth century. The world's oceans contain areas known as "gyres," where the winds and currents cause the sea waters to rotate in immense spiral vortices, and these gyres have swept the plastic debris floating in the oceans into immense pools. The largest of these pools is the Pacific Trash Vortex, located between Southeast Asia and the west coast of North America. By the year 2014, the Pacific Trash Vortex had become larger than the continental United States.

And yet—the biosphere has demonstrated tremendous resilience in the face of man-made pollution. Waterways that were once severely polluted have bounced back with new wildlife after their polluted waters were cleaned up. Air quality in Los Angeles, once the smog capital of the world, has improved dramatically in recent years.[11] And while pollution in China, India, and other developing nations remains a serious problem, experience shows that pollution can be stopped, and its effects reversed, when people are willing to bear the public and private costs that ending pollution entails.

The Disappearing Forests

Man-made deforestation actually began in prehistoric times. Deforestation in the European river valleys first began nearly ten thousand years ago, when the *Linearbandkeramik* and other early Neolithic peoples used polished stone axes to cut down the forests that grew along the banks of the major rivers and to clear the land for agriculture.

Deforestation accelerated during the third millennium BC, when the urban societies of the Fertile Crescent began cutting down the forests of the ancient Middle East to provide timber for constructing the temples, palaces, and seagoing ships that accompanied the rise of urban civilization. The Cedars of Lebanon of Biblical fame were decimated to build seaworthy ships for the Phoenicians and Greeks who brought civilization to the lands bordering the Mediterranean Sea, and the alluvial lowlands of eastern China were gradually stripped of their forests as agriculture spread through the great valleys of the Yellow, Yangtze, and Pearl rivers.

The deforestation of northern Europe was accelerated by the growing use of iron and steel during the late Middle Ages, when entire forests were cut down to provide charcoal for the blast furnaces that were used for smelting iron ore. When the steamboat was invented in the early nineteenth century, the forests that grew along the banks of the Mississippi River in the United States were largely destroyed by the voracious consumption of trees to feed the paddle-wheel steamships' wood-burning fireboxes. In fact, most of the land east of the Mississippi River in North America was covered with forests before the arrival of the Europeans. Today, most of these forests have long since been cut down and converted into farmland.

Until the twentieth century, the tropical rainforests of South America, Southeast Asia, and Africa remained largely intact, but since the year 1900 a significant portion of the world's rainforests have succumbed to the woodsman's axe. In Brazil alone, nearly 300,000 square miles of rainforest—an area almost twice the size of California—have disappeared since 1970. At the present time, more than twenty thousand square miles of tropical forest are disappearing each year, and if this rate of deforestation continues, the rainforests will have largely disappeared by the end of the next century. The deforestation of tropical rainforests is particularly troubling, because these forests are home to roughly half of all the life forms of the biosphere. This means that more plants and animals become extinct with the loss of each rainforest habitat than with the loss of any other terrestrial ecosystem.

China has embarked upon an ambitious reforestation program that produced a 25 percent increase in the size of its forests by between 1990 and 2005. Yet China currently imports approximately 1.5 billion cubic feet of timber annually, much of it from tropical rainforests. This is more than half of all the timber that is shipped into the global timber market.[12] Imports of rosewood, for example, a tropical wood highly valued for manufacturing furniture, increased 1,500 percent in only nine years, from 1.8 million cubic feet in 2003 to 26.5 million cubic feet in 2012.[13] Nor are other nations blameless in this activity: more than half of all rosewood furniture from China is exported to other nations. Deforestation is not a national problem; it is a global problem.

And yet—not only have many nations begun serious programs of reforestation but small remnants of primeval forest also remain, even in Europe, where they serve as sanctuaries for species that have disappeared from deforested areas. One of the best-preserved of these remnants is the Białowieża Forest, a protected forest park approximately 120 square miles in size that straddles the border between Poland and

Belarus. Although it is smaller in area than many of the world's major cities, the Białowieża Forest contains more living species than any other European habitat. It is home to 117 species of birds and fifty-nine species of mammals, including weasels, pine martens, raccoons, badgers, beavers, otters, foxes, lynx, wolves, wild boar, roe deer, moose, and elk, as well as eight hundred European bison.

The difference between the towering trees and rich bird and animal life that has survived in the primeval forest habitat of the Białowieża Forest and the ragged, skimpy, second-growth forest that has replaced similar habitats in Europe over the last few centuries is dramatic—and sobering.[14] But in time, given proper management and protection, the second-growth forests of Europe will mature, and in the future they may eventually achieve the grandeur of the primeval habitats which once flourished there. And in sub-Saharan Africa, Southeast Asia, North America, and South America, many wilderness areas have been set aside as ecological preserves, where ancient habitats and their many life forms continue to cling to a precarious survival.

The Sixth Mass Extinction

The second most immediate threat to intelligent life, after the risk of thermonuclear war, is the risk of massive environmental destruction and life form extinction. Scientists estimate that at the present time approximately thirty thousand life forms are becoming extinct every year, and the rate of extinctions is steadily rising. Furthermore, once a species of plant or animal becomes extinct, it normally takes millions of years before a new life form can evolve to replace it.

There have actually been five previous mass extinctions in geologic history, all of which occurred long before hominids existed. The first of these occurred roughly 445 million years ago during the Ordovician Period, an age of primitive marine creatures such as trilobites and coral. The second occurred 360 million years ago during the Devonian Period, an age of primitive fishes and the earliest land-based plants and insects. The third occurred 250 million years ago during the Permian Period, an age of primitive reptiles. And the fourth occurred two hundred million years ago during the Triassic Period, an age of large amphibians and marine reptiles.

The Ordovician, Devonian, and Triassic extinctions wiped out 75 to 85 percent of all living species, and during the Permian extinction, approximately 95 percent of all species of plants and animals perished. Most scientists believe that these four extinctions occurred during periods of severe and rapid global climate change. These periods of climate change may have been triggered by irregularities in the earth's orbit around the sun, by episodes of intense and sustained volcanic activity, by variations in solar activity, by collisions with large asteroids, or by any combination of these events.

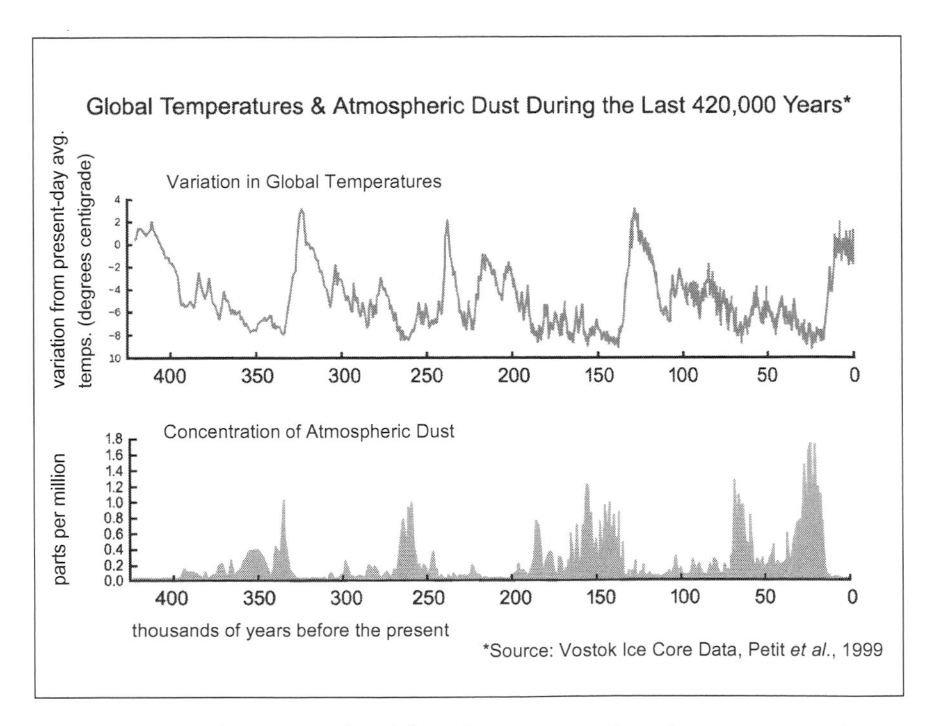

Figure 10.2. *Changes in the global climate are reflected in ice cores taken at the Vostok Research Station in Antarctica. These data show a clear association between levels of atmospheric dust and lowering of global temperatures during the last four ice ages.* Illustration by the author, after Vostok Petit data. Licensed under Creative Commons Attribution-Share Alike 3.0 via Wikimedia Commons.

The fifth and most recent mass extinction in geologic history occurred sixty-five million years ago, at the end of the Cretaceous Period, an age of dinosaurs, flowering plants, modern forms of insects, and early mammals. It is generally believed that the cause of this extinction was the impact of an asteroid more than six miles wide that collided with the earth off the coast of southern Mexico at an estimated speed of 67,000 miles per hour.

Scientists believe that the debris thrown into the atmosphere by this immense collision—and possibly also by volcanic eruptions that may have been set off by the physical shock of the impact—created an immense global dust cloud that spread over the entire world. It is likely that this dust cloud blocked the light from the sun from reaching the earth's surface for several years, resulting in a major cooling of the global climate. The fifth mass extinction brought about the disappear-

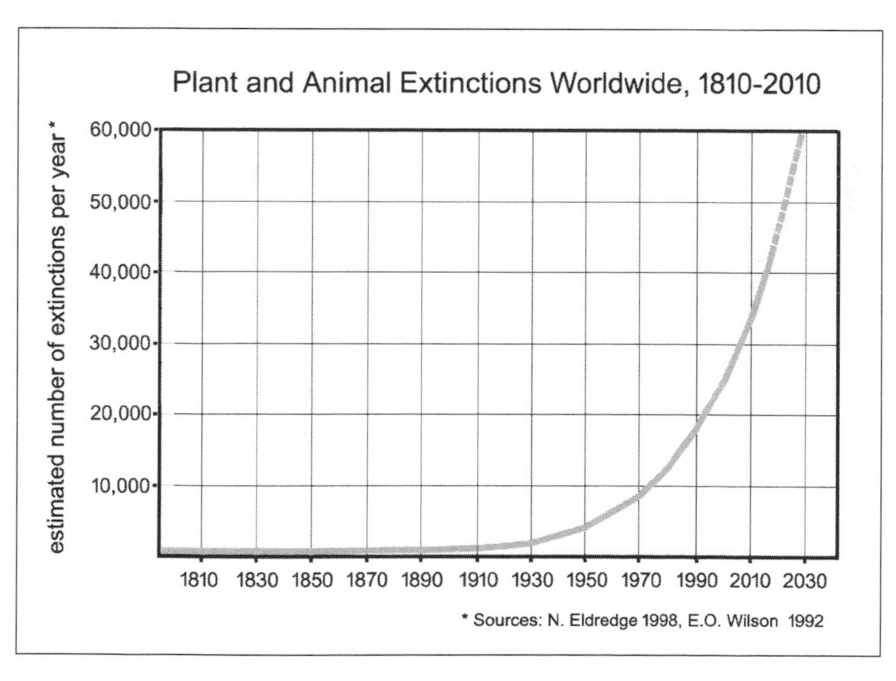

Figure 10.3. *Although the extinction of plant and animal species has been an ongoing process throughout the history of life on Earth, the number of species that have become extinct each year has increased dramatically since the beginning of the twentieth century.*

ance of not only the dinosaurs but also 75 percent of all species of life on Earth.

We are now in the early stages of what may eventually become the sixth mass extinction of life forms in geologic history. But in contrast to the previous five extinctions, this one is the result not of cosmic or geological forces but rather of human activities, including environmental pollution, massive deforestation, the severe depletion of plant and animal populations, and the collapse of terrestrial and marine ecosystems.

Although it is normal for plant and animal species to disappear at a certain rate almost continuously, this normal or "background rate" has been estimated as between ten and one hundred extinctions per year. By contrast, the current rate has been estimated at about thirty thousand extinctions per year. (This is a mid-range estimate by the biologists Edward O. Wilson and Niles Eldredge, but both higher and lower estimates have been published by other scientists.) At present, there is no general agreement about either the rate of extinctions or even about the actual number of living species in the biosphere, but as many as 10 percent of all living species may have become extinct since the beginning of the industrial revolution two hundred years ago.[15]

If the depletion and extinction of living species continues at its present rate, the earth's natural environments will probably experience a cascading series of ecosystem collapses similar to those that occurred inside Biosphere 2. Such ecosystem collapses could produce population explosions of certain insects, plants, fungi, algae, and bacteria; epidemics of new and previously unknown diseases; massive crop failures; and the wholesale starvation of human populations.

Some early warning signs of such ecosystem collapses have already appeared in recent years. These include outbreaks of colony collapse among honeybees, an incurable fungus that causes "white nose syndrome" in bats that have wiped out 95 percent of the population of certain bat colonies, and another incurable fungus that has been linked to population crashes of frogs throughout the world since the 1990s.[16]

Geologists estimate that it took between five million and twenty-five million years for new species to evolve to replace those that were lost in each of the previous mass extinctions. This means that it would take at least five million years for the biosphere's ecological systems to fully recover and regain their previous levels of biodiversity. Five million years ago, our ancestors were quadrupedal apes living in the trees. Five million years from now, we humans will probably have long since evolved into—or been replaced by—another, possibly wiser, form of intelligent life. We can only guess at the ultimate impact this man-made extinction will have on the earth's many fragile ecosystems and on the biosphere as a whole. But as far as the life and history of *Homo sapiens* is concerned, its impact will be permanent and irreversible.

And yet—the current mass extinction is still in its early stages and might still be reversed, and many species of birds and mammals have been brought back from the brink in recent years. The population of whooping cranes has rebounded from its low of twenty-three cranes in 1941 to more than six hundred cranes in 2011. Not long ago, there were only nine California condors still living in the wild, but there are now roughly 175 of these majestic raptors living and breeding in the mountains of California and Mexico (another 175 are in captive breeding programs). And more than ten thousand breeding pairs of bald eagles now live in the lower forty-eight states, where America's national bird was once virtually extinct.

Furthermore, continued advances in DNA sequencing may ultimately enable biologists to recreate some extinct species by combining DNA extracted from museum specimens with living cells from closely related organisms. Humanity has thousands of years of experience with selective breeding while modern science has decades of experience with genetic engineering. Developing new species through such techniques would be much faster than the natural processes of evolution could produce new species. The full story of the sixth mass extinction is still to be written. We may yet hope for a happy ending.

The Perils of Global Climate Change

Of all the threats to the earth's environment that have materialized in recent years, perhaps none has generated as much public concern and urgent calls for action as the threat of global climate change. This concern is based on two simple facts. First, increasing quantities of carbon dioxide are being released into the atmosphere by the combustion of fossil fuels in modern industrial societies. Second, the geological record contains abundant evidence that increases in the concentration of carbon dioxide in the earth's atmosphere have been closely associated with significant periods of warming in global climates.

In the year 1751, roughly three million tons of carbon were being released into the earth's atmosphere each year from the burning of fossil fuels, mostly coal. This amount increased gradually at first, and carbon emissions did not increase to four million tons annually until twenty years later, in 1771. But by 1775, James Watt had perfected his very efficient steam engines, coal had become the premier fuel both for smelting iron ore and for providing mechanical power, and carbon emissions began to accelerate.

In 1781, global carbon dioxide emissions reached five million tons per year, and by 1800 they had grown to eight million tons annually. Over the ensuing hundred years, as the steam engine replaced the windmills and water wheels that had once powered the world's factories, and as the steamship and the railroad revolutionized long-distance transportation, the amount of carbon released into the atmosphere grew from eight million tons in 1800 to 534 million tons in 1900—an increase of 6,675 percent.

This massive growth in carbon emissions was dwarfed again as the automobile age was born and daily life in the twentieth century became increasingly electrified. The revolution in transportation unleashed by the invention of the internal combustion engine created a new and growing demand for gasoline and diesel fuel, and the combustion of petroleum products joined the burning of coal as societies throughout the world became increasingly industrialized. Car-

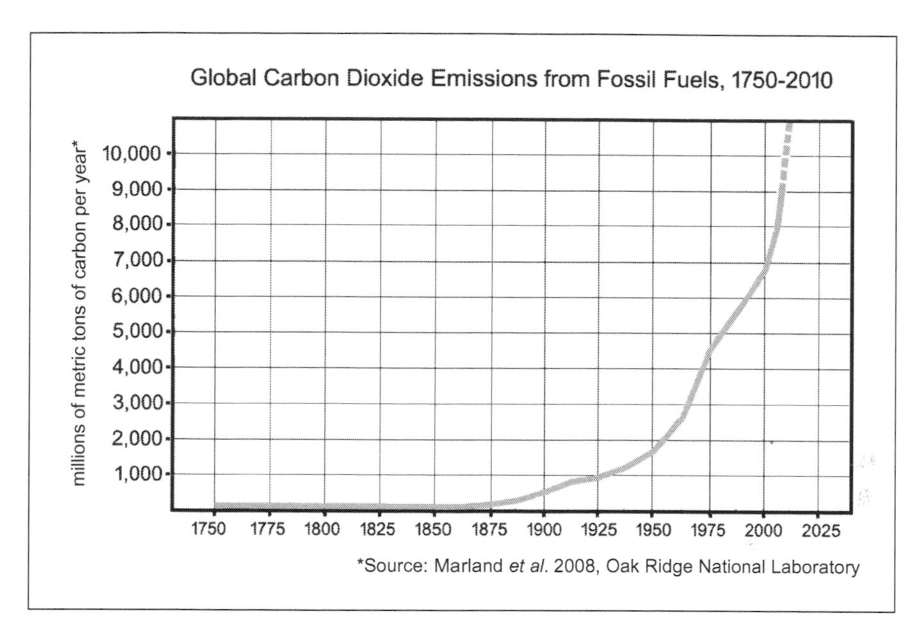

Figure 10.4. *The amount of carbon released into the atmosphere each year in the form of carbon dioxide from burning fossil fuels grew from three million tons in 1750 to fifty-four million tons in 1850 to 1.63 billion tons in 1950. Carbon emissions surpassed nine billion tons per year in 2010 and are projected to exceed fifteen billion tons annually by the year 2025.*

bon emissions from the burning of fossil fuels grew from 534 million tons annually in 1900 to more than nine *billion* tons annually by the year 2010, and it is expected to surpass fifteen billion tons annually by the year 2025.[17]

The burning of coal, oil, and natural gas in increasingly larger quantities is adding carbon dioxide to the earth's atmosphere much faster than the green plants of the world's continents and oceans can absorb this carbon dioxide and use it to synthesize new molecules of living tissue. As a result, the carbon dioxide content of the earth's atmosphere has been steadily increasing, adding to the normal "greenhouse effect" of water vapor and other atmospheric components that keep the earth warm enough to sustain life. Climate scientists are unanimous in their prediction that significant warming of the earth's surface would melt much of the polar ice, raise global sea levels, and increase the frequency

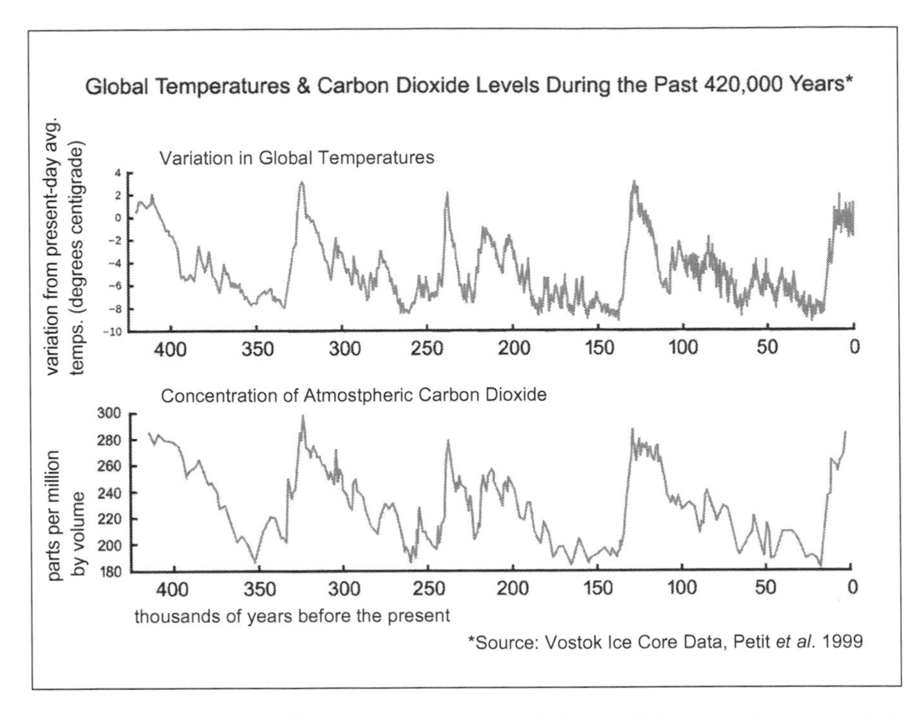

Figure 10.5. *Ice cores from Antarctica's Vostok Research Station show that high concentrations of atmospheric carbon dioxide have been closely associated with increased global temperatures during the past four hundred thousand years.* Illustration by the author, after Vostok Petit data. Licensed under Creative Commons Attribution-Share Alike 3.0 via Wikimedia Commons.

of unusual and dangerous weather, including severe heat waves and storm systems of unprecedented ferocity.

The close association between carbon dioxide levels and average global temperatures has been graphically demonstrated by data from ice cores taken in 1998 at the Vostok Research Station in Antarctica, located eight hundred miles from the South Pole. Analysis of the Vostok ice cores revealed that higher levels of atmospheric carbon dioxide and methane gas—the two major "greenhouse gases" that trap the heat of the sun in the earth's atmosphere—have been closely associated with higher average global temperatures for hundreds of thousands of years.

And yet—global climate change is hardly a new phenomenon in the earth's history. Periods of global warming and global cooling have occurred repeatedly for hundreds of *millions* of years, not only before hominids existed but also before primates, mammals, birds, dinosaurs, reptiles, amphibians, or even fish existed. In fact, most scientists suspect that the five previous extinctions—all of which took place before any modern species of monkeys or apes appeared—were caused primarily by major changes in global climates.

The last ice age ended eleven thousand years ago, just before Neolithic societies began to practice agriculture. Since then, the earth has been in one of the warm periods known as an "interglacial" that have occurred repeatedly over the past million years. While the current warm period has lasted somewhat longer than most previous interglacials, a much more pronounced period of global warming occurred roughly four hundred thousand years ago. During this period, known as Marine Isotope Stage 11, or "MIS 11,"[18] global sea levels rose as much as thirty feet above their present levels, and most of the Greenland Ice Sheet collapsed.[19] Yet the extremely warm period of MIS 11 was eventually followed by new ice ages as severe as any in recent geologic history.

During most of the past million years or so, cooling periods culminating in severe ice ages have lasted for roughly ninety thousand years, and these have been interrupted by warm interglacials that have each lasted for roughly ten thousand years.[20] This suggests that, given the earth's recent climate history, we should be due for the onset of another ice age very soon. That is, we *would have been due* for the onset of another ice age, if it were not for the fact that the concentration of carbon dioxide in the atmosphere is now higher than it has been in any time during the past million years.

During certain periods in the very distant past, the concentration of carbon dioxide in the atmosphere has been as much as fifteen times higher than it is at the present time. In fact, the concentration of atmospheric carbon dioxide peaked at approximately 6,000 parts per million (ppm) around five hundred million years ago and peaked again at roughly 2,500 ppm around two hundred million years ago.

During the past million years, however, carbon dioxide concentrations have not risen above 300 ppm during interglacial periods or fallen below 180 ppm during glacial periods. Yet in the short space of the past two hundred years, the use of fossil fuels by hominids has raised carbon dioxide levels to 400 ppm—33 percent above their previous maximums—and this increase is continuing to accelerate. This means that the current increase in atmospheric carbon dioxide will not only delay the onset of the next ice age but might actually prevent another ice age from occurring altogether.

It is important to recognize that the perils of global climate change are not limited only to the effects of global warming. In fact, if another ice age similar to the most recent one were to return, massive ice sheets hundreds of feet thick and weighing hundreds of billions of tons would slowly descend from the polar regions, crushing everything in their path. Although this process would take tens of thousands of years—several times longer than the entire history of human civilization—the destructive effects of a new ice age would be unprecedented.

In North America, the return of the polar ice sheets would completely destroy the entire nation of Canada and descend into the continental United States, burying most of Minnesota, all of Wisconsin, Michigan, and Ohio, the northern regions of Illinois, Indiana, and Pennsylvania, and all of New England as far south as New York City. The cities of Vancouver, Calgary, Winnipeg, Toronto, Ottawa, Montreal, Quebec, Seattle, Minneapolis, St. Paul, Chicago, Cleveland, Columbus, Buffalo, Albany, Portland, Boston, Hartford, and New York—as well as thousands of neighboring cities and towns—would be crushed under hundreds of feet of ice. In Europe, all of Scandinavia, most of the British Isles, and large portions of Germany, Poland, and Russia would be completely obliterated, and the cities of Dublin, Belfast, Glasgow, Edinburgh, Oslo, Stockholm, Copenhagen, Helsinki, Berlin, Warsaw, and St. Petersburg—as well as thousands of other neighboring cities and towns—would all be literally wiped off the map.

In addition, there would be a severe loss of agricultural land throughout the world. Barren, treeless tundra and permafrost would extend southward from the polar ice sheets into the middle of the continental US as far south as New Mexico, and the tundra would spread across all of Europe as far south as Spain and Portugal. South of the tundra, evergreen forests would cover southern Europe, most of California, and all but the southernmost regions of the United States. Neither the tundra nor the evergreen forest environments are capable of supporting human agriculture.

Furthermore, it would be difficult or impossible to move agricultural production farther south, because the massive depletion of atmospheric water would cause the world's deserts to grow significantly larger than they are today. The Sahara Desert would expand until it covered more than half of the African continent, and the deserts of Central Asia would expand from the Caspian Sea north of present-day Iran all the way across Asia to the Pacific Ocean. Meanwhile, the tropical rainforests of Central Africa and the Amazon Basin would shrink to a fraction of their present size. And since great quantities of the earth's water would be locked up in the form of polar ice, sea levels throughout the world would drop to three hundred feet below their present levels, leaving every port city on Earth stranded dozens—or hundreds—of miles from the open sea.

These are the actual conditions that prevailed at the time of the Last Glacial Maximum only eighteen thousand years ago—not very long ago even in the relatively short history of the human species. Eighteen thousand years ago, the cultures of the Upper Paleolithic were already using sophisticated forms of symbolic communication—including the art, music, language, petroglyphs, and narrative stories invented by fully modern humans—and the Natufians and their contemporaries were already settling in the permanent villages that would eventually produce the agricultural revolution.

While we are justly concerned about the dangers of global warming and the destructive effect it would have on our contemporary societies, we should also recognize that another period of severe global cooling—and the return of another ice age—would actually pose a much greater threat to human civilization. In fact, it is entirely possible that the release of carbon dioxide into the atmosphere by industrial societies will actually *prevent* the return of another ice age that might have otherwise returned in the near future. And because higher levels of carbon dioxide bring about significant changes in the chemistry of the oceans, the effects of global warming could persist almost indefinitely and might prevent the return of more ice ages for hundreds of thousands of years into the future.[21]

We must also acknowledge that a sustained campaign among the world's most developed nations to limit or reduce the emission of greenhouse gases, especially carbon dioxide, has caused carbon dioxide emissions to level off in these nations in recent years. In fact, the continued increase in the size of world carbon dioxide emissions is now due mainly to the growth of populations—and affluence—in the world's developing nations, especially China and India.

Moreover, modern society has made significant progress in developing new carbon-free sources of energy, such as solar and wind power. We should not forget that nearly a century passed after the development of Newcomen's first "atmospheric" steam engine in 1712 before safe and practical reciprocating steam engines were being mass produced for the commercial market, and that it took seventy-five years of experimentation before a safe and reliable internal combustion engine was ready for market.

Finally, although new techniques have been developed for extracting natural gas by hydraulic fracturing and the mining of tar sands, the supply of fossil fuels is not unlimited. Within a few hundred years at most, humanity will have to end its dependence on fossil fuels and will be forced to replace fossil fuels with alternative sources of energy. And when that day finally arrives, modern science and technology will have probably spent many decades searching for practical ways of extracting excess carbon dioxide from the atmosphere, while taking care not to extract so much that we inadvertently trigger another ice age.

Given the past history of our species' incredible technological achievements, it is entirely possible that humanity will eventually succeed in achieving both of these goals. Above all, the phenomenon of global climate change is vivid evidence that the future of life on Earth now depends on the actions of humanity. For better or worse, our species has taken possession of this planet and made it our own.

Planet of the Hominids

The earth's human population is now growing faster than at any previous period in history. The development of scientific agriculture and the mechanization of farming has made possible the feeding of ever larger populations, while modern advances in the effectiveness of medical treatments have steadily reduced human death rates. As the last ice age slowly came to an end and the technology of symbolic communication unleashed the power of cultural evolution, the population of human hunters and gatherers gradually increased until it had grown to a total of roughly five million people.

With the adoption of agriculture and the rise of urban civilization, the human population grew 3,000 percent—to 150 million people—between 5000 BC and 1 AD. Even with the fall of the Roman Empire and the onset of the Dark Ages, the world's population doubled again to

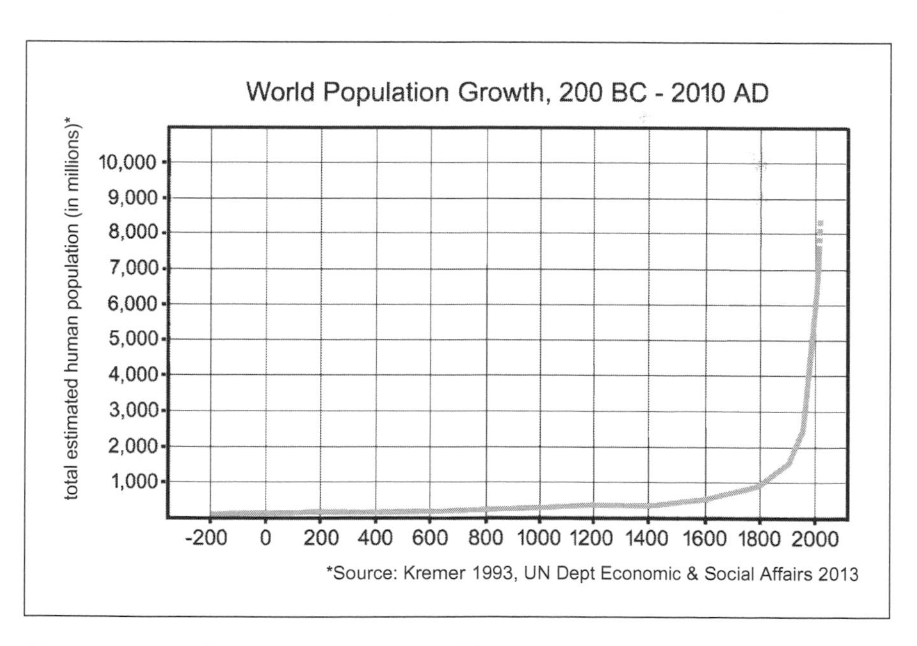

Figure 10.6. *The global population of human hunters and gatherers stood at roughly five million people in 5000 BC and grew to 150 million over the next five thousand years. But when the industrial revolution began in 1800, the human population exploded, increasing from less than one billion to more than six billion people in slightly more than two hundred years.*

300 million by 1100 AD. And during the Middle Ages, in spite of the tremendous loss of life during the massacres of the Mongol invasions in the thirteenth century and the pandemics of the plague or Black Death in the fourteenth century, the world's human population doubled again, to approximately 600 million people, by the year 1600 AD.

When the industrial revolution began in 1800, however, the human population truly exploded, increasing 600 percent in two hundred years from less than one billion to more than six billion people. And if this population explosion continues at its present rate, there will eventually no longer be enough agricultural land on Earth to feed all the people who will inhabit this planet. During the past fifty thousand years—which represent only the most recent 1 percent of hominid history—the earth's human population has increased *seven thousand times* over its original size and is currently increasing by approximately *a quarter of a million people per day.*

And yet—since the 1960s, birth rates have been falling all over the world. In fact, by 2014, birth rates had fallen below the "replacement rate" of 2.1 births per woman in nearly half of the world's nations. The birth rate in Europe as a whole had fallen to 1.5 births per woman by 2005, and the population of Europe is now expected to decline 14 percent between now and the end of the twenty-first century.[22] As human societies have become affluent enough to provide their elderly populations with a decent standard of living, the ancient human practice of having large numbers of children to ensure a means of support in old age is no longer necessary.

Moreover, as the great mass of humanity has made the transition from an agricultural to an urban existence, the economic benefits that children once provided to agricultural families have largely disappeared. In fact, children have actually become a net economic drain on the fortunes of an urban married couple, whose disposable income is significantly reduced by the additional costs of food, clothing, shelter, medical care, and education that each child requires.

Having children does not contribute to the myriad opportunities that modern people enjoy: to be entertained, to travel, to dine out, and to pursue their professional careers and their personal interests. Children are more likely to interfere with such activities than to make it easier for their parents to pursue them. For all of these reasons, birth rates in the most affluent societies have declined to the point where some of the world's richest and most advanced nations, including Germany, Japan, and South Korea, are already experiencing population declines.

This does not mean that people will cease having children. The psychological satisfactions of bearing and raising children—and the connection that children provide to the future and to succeeding generations—ensures that women and men will never stop wanting to have children. More than at any other time in human history, the children born in today's world are the offspring of people who want children and who value the experience of parenthood for its own sake. And while there will always be individuals who choose not to have children, these individuals will pass on neither their DNA nor their cultural attitudes and beliefs to succeeding generations. This fact alone will ensure that humanity will never stop reproducing.

Some economists have bemoaned these stagnating or falling birth rates, seeing this as a harbinger of economic and cultural decline. But this is a short-sighted point of view. Endless population growth is not possible; the earth's biosphere is finite in size; other planets do not offer realistic opportunities for human colonization. In the final analysis, more people will mean less space and a smaller share of the available resources for each person. Less people will mean more space and ultimately a greater share of the available resources for each person.

The current threats to the biosphere are more than the consequences of human technologies. They are also the expression of *Homo sapiens*' animal nature. For all of its technological prowess, our species is still motivated by ancient animal instincts, including the drive to expand and multiply to the limits of the possible. This kind of elemental self-interest is to be expected of any life form. But in our desire to provide ourselves with ever-increasing amounts of progeny, food, clothing, shelter, energy, weapons, and material possessions, we have too often overlooked the needs of other forms of plant and animal life. After all, *Homo sapiens* is the only species capable of both comprehending and addressing the needs of all other life forms. It is therefore incumbent upon us to use our higher consciousness not only for our own well-being but also for the greater well-being of all the many species that share the biosphere with us.

In the final analysis, the dramatic increases in the human population have occurred because *Homo sapiens* has effectively taken full possession of this planet and its resources—and, in the process, has become entirely responsible for its future. Although we may not have consciously intended to do so, we have in effect converted the earth into the "planet of the hominids." As a result, it is we hominids who must manage the earth wisely—not only for our own benefit but for the benefit of all living things, now and in the future.

Beyond the Nation-State

The world today is made up of approximately two hundred nations that among them claim all the land area of our planet except the continent of Antarctica.[23] Although this state of affairs seems perfectly natural to us, our familiar world of nation-states did not begin to take shape until after the year 1800. In fact, until the railroad and the telegraph became commonplace in the nineteenth century, the nation-state as a distinct political and cultural entity was actually a rarity among human societies. Of the two-hundred-odd nation-states that presently rule the world, no less than 177 of them were created *after* the beginning of the twentieth century.[24]

Yet in spite of their proclamations of independence and national sovereignty, the nation-states of the present era have never been entirely independent. World trade, which grew rapidly throughout the nineteenth century, has long since created a vital economic integration among the world's nations. It is a fact of life that by the dawning of the twentieth century a global economy had already been created in which no single nation-state or economic region could truly prosper without maintaining robust economic relationships with many other nations.

The stark reality of this fact was driven home in spectacular fashion when, in June of 1930, the Smoot-Hawley Tariff Act was passed by the US Congress in a misguided attempt to protect American industry and agriculture from foreign competition during the economic downturn that began in 1929. But when other countries—especially Canada, America's most important trading partner—retaliated by raising their own tariffs on American goods, the entire system of global trade went into rapid decline. US imports from Europe tumbled from $1.3 billion to $390 million between 1929 and 1932, while US exports to Europe shrank from $2.3 billion to $784 million during the same three years. In fact, world trade as a whole declined by two-thirds between 1929 and 1934, plunging all of the world's industrialized countries into the economic doldrums of the Great Depression.

The emergence of the nation-state as the dominant form of human society was the single greatest act of social fusion in the history of humanity. Never before had millions of people, living in thousands of distinct geographical communities, regarded themselves as brethren belonging to the same ethnic group. It is worth noting that even as the world's human population has grown from less than one million people to seven billion and counting, the number of independent human societies has actually been shrinking.

There is no reason to suppose that this long history of social fusion has entirely run its course, or that humanity will remain permanently divided into two hundred independent nations. Over and over again, significant advances in the technologies of interaction have led to social fusions that created much larger human groups than those that formerly existed. Since the invention of information technology has now created new and more powerful forms of transportation and communication, it is reasonable to expect that humanity is destined to experience yet another great process of social fusion in the not-too-distant future. In fact, the foundations of humanity's first truly global civilization are already in place, and if we look carefully, we can already see the tentative outlines of a future world taking shape beyond the boundaries of the world's nation-states.

The Birth of Global Civilization

The seeds of global civilization are many, and some of them are very old. They begin at least as long ago as the time of Jesus of Nazareth, who preached that every human being was equal in the sight of God and that by accepting Jehovah as the only true deity, anyone, no matter what his or her culture of origin, could join the fellowship of the faithful. Two thousand years later, the followers of Jesus number more than two billion and constitute the single largest religious group on Earth.

The Romans practiced their own secular version of cultural acceptance and human equality. They granted the privileges of Roman citizenship to favored allies, and they built roads, public baths, and elaborate fresh water systems not only for their own comfort and safety but also for the well-being of their subject peoples. They filled the granaries of their colonies to keep their subject populations from starvation; they erected grand stadiums and coliseums for the entertainment of all; and for the most part they granted the people they conquered the freedom to pursue their own cultural traditions and worship their own religions. In fact, it was the tolerance of the Romans toward other cultural traditions, combined with their sense of responsibility for the well-being of everyone living under their dominion, that was largely responsible for the stability of Roman society and the longevity of Roman rule throughout the ancient world.

In the centuries that followed the fall of Rome, the religions of Hinduism, Zoroastrianism, Christianity, Buddhism, Confucianism, and Islam spread far and wide among diverse cultures and across immense geographical areas. These faiths were all built on the principle of cultural inclusion, and it was their inclusionary philosophies that enabled them to absorb new populations and new ethnicities, transcend cultural boundaries, and become true world religions.

We should not forget that these great inclusionary religions arose and prospered at a time when writing was practiced only by a few members of the elite, when riding horseback was the fastest means of transportation, when roads between cities were few and far between, and when the sailing ship was a small wooden vessel that was as likely to be propelled by oars as by the wind. This alone is an eloquent testament to the power of their philosophies and to the magnetism of their founders, all of whom preached the doctrine of multicultural fusion. The great success of these ancient religions suggests that a similar process of cultural fusion will probably be our species' ultimate destiny.

Two conditions that have emerged in recent decades are now driving humanity toward this ultimate fusion. The first is the rapidly growing threat to the planetary environment and to all human societies. This global threat cannot be resolved by any single nation or regional group of nations. Instead, all nations must work together to resolve it. The second condition is the means—made possible by digital technologies—for rapid, efficient, and affordable interaction among the inhabitants of all the various societies and cultures scattered across the globe. The potential synergies between these two conditions should not be underestimated. In fact, evidence that this ultimate process of fusion is already under way is revealed by the emergence of key social and cultural elements of humanity's first global civilization.

The customs and traditions surrounding the preparation of food—considered by anthropologists to be one of the most distinctive expressions of ethnic and cultural identity—have been undergoing an unprecedented process of fusion. Culinary traditions from Europe, the Americas, the Middle East, Central Asia, and the Far East have spread far and wide from their cultures of origin and have become popular favorites throughout the world.

Music, another distinctive expression of ethnic and cultural identity, is undergoing a process of globalization that began with the worldwide diffusion of European classical music and that has now been joined by the spread of musical traditions from all over the globe. Musical harmonies, rhythms, and instrumental styles from a myriad of ethnicities have migrated far from their cultures of origin and have developed their own robust global audiences. In fact, both food and music have begun to merge the traditions of different cultures into new forms of "fusion food" and "fusion music."

The cultural fusion that is taking place in the spheres of food and the arts is equally apparent in the material spheres of science, commerce, and industry. Consumer brands from numerous industries—including foods, beverages, automobiles, jewelry, perfumes, watches, clothing, and electronics—have achieved global recognition and now enjoy the fierce loyalty of hundreds of millions of customers all over the world. And the rise of multinational corporations—some of which have larger economies than most nations—has knitted the global economy together more tightly than ever before.

The pursuit of research and development in science and technology had an international character from its earliest beginnings and has long since become a global enterprise. More than half of the Nobel Prizes in physics, chemistry, and medicine awarded during the last fifty years were bestowed upon collaborating scientists based in different nations.[25] In fact, our contemporary culture of science and technology is one of the strongest and most vibrant elements of this emerging global civilization.

At the same time, the world's institutions of higher learning are slowly becoming global entities, with both faculties and student bodies increasingly composed of international and multicultural populations. Finally, the number of international organizations has multiplied exponentially during the modern era. In 1910, there were approximately two hundred non-governmental organizations (NGOs) that were international in character. By 2010, this number had grown to nearly *sixty thousand.* And the growth of international governmental organizations (IGOs) has been equally as great.

Yet even as humanity has fused into about two hundred nation-states, all of the traditional human groups that were created in former transformations—the countless thousands of tribes, villages, towns, cities, and regions that constitute the fabric of human society—have continued to maintain their separate identities and to govern their appropriate spheres of human life. This alone should reassure us that the nation-states of the world are in no danger of disappearing. The rise of global civilization is destined not to replace the world's nations but rather to facilitate a sense of common purpose among them. Global institutions are, in the final analysis, networks of people, and among the most extraordinary of human behaviors is the ease with which human beings can simultaneously identify with multiple groups on multiple levels.

Finally, history has repeatedly shown that nothing brings people together so much as a common enemy. Nearly every instance of social and cultural fusion, in which smaller social units have successfully merged into larger ones, has been driven, at least in part, by confrontations with a hostile adversary. When the world's forests and primeval habitats have been reduced to a fraction of their original size, when the ocean's resources have been largely depleted, and when even the miracles of scientific agriculture are no longer able to feed the expanding population of humans, the "enemy" of human civilization will be nothing more nor less than the existential threats to the earth's biosphere produced by human technologies. As these threats continue to multiply, humanity will be forced to choose between continuing on the pathways that lead ultimately to depletion, extinction, and annihilation or—by cooperating on a global scale—turn to new pathways that lead to restoration, reconciliation, and a sustainable relationship between ourselves and the biosphere.

We should not expect this transformation into a global civilization to be realized either quickly or easily. The effects of information technology did not begin to be felt until the middle of the twentieth century, and the deep integration of digital technologies into the processes of contemporary finance, trade, communication, and transportation did not take place until even later. Since we are now at the very beginning of this latest transformation, we cannot truly anticipate the changes in human life and society that it is destined to produce. But the history of past metamorphoses suggests two conclusions. First, that human life and society will be transformed no less by information technology than it has been by the key technologies of former ages. And second, that another great fusion of human society lies ahead.

♦

We who are alive in the twenty-first century stand at the pinnacle of human achievement, even as we contemplate a looming catastrophe of our own making. Yet the genius of humanity shows no sign of coming to an end. Throughout our long history, we have demonstrated an uncanny capacity to rise to the challenges before us and to overcome the obstacles in our path.

During all of our unique and remarkable history, we hominids have used technology to transform ourselves, and in the process, we have transformed the world. The metamorphoses in human life and society wrought by weapons, fire, clothing, shelter, symbolic communication, agriculture, urban civilization, precision machinery, and information technology have propelled us into a new and unique relationship with the biosphere that originally gave us life. We have become its stewards; it has become our responsibility; it is we who must now decide its fate.

Acknowledgments

I would like to express my sincere gratitude to those who contributed in significant ways to the successful completion of *Unbound.*

Early in the process of developing the manuscript, my anthropologist colleagues Jack M. Potter, Robert Bates Graber, Richard Robbins, and Anek R. Sankhyan provided both valuable advice and gracious endorsements that were of great help in finding a quality publisher. My son Guy Currier diligently read an early draft and responded with copious feedback. My son Chad Currier and my daughter Rebecca Meyer offered their constant encouragement throughout the long months of writing the manuscript, and my son-in-law Christopher Meyer helped to rescue my precious files from oblivion when my hard disk crashed. My friends Richard Foley, Alicia Larsson, and Terry Lee Wilder all read the completed first draft and provided timely and useful feedback.

Several individuals and organizations kindly granted permission to reprint many of the best illustrations in *Unbound,* including Frans Lanting, John Reader, Richard Dutton, Mike Storey, Elsevier's *Journal of Human Evolution*, Skullduggery, Inc., the Science Outreach Program at the University of Canterbury in New Zealand, the Florida Center for Instructional Technology, and the Centers for Disease Control and Prevention's Laboratory Identification of Parasitic Diseases.

Above all, my heartfelt thanks goes to three remarkable people, without whom this book might never have seen the light of day. My agent, Roger Williams, never doubted the potential of this book project, never wavered in his confidence and optimism about its future, and worked tirelessly to secure a fine contract with an excellent publisher. My editor Cal Barksdale's meticulous reading of the final manuscript, numerous valuable editorial suggestions, and pivotal role in the development of the book's final title and physical design all made *Unbound* a far better book than it would have been without him. Finally, I cannot give enough credit to my life partner, Cara L. Keith, for her enormous contributions to this work. She read the entire manuscript several times and—with her sharp editor's eye and her uncompromising sensitivity to anything superfluous or unreasonable—contributed in countless ways to the focus and clarity of each chapter. Her inexhaustible confidence kept my spirits afloat when the going was rough, and it is to her wisdom, love, and support that I owe my greatest debt.

Notes

Introduction

1. This quotation is typically attributed to the German philosopher Arthur Schopenhauer, but it does not appear in any of Schopenhauer's published works. What Schopenhauer actually wrote was: "*Der Wahrheit ist allezeit nur ein kurzes Siegesfest beschieden, zwischen den beiden langen Zeiträumen, wo sie als Paradox verdammt und als Trivial gering geschätzt wird.*" ("Truth is allowed only a brief celebration of victory, between two long periods, during which it is condemned as paradoxical and disparaged as trivial.") From *Die Welt als Wille und Vorstellung* (*The World as Will and Representation*), first published in Germany in 1818.
2. See Wildman, Derek E., Monica Uddin, Guozhen Liu, Lawrence I. Grossman, and Morris Goodman, 2003, "Implications of Natural Selection in Shaping 99.4% Nonsynonymous DNA Identity between Humans and Chimpanzees: Enlarging Genus Homo." *Proceedings of the National Academy of Sciences of the United States of America* 100: 7181–7188.

Chapter 1

1. There is, of course, facial hair above and below the lips in the form of mustaches and beards among the adult males in most human populations. However, the presence or absence of this facial hair varies greatly. In some human societies, such as the San Bushman of Africa's Kalahari Desert and the Yanomamö of South America's Amazon Basin, adult males typically have little or no facial hair. Some non-human primates also have facial hair: the male orangutan often grows a beard, and both male and female patas monkeys typically have well-developed mustaches and beards in adulthood.
2. Some baboon species (*e.g.*, the hamadryas baboon of Ethiopia) that inhabit open, exposed environments have a flexible, multi-layered social structure, in which several harems combine to form "clans" that forage

together during the day and then combine with other clans to share a common sleeping area at night. This enables the baboons, while maintaining their pattern of forming exclusive harems, to form very large groups of up to 750 individuals that are effective defenses against predators during the hours of darkness.

3. In the modern world, ethnic identities are not as uniformly strong as they are in traditional societies. Some people tend to remain faithful to their cultural origins, while others—especially the young—prefer to embrace innovation and change. But the tremendous increase in contact among the world's cultures made possible by industrial technologies of interaction and later by the invention of digital technologies—combined with the hugely accelerated pace of technological and cultural change—has in recent years caused many ethnic traditions to lose their former power and importance, much to the despair of the older generations.

Chapter 2

1. For a detailed description of the discovery of the Laetoli footprints, see John Reader, *Missing Links: In Search of Human Origins* (Oxford: Oxford University Press, 2011).
2. As I have previously explained in greater detail in the introduction, for many decades, the familiar term "hominids" was universally used by anthropologists, paleontologists, and all other scientists to refer to the various species of prehistoric and modern humans. But "hominids" fell out of favor during the 1990s, when advances in DNA analysis showed that gorillas and chimpanzees are genetically closer to humans than had been previously assumed. As a result of these findings, all of the great apes were reclassified and grouped together into the family *Hominidae*. In the years since this reclassification took place, anthropologists have generally favored using the term "hominins"—on the theory that it identifies humanity as belonging not to the family *Hominidae* but rather to the sub-family *Homininae* and/or the tribe *Hominini*. However, since the *Hominini* includes chimpanzees and the *Homininae* includes both chimpanzees and gorillas—both of which are quadrupedal apes that are not human and are not part of the human family tree—I have decided, with apologies to my fellow anthropologists, to use the more familiar term "hominid" in this book, since it is addressed to the general reader as well as to the scholar and scientist.
3. A. L. Kroeber, *Anthropology* (New York: Harcourt, Brace and Company, 1948), 79. In the same work, Kroeber also asserts that culture "is that

which the human species has and other social species lack" (p. 253)—another "prevailing opinion" among scientists of the day that has since been completely discredited.

4. The Lomekwi 3 find consists of 149 stone artifacts showing clear evidence of having been "worked" by hand and containing characteristic fractures that do not occur from natural processes. Many of these artifacts are quite large and heavy, averaging more than six pounds each. Some of them were apparently used as hammers, others as anvils, and the purpose of others is still unknown. See Sonia Harmond, Jason E. Lewis, Craig S. Feibel, Christopher J. Lepre, Sandrine Prat, Arnaud Lenoble, Xavier Boës, Rhonda L. Quinn, Michel Brenet, Adrian Arroyo, *et al.*, "3.3-million-year-old stone tools from Lomekwi 3, West Turkana, Kenya." *Nature* 521 (2015), 310–315.
5. Even the kangaroo (which is not a placental mammal but rather a marsupial), with its specialized ability to stand comfortably on its hind legs and to run across the landscape with great speed by jumping with its hind legs, must get down on all fours when it walks. And the kangaroo is burdened by an immensely heavy tail, required to counterbalance its forequarters. Even when the kangaroo is standing erect, its spine is held not in a vertical position but rather at an angle, halfway between vertical and horizontal.
6. See P. S. Rodman and H. M. McHenry, "Bioenergetics and the Origin of Hominid Bipedalism," *American Journal of Physical Anthropology* 52, (1980), 103-106.
7. See Tim D. White et al., "*Ardipithecus ramidus* and the Paleobiology of Early Hominids," *Science* 326, no. 5949 (2009), 64, 75–86.
8. See C. Owen Lovejoy, "The Origin of Man," *Science* (new series) 211, no. 4480 (1981), 341–350.
9. See R. W. Newman, "Why Man Is Such a Sweaty and Thirsty Naked Animal: A Speculative Review," *Human Biology* 42 (1970):12–27, and Peter E. Wheeler, "The Evolution of Bipedality and Loss of Functional Body Hair in Humans," *Journal of Human Evolution* 13 (1984), 91–98 .
10. See N. Jablonski, and G. Chaplin, "Origin of Habitual Terrestrial Bipedalism in the Ancestor of the *Hominidae*," *Journal of Human Evolution* 24 (1993), 259–280.
11. See Kevin D. Hunt, "The Evolution of Human Bipedality: Ecology and Functional Morphology," *Journal of Human Evolution* 26 (1994), 183–202.
12. See Ralph L. Holloway, "Tools and Teeth: Some Speculations Regarding Canine Reduction," *American Anthropologist* 69(1) (1967): 63–67.

For a different perspective, see Sherwood Washburn and R. Ciochon, "Canine Teeth: Notes on Controversies in the Study of Human Evolution," *American Anthropologist* 76, no. 4 (1974), 765–784.

13. Gen Suwa noted that the canine teeth of Ardipithecus had become significantly reduced when compared with those of its likely ancestors. "In the hominid precursors of *Ar. ramidus*, the predominant and cardinal evolutionary innovations of the dentition were reduction of male canine size and minimization of its visual prominence . . . .The fossils now available suggest that male canine reduction was well underway by six million years ago." See Gen Suwa et al., "Paleobiological Implications of the *Ardipithecus ramidus* Dentition," *Science* 326, no. 5949 (2009), 94–99.
14. For a detailed explanation of this hypothesis addressed to a scholarly audience, see Richard L. Currier, "Canine Teeth and Lethal Weapons: Was the Fabrication of Wooden Spears and Digging Sticks by Human Ancestors Responsible for the Evolution of Bipedal Locomotion?" Available at: http://www.richardlcurrier.com/articles/canine-teeth-and-lethal-weapons.html.
15. Charles Darwin, *The Descent of Man* (New York: Penguin Books, 2007), 90–91.
16. C. Loring Brace, *The Stages of Human Evolution*, 5th ed (New York: Prentice-Hall, 1995), 130–131.
17. Graber, Robert Bates, Randall R. Skelton, Ralph M. Rowlett, Ronald Kephart, and Susan Love Brown, *Meeting Anthropology Phase to Phase: Growing Up, Spreading Out, Crowding In, Switching On* (Durham, NC: Carolina Academic Press, 2000), 90.
18. Monogamy is common among birds, because once the female has laid the eggs, either parent can perform all the remaining tasks required to care for the offspring, from sitting on the eggs to keep them warm to protecting and feeding the nesting chicks. Thus it stands to reason that two birds would generally have a better chance of successfully raising a nest of offspring than would a single bird. But monogamy is rare among mammals, including primates, among which the care and feeding of the young is primarily or exclusively the burden of the female, the giver of milk (although among a few monogamous primate species, the father tends to become the preferred beast of burden). There are several monogamous primates, including gibbons, siamangs, titi monkeys, and marmosets, but in every one of these species the primary group consists of a single breeding pair.

Humans are the only primates that form monogamous families that remain integrated within larger groups of cooperating adults.

Chapter 3

1. See Charles K. Brain and Andrew Sillen, "Evidence from the Swartkrans Cave for the Earliest Use of Fire," *Nature* 336 (1989), 464–466.
2. In *The Spark That Ignited Human Evolution* (Albuquerque: University of New Mexico Press, 2009), 173, Frances Burton, an anthropologist at the University of Toronto, writes: "It is hard to identify why the Ancestor [the earliest hominid] first approached fire. It surely must have occurred, however, somewhere between 7 and 10 million years ago." On the other hand, Steven R. James, an anthropologist at the California State University at Fullerton, wrote in 1989 that "there are no actual hearths until the appearance of the Neandertals . . . at the end of the Middle Pleistocene [approximately 130,000 years ago]. Much of the evidence [for the human use of fire] prior to this time is equivocal, and natural processes may explain it." See Steven R. James, "Hominid use of Fire in the Lower and Middle Pleistocene: A Review of the Evidence," *Current Anthropology* 30, no. 1 (1989), 1–11.
3. It is no accident that those species of non-human primates who have adapted to a terrestrial existence—including baboons, vervets, and patas monkeys—never venture far from the safety of the trees and high cliffs that they use as sleeping places when darkness falls. Even the chimpanzee, hardly a terrestrial species, always sleeps in the trees, even though, due to its large size, it must construct a sleeping platform out of tree branches that is strong enough to support its weight. Only the massive gorilla, which lives in mountain habitats where large predators are rare, and is strong enough to overpower most predators, sleeps on the ground at night.
4. The various emerging humans that have been given their own status as separate "species" to date include *Homo habilis*, *Homo ergaster*, *Homo rudolfensis*, *Homo erectus*, *Homo antecessor*, *Homo heidelbergensis*, *Homo rhodesiensis*, and *Homo floresiensis*. Some of them, such as *H. habilis* and *H. ergaster*, are very ancient and have long been extinct, while others, such as *H. floresiensis*, may have survived until as recently as twelve thousand years ago. Still others, especially *Homo heidelbergensis*, were so advanced that they may not have been emerging humans at all but rather early forms of modern humans. Precisely how all of the many fossil remains of the various emerging humans should be classified continues to be the

subject of lively debate among prehistorians and will doubtless remain so for a long time to come.

5. For a fascinating discussion of the probable antiquity of *Homo erectus,* see Christopher Wills, *Children of Prometheus: The Accelerating Pace of Human Evolution,* 164–171.
6. For a full explanation of this hypothesis, see Richard Wrangham, *Catching Fire: How Cooking Made Us Human* (New York: Basic Books, 2009), 98–102. Wrangham further elaborated this argument in a publication with Rachel Carmody in 2010. See Richard Wrangham and Rachel Carmody, "Human Adaptation to the Control of Fire," *Evolutionary Anthropology* 19 (2010), 187–199. See especially 190–191 and 196.
7. In captivity, however, the chimpanzee has demonstrated an uncanny degree of casualness in handling and using fire—complete with the ability to smoke cigarettes and cigars, a behavior that is described in detail in A. S. Brink, "The Spontaneous Fire-Controlling Reactions of Two Chimpanzee Smoking Addicts," *South African Journal of Science* 53 (1957), 241–247. One particularly precocious chimpanzee was even capable of gathering dry twigs together, lighting them with a cigarette lighter, adding more twigs to this little fire, and toasting a marshmallow. See Frances D. Burton, *Fire: The Spark That Ignited Human Evolution* (Albuquerque: University of New Mexico Press, 2009).
8. Several of the more primitive primate species, however, are nocturnal in nature. Aside from a few species of "night monkeys" (*aotidae*) that live in Central and South America, most of these nocturnal primates belong to a more ancient and more primitive suborder of primates called "prosimians" (literally, "before monkeys") that include dwarf lemurs, tarsiers, bush babies, galagos, lorises, and pottos. As a rule, the nocturnal prosimians tend to be rare, relatively small-brained, and solitary. Most of them survive in small numbers in the more remote habitats of Africa and Madagascar.
9. The release of melatonin is also governed by the normal daily pattern of wakefulness and sleep, a phenomenon known as the "circadian rhythm."
10. With apologies to the French anthropologist Claude Lévi-Strauss, who published *The Raw and the Cooked: Mythologiques, volume 1* in 1964, a classic work that explored the food mythologies of numerous societies. In it, Lévi-Strauss concluded that all human cultures distinguish between food in its natural state ("the raw") and food that has been processed by human activity ("the cooked"). He further theorized that the process

of cooking was viewed in all cultural mythologies as the crossing of the boundary that exists in the human mind between nature and culture.

11. See Francesco Berna, Paul Goldberg, Liora Kolska Horwitz, James Brink, Sharon Holt, Marion Bamford, and Michael Chazan, "Microstratigraphic Evidence of in situ Fire in the Acheulean Strata of Wonderwerk Cave, Northern Cape Province, South Africa," *Proceedings of the National Academy of Sciences of the United States of America* 109, no. 20 (2012), 1215–1220.
12. See Victoria Wobber, Brian Hare, and Richard Wrangham, "Great Apes Prefer Cooked Food," *Journal of Human Evolution* 55 (2008), 343–348.
13. The difficulties that humans encounter when they attempt to live on a diet of raw foods are described in detail in Richard Wrangham's *Catching Fire: How Cooking Made Us Human*, 15–36.
14. See Leslie C. Aiello and Peter Wheeler, "The Expensive-Tissue Hypothesis: The Brain and the Digestive System in Human and Primate Evolution," *Current Anthropology* 36, no. 2 (1995), 199–221.
15. It should be noted that Aiello and Wheeler argued that the increase in brain size between the early hominids and the emerging humans was most likely due to the addition of substantial amounts of meat in the hominid diet, and they assumed that cooking was invented much later, closer to 250,000 years ago. But Wrangham makes a convincing case that cooking was indeed responsible for the first and most dramatic increase in brain size, and his argument has been strengthened by the steady accumulation of evidence that hominids were already using fire on a regular basis well before one million years ago. See Richard Wrangham, *Catching Fire: How Cooking Made Us Human*, 96–127.
16. Just as the human embryo has a tail and other archaic structures in an early phase of its development, the human fetus may have a considerable amount of fine hair all over its body before it is born. This fetal hair, or lanugo, is typically shed in the womb several weeks before birth.
17. In *The Palaeolithic Settlement of Asia* (Cambridge: Cambridge University Press, 2009), Robin Dennell explains in detail how changes in geology and climate resulted in a huge expansion of grasslands in Asia and Africa and that by three million years ago, the grasslands of the two continents had merged into a vast belt of savannah habitats that stretched uninterrupted from West Africa to northern China. Dennell theorized that this unbroken belt of grasslands provided a pathway for the migration

of hominids north out of Africa and eastward all the way to the Pacific shores of East Asia.

Chapter 4

1. The Schöningen javelins described in chapter 2 provide one of the rare exceptions. These wooden spears, fortuitously preserved in the highly acidic environment of German peat bogs for four hundred thousand years, showed that the manufacture of wooden artifacts by emerging humans was already in an advanced state by the beginning of the Middle Paleolithic approximately three hundred thousand years ago. Yet of all the billions of wooden artifacts that were doubtless created by the hominids who roamed the earth for several million years, only these four objects, some wooden artifacts from Bilzingsleben in Germany, and a single spear point from Clacton, England, have been found that date from before fifty thousand years ago.
2. Among the rare exceptions are the decorative beads that some prehistoric humans fashioned from seashells beginning about eighty-five thousand years ago.
3. While gold, silver, and copper were used extensively throughout the Americas in pre-Columbian times, these metals were used primarily for the manufacture of jewelry and other objects that functioned as symbols of wealth and status. The techniques we usually associate with metallurgy and that characterized the technologies of the Bronze Age and Iron Age cultures in Europe and Asia—the smelting of ores, casting of metal into shapes, mixing of molten metals to create alloys, and the tempering of finished metal objects—were rudimentary in pre-Columbian America and were not central to their technologies. In fact, in some of the most advanced of these civilizations—such as the Mayan civilization of Guatemala and the Yucatán Peninsula—metal artifacts played only a marginal role and were not essential components of their military, political, or economic activities.
4. See Colin Groves and Jordi Sabater-Pi, "From Ape's Nest to Human Fix-Point," *Man* 20, no. 1 (1985), 22–47. "Anyone however slightly familiar with Great Apes in their natural environment," the article begins, "will have been struck by their elaborate nests: their ubiquity, the regularity of their construction, [and] the skill required to make them."
5. A few instances of nest-sharing between adults have been observed between consort pairs of bonobos, which occasionally share sleeping nests with their partners. But juvenile gorillas and chimpanzees as young as one year of age have been observed beginning to construct their own

nests, and juveniles older than five years of age regularly construct their own nests and use them to sleep apart from their mothers.

6. Groves and Sabater-Pi, 1985, 38.
7. This was the time period during which the inhabitants of Schöningen, less than one hundred miles to the north of Bilzingsleben, were making their famous hardwood javelins and using them for hunting wild horses.
8. For an excellent argument in support of the view that Neandertals must have used dwellings and clothing in spite of the near total lack of archeological evidence, see Mark J. White, "Things to Do in Doggerland When You're Dead: Surviving OIS3 at the Northwestern-Most Fringe of Middle Palaeolithic Europe," *World Archaeology* 38 (2006), no. 4, 547–575. For a rebuttal of the hypothesis that the stone circles of Terra Amata were in fact the remains of prehistoric habitations, see Paola Villa, *Terra Amata and The Middle Pleistocene Archaeological Record of Southern France* (Berkeley: University of California Press, 1983).
9. See Aiello and Wheeler, "Neandertal Thermoregulation and the Glacial Climate" in Tjerd van Andel and William Davies, eds., *Neandertals and Modern Humans in the European Landscape of the Last Glaciation: Archaeological Results of the Stage 3 Project*, (Cambridge: McDonald Institute for Archaeological Research, 2003).
10. Remains from Paleolithic sites in eastern Europe indicate that as long as twenty-seven thousand years ago, prehistoric people had already developed sophisticated techniques of weaving and sewing. They made cordage, knotted netting, wicker basketry, and woven and twilled textiles sewn together with bone needles. In fact, the advanced state of weaving and sewing that was already evident by twenty-seven thousand years ago indicates that these technologies had actually originated much farther back in time. Finally, many of the Venus figurines recovered from these time periods are inscribed with patterns and designs suggesting that the people of those ancient societies were customarily wearing woven, tailored clothing.
11. Since women in preindustrial societies typically bore several children during their lifetimes, it can be estimated that as many as 10 percent of all human mothers died in childbirth throughout all but the most recent decades of human history. This high rate of mortality in human childbirth still exists throughout much of the developing world, and it continues to be the norm in much of sub-Saharan Africa.

12. See Vincent Balter and Laurent Simon, "Diet and Behavior of the Saint-Césaire Neandertal Inferred from Biogeochemical Data Inversion," *Journal of Human Evolution* 51 (2006), 329–338.

Chapter 5

1. The spear-thrower or atlatl is a short stick with a cupped or hooked end that is held in the throwing hand, with the hooked end fitted into the butt end of the spear, and used to increase the effective length of the hunter's throwing arm. The use of a spear-thrower makes it possible for a spear to be hurled with significantly greater force and speed than is possible when a spear is thrown while held only in the hand. A light spear or javelin launched with a spear-thrower can attain speeds of 150 miles per hour and has enough force to pass entirely through the body of an antelope when thrown from a distance of less than twenty-five feet. Spear-throwers first appeared about thirty thousand years ago and were still in use until modern times by hunter-gatherers all over the world.
2. The term "petroglyph" was originally coined in the eighteenth century by the early investigators of cave art, who combined two Greek words: *petros*, meaning "stone," and *glyphē*, meaning "carving." The meaning of this term has since been extended to mean any man-made design on a rock or stone surface, whether carved or painted, either out in the open or on the walls of caves.
3. See Carl C. Swisher, III, W. J. Rink, S. C. Antón, H. P. Schwarcz, Garniss H. Curtis, and A. Suprijo Widiasmoro, "Latest *Homo erectus* of Java: Potential Contemporaneity With *Homo Sapiens* in Southeast Asia." *Science* vol. 274, no. 5294 (1996), 1870–1874.
4. *Homo floresiensis* stood only three feet tall and had a brain no larger than the typical chimpanzee brain. But the anatomy of its brain case indicates that its prefrontal lobes—the area of the brain devoted to conscious thought—may have been more highly developed than the corresponding regions of the brains of early hominids. This suggests that *Homo floresiensis* may have been the offshoot of a *Homo erectus* population that had become isolated on Flores Island when the sea levels rose during a warming period in the global climate. See P. T. Brown, et al., "A New Small-Bodied Hominin from the Late Pleistocene of Flores, Indonesia." *Nature* 431, no. 7012 (2004), 1055–1061. See also M .J. Morwood and W. L. Jungers, "Conclusions: Implications of the Liang Bua Excavations for Hominin Evolution and Biogeography," *Journal of Human Evolution* 57 (2009),

640–648, and Leslie C. Aiello, "Five Years of *Homo floresiensis*," *American Journal of Physical Anthropology* 142, issue 2 (2010), 167–179.

5. For a detailed discussion of how anatomically modern humans were responsible for the decline and eventual extinction of the Neandertals, see Pat Shipman, *The Invaders: How Humans and Their Dogs Drove Neanderthals to Extinction* (Cambridge, MA: Harvard University Press, 2015).

6. The *Paleolithic* or "Old Stone Age" technically refers to the entire span of time during which the hominids evolved from the earliest bipedal forms such as *Australopithecus* to the first anatomically modern humans, because the remains of stone tools were by far the most common artifacts found in prehistoric habitation sites throughout this long period.

   The Paleolithic is usually subdivided into three eras based on tool types: 1) the *Lower Paleolithic*, which begins roughly three million years ago with the manufacture of the first crude Oldowan pebble tools by the early hominids and continues throughout the 1.5-million-year period during which the emerging humans such as *Homo erectus* made their more finely worked Acheulean hand axes; 2) the *Middle Paleolithic*, which begins approximately 250,000 years ago and is generally associated with the Neandertals and their Mousterian flake tools; and 3) the *Upper Paleolithic*, which begins roughly fifty thousand years ago and corresponds to the time of the anatomically modern humans such as the Cro-Magnons, who made the finest and most varied stone tools from long, thin blades of fine-grained stone such as flint and obsidian.

   The term "Paleolithic" is useful here because it distinguishes the societies of the nomadic hunters and gatherers from those of the *Neolithic* or "New Stone Age," which begins shortly before the development of agriculture. During the Neolithic, the larger stone tools are increasingly made by grinding and polishing rather than by flaking, while the smaller stone tools made by flaking—generally called microliths—became very small and specialized. In the later years of the Neolithic, the use of pottery became widespread, and the remains of broken pottery eventually became the most common type of artifact in prehistoric habitation sites. In between these clearly defined periods, there is a somewhat vague middle ground called the *Mesolithic* or "Middle Stone Age." This is sometimes used as a catch-all term for prehistoric cultures that do not fit neatly into the categories of either Paleolithic or Neolithic.

   I have used the term "Upper Paleolithic" extensively in this chapter to refer to the period when anatomically modern humans appeared in

Europe roughly fifty thousand years ago—and who lived by nomadic hunting and gathering—in order to distinguish them from the food-producing societies of the Neolithic that first began to appear fifteen thousand years ago and that will be described in detail in chapter 6.

7. The anatomically modern humans of the Upper Paleolithic were the first hominids to show clear evidence of distinct cultural traditions. Although many different Upper Paleolithic cultures have been identified in different areas of the world, those of Europe are the best known and have been the most carefully studied. While most of these cultures overlap in time with each other, they are usually organized in a rough chronological order. The oldest of these cultures was that of the Châtelperronians, who lived roughly thirty-three thousand years ago. They were followed by the Aurignacians, Gravettians, and Solutreans. The most recent culture was that of the Magdalenians, who survived in Europe until approximately eleven thousand years ago.

8. These include the grotto of Chabot in Gard in 1878, the cave of La Mouthe near Les-Eyzies-de-Tayac in 1895, the caves of Les Combarelles and Font-de-Gaume at Les Eyzies in 1901, Marsoulas Cave in the Pyrenees in 1902, and La Calevie and Bernifal caves in the Vézère Valley in 1903.

9. I am indebted to Robert Bates Graber for this observation. See Bronislaw Malinowski, *Magic, Science, and Religion* (Garden City, NY: Doubleday & Company, 1954), 30–31.

10. *Homo heidelbergensis* was a large-brained emerging human so similar to the more advanced forms of *Homo erectus* that many scientists consider them to be essentially the same species. It was originally identified from a fossil jaw found near Heidelberg, Germany in 1907, and other remains of this hominid have since been found in Africa and western Asia. *Homo heidelbergensis* is the evolutionary branch of *Homo erectus* that is most likely to have been ancestral to the Neandertals as well as to anatomically modern humans.

11 See John Feliks, "The Graphics of Bilzingsleben: Sophistication and Subtlety in the Mind of *Homo erectus*," *Proceedings of the XV UISPP World Congress*, Oxford: BAR International Series, 2006. Available at: http://www-personal.umich.edu/~feliks/graphics-of-bilzingsleben/index.html. See also: John Feliks, "The Golden Flute of Geissenklösterle: Mathematical Evidence for a Continuity of Human Intelligence as Opposed to Evolutionary Change Through Time," *Aplimat—Journal of Applied Mathematics* 4, no. 4 (2011), 157–162.

12. This argument is set forth by Iain Davidson and William Noble in "The Archaeology of Perception: Traces of Depiction and Language." *Current Anthropology* 30, no. 2 (1989), 125–156.
13. See Leslie C. Aiello and Robin I. M. Dunbar, "Neocortex Size, Group Size, and the Evolution of Language," *Current Anthropology* 34, no. 2 (1993), 184–193.
14. From *Alice in Wonderland*, by Lewis Carroll.

Chapter 6

1. For the "oasis theory," see V. Gordon Childe, "Chapter V: The Neolithic Revolution" in *Man Makes Himself* (New York: Oxford University Press, 1936). For the "hilly flanks" theory, see Robert J. Braidwood, "The Agricultural Revolution," *Scientific American* 203 (1960), 130–148. For the "demographic" or "population pressure" theory, see Lewis R. Binford, "Post-Pleistocene Adaptations," in Sally R. Binford and Lewis R. Binford, *New Perspectives in Archaeology* (Chicago: Aldine Publishing Company, 1968), 313–342. See also Kent Flannery, "The Origins of Agriculture," *Annual Review of Anthropology* 2 (1973), 271–310, and Mark N. Cohen, *The Food Crisis in Prehistory: Overpopulation and the Origins of Agriculture* (New Haven: Yale University Press, 1977). For the "co-evolutionary" theory, see David Rindos, *The Origins of Agriculture: An Evolutionary Perspective* (Orlando, Florida: Academic Press, 1984). For the "competitive feasting" theory, see Brian Hayden, "Nimrods, Piscators, Pluckers, and Planters: The Emergence of Food Production," *Journal of Anthropological Archaeology* 9, no. 1 (1990), 31–69. For the "hospitable climate" theory, see Peter J. Richerson, Robert Boyd, and Robert L. Bettinger, "Was Agriculture Impossible during the Pleistocene but Mandatory during the Holocene?," *American Antiquity* 66, no. 3 (2001), 387–411.
2. Graeme Barker, *The Agricultural Revolution in Prehistory* (Oxford: Oxford University Press, 2006), 383.
3. The plump yellow grain that grows in large cobs on tall fleshy plants is known throughout most of the world as "maize" or "mealie" but is called "corn" in the United States, Canada, Australia, and New Zealand. By contrast, the word "corn" is used throughout most of the world for the cereal grains such as wheat, rye, and barley.
4. The horse was domesticated about six thousand years ago in the steppes, or grasslands, of present-day Kazakhstan, nearly a thousand miles north and

east of the Fertile Crescent, where it first became important as a means of transportation. The important story of the domesticated horse will be discussed at length in the next chapter, where we will see how, as one of the key elements in the technologies of interaction, the horse contributed significantly to the emergence of cities and the emergence of urban civilization.

5. Andrew Sherratt, "Climatic Cycles and Behavioural Revolutions: The Emergence of Modern Humans and the Beginning of Farming," *Antiquity* 71, 1997, 277.
6. See R. Alexander Bentley et al., "Community Differentiation and Kinship Among Europe's First Farmers," *Proceedings of the National Academic of Sciences of the United States of America* 109, no. 24 (2012), 9326–9330.
7. See Verrier Elwin, *The Kingdom of the Young* (London: Oxford University Press, 1968), 165–169; Clellan S. Ford and Frank A. Beach, *Patterns of Sexual Behavior* New York: Harper Torchbooks, 1951), 268; William A. Lessa, *Ulithi: A Micronesian Design for Living* (New York: Holt, Rinehart and Winston, 1966), 88; Bronislaw Malinowski, *The Sexual Life of Savages in North-Western Melanesia* (New York: Harcourt, Brace and Company, 1929), 198–200; and Marjorie Shostak, *Nisa: The Life and Words of a ¡Kung Woman* (New York: Vintage Books, 1983), 150–151.
8. See Melvin J. Konner, "Hunter-Gatherer Infancy and Childhood: The !Kung and Others," in Barry S. Hewlett and Michael E. Lamb, eds., *Hunter-Gatherer Childhoods: Evolutionary, Developmental and Cultural Perspectives* (Piscataway, NJ: Aldine Transaction, 2006), 19–64.
9. The purpose of these arrangements was mainly to obtain the allegiance—and often the services—of the prospective son-in-law. However, when these girls reached adulthood, they were usually free to leave their "assigned" husband in favor of some other man who was more to their liking. See Victoria K. Burbank, "Premarital Sex norms: Cultural interpretations in an Australian Aboriginal Community," *Ethos* 15, no. 2 (1987), 226–234.
10. See Napoleon Chagnon, *Yanomamö: The Fierce People,* 2nd ed. (New York: Holt, Rinehart and Winston, 1977), 119–124. See also Kenneth Good, *Into the Heart: One Man's Pursuit of Love and Knowledge among the Yanomama* (New York: Simon & Schuster, 1991), 72–74 and 194–204.

Chapter 7

1. Jared Diamond, *The World until Yesterday* (New York: Viking, 2012), 4–5.

2. In deference to the sensitivities of atheists, agnostics, and non-Christians, scientists and scholars have recently begun using the term "BCE," meaning "Before the Common (or Current) Era," in preference to the traditional "BC" or "Before Christ." For the same reason, they have also begun using the corresponding term "CE," meaning "Common (or Current) Era," instead of the more traditional "AD" or "*Anno Domini*" ("the year of our Lord"). While not meaning to take issue with the logic of this change—but for the convenience of readers who are not be familiar with these terms—I have continued using the more traditional "BC" and "AD" in this book for dates before and after the birth of Christ.
3. These were the Warring States period (476–221 BC), the Sixteen Kingdoms period (304–439 AD), the Five Dynasties and Ten Kingdoms period (907–960 AD), and the Song, Liao, Jin, and Western Xia dynasties period (960–1234 AD).
4. See Robert G. Bednarik, "Seafaring in the Pleistocene," *Cambridge Archaeological Journal* 13, no. 1 (2003), 41–66.
5. Although three hundred miles of open sea separate Timor from the Australian coast today, sea levels during the ice ages were three hundred feet lower than they are now, and large areas of the continental shelves on both sides of the Timor Strait were exposed. The distance between Australia and Timor sixty thousand years ago was approximately fifty miles of open sea.
6. See Robert G. Bednarik, "Crossing the Timor Sea by Middle Palaeolithic Raft," *Anthropos* 95 (2000):37–47, and Robert G. Bednarik, "Seafaring," in Helaine Selin Springer, ed., *Encyclopaedia of the History of Science, Technology, and Medicine in Non-Western Cultures,* 2nd ed., 2008; and Robert G. Bednarik, "The Beginnings of Maritime Travel," *Advances in Anthropology* 4 (2014):209–221.
7. With apologies to David W. Anthony, whose exhaustive study *The Horse, the Wheel, and Language* has illuminated the pivotal role played by the domesticated horse in the histories of both ancient and modern civilizations.
8. See David W. Anthony, *The Horse, the Wheel, and Language: How Bronze-Age Riders from the Eurasian Steppes Shaped the Modern World* (Princeton: Princeton University Press, 2007).
9. See Robin I. M. Dunbar, "Neocortex Size as a Constraint on Group Size in Primates," *Journal of Human Evolution* 20 (1992), 469–493.

Chapter 8

1. Daniel J. Boorstin, *The Discoverers: A History of Man's Search to Know His World and Himself* (New York: Vintage Books, 1985), 62–63.
2. Peter Hutchinson, "Magazine Growth in the Nineteenth Century," *A Publisher's History of American Magazines* (2008), 1. Available at: http://www.themagazinist.com/Magazine_History.html.
3. J. V. Thirgood, "The Historical Significance of Oak," in *Oak Symposium Proceedings: 1971 August 16–20* (Upper Darby, PA: US Department of Agriculture, Forest Service, Northeastern Forest Experiment Station), 9.
4. Ibid., 10.
5. Nicéphore Niépce is better known as one of the first inventors of photography. His "View from the Window at Le Gras," taken in 1827, is the world's oldest surviving photograph.
6. Kenneth Chase, *Firearms: A Global History to 1700.* (Cambridge: Cambridge University Press, 2003), 25.
7. Warren D. Devine, "From Shafts to Wires: Historical Perspective on Electrification," *The Journal of Economic History* 43 (1983), 349.
8. Perhaps this is why many behavioral scientists in the twentieth century embraced the preposterous idea that humans are the only animals capable of reason and emotion—as if such mental activities had emerged full-blown for the first time in the history of life on Earth with the appearance of *Homo sapiens.* Such views—which were never supported by credible scientific evidence—could only have been taken seriously by someone who had never lived with an intelligent animal such as a dog, cat, horse, pig, monkey, or parrot. People who have lived intimately with such animals are well aware that they are capable of both reason and a wide range of emotions.
9. In 1935, there were nearly seven million farms in the United States. By the year 2007, the number of farms had declined to 2.2 million. In that year, the largest 188,000 farms accounted for 63 percent of all agricultural products sold, while the smallest 2,012,000 farms accounted for only 37 percent of all agricultural products sold (see US Department of Agriculture, "2007 Census of Agriculture").

Chapter 9

1. See Jonathan Fildes, "Campaign builds to construct Babbage Analytical Engine," *BBC News*, October 14, 2010, and John Graham-Cumming,

"Let's Build Babbage's Ultimate Mechanical Computer," *New Scientist* 2791, December 23, 2010.

2. See M. H. Weik, "Computers with Names Starting with E through H," *A Survey of Domestic Electronic Digital Computing Systems,* 1955.
3. Moore's exact words in 1965 were as follows: "The complexity for minimum component costs has increased at a rate of roughly a factor of two per year (see graph). Certainly over the short term this rate can be expected to continue, if not to increase. Over the longer term, the rate of increase is a bit more uncertain, although there is no reason to believe it will not remain nearly constant for at least ten years. That means by 1975, the number of components per integrated circuit for minimum cost will be 65000." Gordon E. Moore, "Cramming More Components onto Integrated Circuits." *Electronics* 38, no. 8 (1965), 114–117.
4. ENIAC contained 17,468 vacuum tubes, each of which was equivalent to a modern transistor. Since the 15-Core Xeon E7 v2 microprocessor contains 4.31 billion transistors, it would require the equivalent of 4,310,000,000 divided by 17,468 or 246,737 ENIACs to equal the processing power of one 15-Core Xeon E7 v2 microprocessor. Each ENIAC was 100 feet long, thus 246,737 ENIACs laid end to end would measure 24,673,700 feet or 4,673 miles in length. The weight of each ENIAC was thirty tons or 60,000 pounds. Thus, 246,737 ENIACs would weigh 7,402,110 tons—equal to the weight of sixty-six Nimitz-class supercarriers weighing 112,000 tons each. The ENIAC cost $500,000 to construct, which in 2014 dollars amounts to approximately $6 million. Thus, 246,737 ENIACs would have cost $1,480,422,000,000 to build in 2014.
5. Since the best vacuum tube computers of the 1950s averaged a maximum consecutive running time of ten hours (or thirty-six thousand seconds) before experiencing a vacuum tube failure, a computer as large as 246,737 ENIACs would probably run for only thirty-six thousand seconds divided by 246,737, or one-seventh of a second, before one of its 4,310,000,000 vacuum tubes would fail.
6. See Jaxon Van Derbeken, Demian Bulwa, and Erin Allday, "SF Plane Crash: Crew Tried to Abort Landing," *San Francisco Chronicle*, July 8, 2013.
7. See *International Technology Roadmap for Semiconductors, 2010, Overall Technology Roadmap Characteristics. 2010 Update*, 8–14.

8. See John Hultsman and William Harper, "The Problem of Leisure Reconsidered," in *Journal of American Culture* 16, issue 1 (2004), 47–54.
9. See Edgardo Sica, "International Tourism: A Driving Force for Economic Growth of Commonwealth Countries," *The Commonwealth Finance Ministers Meeting 2007.*
10. United Nations, Department of Economic and Social Affairs, Population Division, *International Migration Report 2013*, 11–17.
11. See Lewis et al., eds., *Ethnologue: Languages of the World,* 17th ed., 2014.

Chapter 10

1. Frank Borman, *Countdown: An Autobiography* (New York: Silver Arrow Books, 1988), 212.
2. Nancy Atkinson, "A Conversation with Jim Lovell, Part 2: Looking Back," *UniverseToday.com*, September 27, 2010. Available at: http://www.universetoday.com/74396/a-conversation-with-jim-lovell-part-2-looking-back/. Accessed 6/12/14.
3. Due to their small size and immense distance from Earth, no planet outside of our own solar system has ever been observed directly. Instead, their existence is inferred from "wobbles" in the stars themselves, caused by the gravitational pull of the planets revolving around them.
4. Light travels at the speed of 670,616,629 miles per hour. Multiplied by 24 hours in a day, this is 16,094,799,096 miles per day. Multiplied by 365 days in a year, this is 5,874,601,670,040 miles per year. Sixteen light-years is thus 93,993,626,720,640 miles. At 450,000 miles per hour, it would require 208,874,726 hours, which equals 8,703,114 days or 23,844 years to cover the distance from the earth to Gliese 832 c.
5. The term "biosphere" was coined by the Austrian geologist and paleontologist Eduard Suess in 1875 to describe the total mass of living things that inhabit the surface of the earth. Since the environments that support life are all found only on the earth's surface, the physical mass of living organisms on our planet takes the shape of a sphere—hence the term "biosphere." The same logic was used in the 1600s, when the Greek word *atmos,* meaning "vapor," was combined with the earth's spherical shape to create the word "atmosphere."
6. Biosphere 1 was not an earlier version of this experiment but was actually the Biospherians' name for the earth's natural ecosystem, or biosphere.

7. *Paratrechina longicornus,* the "longhorn crazy ant," is one of the most common species of ants and is found in human habitations throughout the world. Its name is derived from its long antennae and its habit of running erratically at high speeds in all directions.

8. Joel E. Cohen and David Tilman, "Biosphere 2 and Biodiversity: The Lessons So Far," *Science* 274, no. 5290, November 15, 1996, 1151.

9. Although Israel has never publicly acknowledged the existence of its nuclear weapons program, it has never denied its existence either. In 2002, Robert S. Norris and his colleagues described the Israeli nuclear program as follows:

   > After Egyptian President Gamal Abdel Nasser closed the Straits of Tiran in 1953, Israeli Prime Minister David Ben Gurion began development of nuclear weapons and other unconventional munitions. His protégé, Shimon Peres, played a central role in securing an agreement with France in 1956 for a nuclear research reactor. Physicist Ernst David Bergmann, director of the Israeli Atomic Energy Commission, provided early scientific direction. . . . With French assistance, Israel built a nuclear weapons facility at Dimona in the Negev desert. The Dimona site has a plutonium/tritium production reactor, an underground chemical separation plant, and nuclear component fabrication facilities.

   "[I]t is generally accepted by friend and foe alike," the authors concluded, "that Israel has been a nuclear state for several decades." See Robert S. Norris, William M. Arkin, Hans M. Kristensen, and Joshua Handler, "Israeli Nuclear Forces, 2002," *Bulletin of the Atomic Scientists* 58, no. 5 (2002), 72.

10. A summary of world nuclear weapons stockpiles is available at: http://www.ploughshares.org/world-nuclear-stockpile-report. Accessed 6/3/14.

11. In 1976, there were 102 Stage 1 smog alerts in the Los Angeles Basin, while in 1998 there were only twelve Stage 1 alerts. See "Pollution in Los Angeles County," *RabbitAir,* 2014. Available at: http://www.rabbitair.com/pages/pollution-in-los-angeles-county. Accessed 6/18/2014.

12. See William Laurance, "China's Appetite for Wood Takes a Heavy Toll on Forests," *Yale Environment 360*, November 17, 2011.

13. Huang Wenbin and Sun Xiufang, "Tropical Hardwood Flows in China: Case Studies of Rosewood and Okoumé," *Forest Trends*, December 2013, 5.

14. For an eloquent description of the *Białowieża Puszcza* (pronounced "bialoVIESKa PUSHta"), see chapter 1, "A Lingering Scent of Eden," in *The World Without Us* by Alan Weisman, 9–16.

15. See Edward O. Wilson, *The Diversity of Life* and Niles Eldredge, *Life in the Balance: Humanity and the Biodiversity Crisis.* For an excellent article explaining the difficulty of accurately estimating the rates of extinction, see Vânia Proença and Henrique Miguel Pereira, "Comparing Extinction Rates: Past, Present, and Future," in *Encyclopedia of Biodiversity,* vol. 2, 167–176.
16. The epidemic of infestation in frog populations has been caused by several strains of the fungus *Batrachochytrium dendrobatidis* or "BD," and is especially troubling since it has affected frogs on every continent and has already been linked to the extinction of dozens of species. Nevertheless, biologists have reported limited success in helping some species of frogs to develop an immunity to BD. See Carl Zimmer, "Hope for Frogs in Face of a Deadly Fungus," *New York Times* July 9, 2014.
17. The annual extent of global carbon emissions have been painstakingly compiled by the Carbon Dioxide Information Analysis Center at the Oak Ridge National Laboratory. For details on how this information is compiled, see G. Marland, T. A. Boden, and R. J. Andres, "Global, Regional, and National Fossil Fuel CO2 Emissions," in *Trends: A Compendium of Data on Global Change* (Oak Ridge, Tennessee: Carbon Dioxide Information Analysis Center, Oak Ridge National Laboratory, US Department of Energy). Available at: http://cdiac.ornl.gov/trends/emis/overview. Detailed annual emissions data since 1751 is available at: http://cdiac.ornl.gov/trends/emis/tre_glob.html.
18. Marine isotope stages (MIS), also known as "oxygen isotope stages (OIS)," refer to periods in the earth's geological history when global temperatures became either warmer or cooler than they had been during the preceding time period. These stages are determined by measuring the ratio of two isotopes of oxygen—oxygen-16 and oxygen-18—in the shells of marine organisms that were buried at different geologic times in the sea floor. As the oceans become warmer, the proportion of oxygen-18 in the shells decreases; as the oceans become cooler, the proportion of oxygen-18 increases. There have been twenty-two of these stages during the past million years, reflecting the alternating states of global warming and global cooling as the ice ages have advanced and retreated. Odd-numbered stages, including MIS 1 (the past eleven thousand years) and MIS 11 represent the warm periods, while even numbered stages represent cold periods, including all of the major ice ages that have occurred over the past million years.

19. See William R. Howard, "Palaeoclimatology: A Warm Future in the Past," in *Nature* 388 (1997), 418–419, and Alberto V. Reyes, Anders E. Carlson, Brian L. Beard, Robert G. Hatfield, Joseph S. Stoner, Kelsey Winsor, Bethany Welke, and David J. Ullman, "South Greenland Ice-Sheet Collapse during Marine Isotope Stage 11," *Nature* 510 (2014), 525–528.
20. Climate scientists are still struggling to understand exactly why the several most recent ice ages have begun with uncanny regularity every one hundred thousand years, but the prevailing consensus is that this phenomenon is related to irregularities in the earth's orbit around the sun. See John Imbrie, A. Berger, E. A. Boyle, S. C. Clemens, A. Duffy, W. R. Howard, G. Kukla, J. Kutzbach, D. G. Martinson, A. Mcintyre et al., "On the Structure and Origin of Major Glaciation Cycles: 2. The 100,000-Year Cycle," *Paleoceanography* 8, no. 6 (1993), 699–735.
21. See Toby Tyrrell, John G. Shepherd, and Stephanie Castle, "The Long-Term Legacy of Fossil Fuels," in *Tellus B,* vol. 59 (2007), 664672, and also Fred Hoyle and Chandra Wickramasinghe, "On the Cause of Ice-Ages," *Cambridge-Conference Network*, July 1999. Available at: http://abob.libs.uga.edu/bobk/ccc/ce120799.html.
22. See United Nations Department of Economic and Social Affairs, Population Division, *World Population Prospects: The 2012 Revision* (New York: United Nations, 2014).
23. The exact number of nation-states is not only frequently changing but is also subject to interpretation. For example, thirty new nations came into being in the seven years between 1956 and 1963, when the colonies of Africa achieved independence from Europe and became independent nations. And the total number of nation-states increased again by five during the 1990s, when a single nation-state, the former Republic of Yugoslavia, fissioned into six independent nations: Serbia, Croatia, Slovenia, Macedonia, Montenegro, and Bosnia-Herzegovina. Moreover, it can sometimes be difficult to determine whether a given political and geographical entity is in fact a nation-state in its own right. For example, is Palestine a nation—as the Palestinians themselves asserted in their Declaration of Independence in 1988—or is it a United Nations mandate without any status as an independent nation, as the Israeli government currently maintains? There is no consensus on this point.

24. See Philip G. Roeder, *Where Nation-States Come From: Institutional Change in the Age of Nationalism* (Princeton: Princeton University Press, 2007), 10.

25. Of the 150 Nobel Prizes awarded in physics, chemistry, and medicine during the fifty years beginning in 1964 and ending in 2013, eighty-two, or 55 percent, were awarded to international teams. See Wikipedia, *List of Nobel Laureates,* 2014.

Bibliography

Agustí, Jordi, and David Lordkipanidze. "How 'African' Was the Early Human Dispersal Out of Africa?" *Quaternary Science Reviews* 30, no. 11–12 (2011): 1338–1342.

Aiello, Leslie C. "Five Years of *Homo floresiensis." American Journal of Physical Anthropology* 142, Issue 2 (2010), 167-179.

Aiello, Leslie C., and Peter Wheeler. "The Expensive-Tissue Hypothesis: The Brain and the Digestive System in Human and Primate Evolution." *Current Anthropology* 36, no. 2 (1995): 199–221.

———. "Neanderthal Thermoregulation and the Glacial Climate." *Neanderthals and Modern Humans in the European Landscape of the Last Glaciation: Archaeological Results of the Stage 3 Project*, Tjerd van Andel and William Davies, eds. Cambridge: McDonald Institute for Archaeological Research, 2003.

Aiello, Leslie C., and R. I. M. Dunbar. "Neocortex Size, Group Size, and the Evolution of Language." *Current Anthropology* 34, no. 2 (1993): 184–193.

Albert, Rosa M., Ofer Bar-Yosef, Liliane Meignen, and Steve Weiner, eds. "Quantitative Phytolith Study of Hearths from the Natufian and Middle Palaeolithic Levels of Hayonim Cave (Galilee, Israel)." *Journal of Archaeological Science* 30, no. 4 (2003): 461–480.

Alfred, Randy. "Univac gets election right, but CBS balks." *Wired*, November 4, 2008. http://www.wired.com/2010/11/1104cbs-tv-univac-election (accessed June 12, 2014).

Algaze, Guillermo. "Initial Social Complexity in Southwestern Asia: The Mesopotamian Advantage." *Current Anthropology* 42, no. 2 (2001): 199–233.

Allen, John, and Mark Nelson. "Biospherics and Biosphere 2, Mission One (1991–1993)." *Ecological Engineering* 13 (1999): 15–29.

Allen, Robert C. "Agriculture and the Origins of the State in Ancient Egypt." *Explorations in Economic History* 34, no. 2 (1997): 135–154.

Alperson-Afil, Nira. "Continual Fire-Making by Hominins at Gesher Benot Ya'aqov, Israel." *Quaternary Science Reviews* 27 (2008): 1733–1739.

Amsler, Sylvia J. "Ranging Behavior and Territoriality in Chimpanzees at Ngogo, Kibale National Park, Uganda." PhD diss., University of Michigan, 2009.

Anderson, Stephen R. *How Many Languages Are There in the World?* Linguistic Society of America Brochure Series. Washington, DC: Linguistic Society of America.

Anthony, David W. *The Horse, the Wheel, and Language: How Bronze-Age Riders from the Eurasian Steppes Shaped the Modern World.* Princeton, NJ: Princeton University Press, 2007.

Arensburg, B., L. A., Schepartz, A. M. Tillier, B. Vandermeersch, and Y. Rak. "A Reappraisal of the Anatomical Basis for Speech in Middle Palaeolithic Hominids." *American Journal of Physical Anthropology* 83, no 2 (2005): 137–146.

Arkush, Elizabeth N. *Hillforts of the Ancient Andes: Colla Warfare, Society, and Landscape.* Gainesville, FL: University Press of Florida, 2011.

Armstrong, John A. *Nations Before Nationalism.* Chapel Hill, NC: University of North Carolina Press, 1982.

Ashton, Nick, Simon G. Lewis, Simon A. Parfitt, Kirsty E.H. Penkman, and G. Russell Coope. "New Evidence for Complex Climate Change in MIS 11 from Hoxne, Suffolk, UK." *Quaternary Science Review* 27, no. 7–8 (2008): 652–668.

Atkinson, Nancy. "A Conversation with Jim Lovell, Part 2: Looking Back." *Universe Today*, September 27, 2010. http://www.universetoday.com/74396/a-conversation-with-jim-lovell-part-2-looking-back (accessed June 6, 2012).

Ballard, Chris, Richard Bradley, Lise Nordenborg Myhre, and Meredith Wilson. "The Ship as Symbol in the Prehistory of Scandinavia and Southeast Asia." *World Archaeology* 35, no. 3 (2003): 385–403.

Balter, Michael. "Candidate Human Ancestor from South Africa Sparks Praise and Debate." *Science* 328 (2010): 154–155.

Balter, Vincent and Laurent Simon. "Diet and Behavior of the Saint-Césaire Neanderthal Inferred from Biogeochemical Data Inversion." *Journal of Human Evolution* 51 (2006): 329–338.

Barker, Graeme. *The Agricultural Revolution in Prehistory*. New York: Oxford University Press, 2006.

Bar-Yosef, Ofer. "The Natufian Culture in the Levant, Threshold to the Origins of Agriculture." *Evolutionary Anthropology* 6, no. 5 (1998): 159–177.

Barnosky, Anthony D., Nicholas Matzke, Susumu Tomiya, Guinevere O. U. Wogan, Brian Swartz, Tiago B. Quental, Charles Marshall, et al. "Has the Earth's Sixth Mass Extinction Already Arrived?" *Nature* 471, no 7336 (2011): 51–57.

Beaver, S. H. "Coke Manufacture in Great Britain: A Study in Industrial Geography." *Transactions and Papers (Institute of British Geographers)* 17 (1951): 133–148.

Beck, Benjamin B. *Animal Tool Behavior: The Use and Manufacture of Tools by Animals*. New York: Garland Press, 1980.

Bednarik, Robert G., 1997. "The Earliest Evidence of Ocean Navigation." *International Journal of Nautical Archaeology* 26, no. 3 (1997): 183–191.

———. "The 'Australopithecine' Cobble From Makapansgat, South Africa." *South African Archaeological Bulletin* 53 (1998): 4–8.

———. "Beads and Pendants of the Pleistocene." *Anthropos* 96 (2001): 545–555.

———. "The Beginnings of Maritime Travel." *Advances in Anthropology* 4 (2014): 209–221.

———. "Crossing the Timor Sea by Middle Palaeolithic Raft." *Anthropos* 95 (2000): 37–47.

———. "Middle Pleistocene Beads and Symbolism." *Anthropos* 100 (2005): 537–552.

———. "Seafaring in the Pleistocene." *Cambridge Archaeological Journal* 13, no. 1 (2003): 41–66.

———. "Seafaring." *Encyclopedia of the History of Science, Technology, and Medicine in Non-Western Cultures,* 2nd Edition, ed. Helaine Selin. Springer, 2008.

Belfer-Cohen, Anna. "The Natufian Graveyard in Hayonim Cave." *Paléorient* 14, no. 2 (1988): 297–308.

Belfer-Cohen, Anna, and Ofer Bar-Yosef. "Early Sedentism in the Near East: A Bumpy Ride to Village Life," in *Life in Neolithic Farming Communities: Social Organization, Identity, and Differentiation,* ed. Ian Kuijt. New York: Plenum Publishers, 17–37.

Bellomo, Randy V. "Methods of Determining Early Hominid Behavioral Activities Associated with the Controlled Use of Fire at FxJj 20 Main, Koobi Fora, Kenya." *Journal of Human Evolution* 27, no. 1–3 (1994): 173–195.

Bellwood, Peter, and Marc Oxenham. "The Expansions of Farming Societies and the Role of the Neolithic Demographic Transition," in *The Neolithic*

Demographic Transition and its Consequences. New York: Springer, 2008, 13–34.

Bentley, Jerry H. *Old World Encounters: Cross-Cultural Contacts and Exchanges in Pre-Modern Times*. Oxford: Oxford University Press, 1993.

Bentley, R. Alexander, Penny Bickle, Linda Fibiger, Geoff M. Nowell, Christopher W. Dale, Robert E. M. Hedges, Julie Hamilton, et al. "Community Differentiation and Kinship among Europe's First Farmers." *Proceedings of the National Academic of Sciences of the United States of America* 109, no. 24 (2012): 9326–9330.

Benton, Adam. "What Was Neanderthal Clothing Like?" *EvoAnth,* November 13, 2012. http://www.evoanth.net/2012/11/13/what-did-neanderthals-wear/ (accessed December 10, 2013).

Berbesque, J. Colette, Frank W. Marlowe, Peter Shaw, and Peter Thompson. "Hunter–gatherers Have Less Famine Than Agriculturalists." *Biology Letters* 10, no. 1 (2014).

Berger, Lee R., Darryl J. deRuiter, Steven E. Churchill, Peter Schmid, Kristian J. Carlson, Paul H. G. M. Dirks, and Job M. Kibii. "Australopithecus Sediba: A New Species of Homo-Like Australopith from South Africa." *Science* 328 (2010): 195–204.

Berlin, Leslie. *The Man Behind the Microchip: Robert Noyce and the Invention of Silicon Valley*. New York: Oxford University Press, 2005.

Berna, Francesco, Paul Goldberg, Liora Kolska Horwitz, James Brink, Sharon Holt, Marion Bamford, and Michael Chazan. "Microstratigraphic Evidence of *In Situ* Fire in the Acheulean Strata of Wonderwerk Cave, Northern Cape Province, South Africa." *Proceedings of the National Academy of Sciences* 109, no. 20 (2012): 1215–1220.

Bicho, Nuno, Antonio F. Carvalho, Cesar González-Sainz, Jose Luis Sanchidrián, Valentín Villaverde, and Lawrence G. Straus. "The Upper Paleolithic Rock Art of Iberia." *Journal of Archaeological Method and Theory* 14, no. 1 (2007): 81–151.

Binford, Lewis R. "Post-Pleistocene Adaptations" in *New Perspectives in Archaeology,* ed. Sally R. Binford and Lewis R. Binford. Chicago: Aldine Publishing Company, 1968.

Bird, Michael. "Fire, Prehistoric Humanity, and the Environment." *Interdisciplinary Science Reviews* 20, no. 2 (1995): 141–154.

Bjorklund, David F. "The Role of Immaturity in Human Development." *Psychological Bulletin* 122, no. 2 (1997): 153–169.

Black, Brian. "Oil Creek's Industrial Apparatus: Re-Creating the Industrial Process Through the Landscape of Pennsylvania's Oil Boom." *Environmental History* 3, no. 2 (1998): 210–229.

Bocquet-Appel, Jean-Pierre. "The Demographic Impact of the Agricultural System in Human History." *Current Anthropology* 50, no. 5 (2009): 657–660.

———. "Paleoanthropological Traces of a Neolithic Demographic Transition." *Current Anthropology* 43, no. 4 (2002): 637–650.

———. "Testing the Hypothesis of a Worldwide Neolithic Demographic Transition: Corroboration from American Cemeteries." *Current Anthropology* 47, no. 2 (2006): 341–365.

———. "When the World's Population Took Off: The Springboard of the Neolithic Demographic Transition." *Science* 333, no. 6042 (2011): 560–561.

Boesch, Christophe, and Hedwige Boesch. "Hunting Behavior of Wild Chimpanzees in the Tai National Park." *American Journal of Physical Anthropology* 78: 547–573.

———. "Optimisation of Nut-Cracking with Natural Hammers by Wild Chimpanzees." *Behaviour* 83, no. 3–4 (1983): 265–286.

———. "Tool Use and Tool Making in Wild Chimpanzees." *Folia Primatologica* 54 (1990): 86–99.

Boesch, Christophe, and Hedwige Boesch-Achermann. *Chimpanzees of the Tai Forest: Behavioral Ecology and Evolution.* New York: Oxford University Press, 2000.

Boesch, Christophe and Michael Tomasello. "Chimpanzee and Human Cultures." *Current Anthropology* 39, no. 5 (1998): 591–614.

Boorstin, Daniel J. *The Discoverers: A History of Man's Search to Know His World and Himself.* New York: Vintage Books, 1985.

Borman, Frank. *Countdown: An Autobiography.* New York: Silver Arrow Books, 1988.

Brace, C. Loring. *The Stages of Human Evolution.* 5th ed. Upper Saddle River, NJ: Prentice Hall, 1995.

Braidwood, Robert J. "The Agricultural Revolution." *Scientific American* 203 (1960): 130–148.

Brain, Charles K. "Raymond Dart and our African origins," in *A Century of Nature: Twenty-One Discoveries that Changed Science and the World,* Laura Garwin and Tim Lincoln, eds. Chicago: University of Chicago Press, 3–9.

Brain, Charles K., and A. Sillen. "Evidence from the Swartkrans Cave for the Earliest Use of Fire." *Nature* 336 (1989): 464.

Brainard, George C., John P. Hanifin, Jeffrey M. Greeson, Brenda Byrne, Gena Glickman, Edward Gerner, and Mark D. Rollag. "Action Spectrum for Melatonin Regulation in Humans: Evidence for a Novel Circadian Photoreceptor." *Journal of Neuroscience* 21, no. 1 (2001): 6405–6412.

Brantingham, P. Jeffrey. Review of *The Palaeolithic Settlement of Asia* by Robin Dennell. *Geoarchaeology: An International Journal* 25, no. 5 (2009): 668–670.

Brink, A. S. "The Spontaneous Fire-Controlling Reactions of Two Chimpanzee Smoking Addicts." *South African Journal of Science* 53 (1957): 241–247.

Broad, William J. "Paradise Lost: Biosphere Retooled as Atmospheric Nightmare." *New York Times,* November 19, 1996. http://www.nytimes.com/1996/11/19/science/paradise-lost-biosphere-retooled-as-atmospheric-nightmare.html (accessed June 12, 2014).

Brown, Kyle S., Curtis W. Marean, Andy I. R. Herries, Zenobia Jacobs, Chantal Tribolo, David Braun, David L. Roberts, Michael C. Meyer, and Jocelyn Bernatchez. "Fire as an Engineering Tool of Early Modern Humans." *Science* 35, no. 5942 (2009): 859–862.

Brown, P. T. Sutikna, M. J. Morwood, R. P. Soejono, Jatmiko, E. Wayhu Saptomo, and Rokus Awe Due. "A New Small-Bodied Hominin from the Late Pleistocene of Flores, Indonesia." *Nature* 431, no. 7012 (2004): 1055–1061.

Bryce, Trevor. "The Last Days of Hattusa: The Mysterious Collapse of the Hittite Empire." *Archaeology Odyssey* 8, no. 1 (2005).

Bulfinch, Thomas. *The Age of Fable, or Stories of Gods and Heroes.* Boston: Sanborn, Carter, and Bazin, 1856.

Bunn, Henry T., Ellen M. Kroll, Stanley H. Ambrose, Anna K. Behrensmeyer, Lewis R. Binford, Robert J. Blumenschine, Richard G. Klein, Henry M. McHenry, Christopher J. O'Brien, and J. J. Wymer. "Systematic Butchery by Plio/Pleistocene Hominids at Olduvai Gorge, Tanzania." *Current Anthropology* 27, no. 5 (1986): 431–452.

Burbank, Victoria K. "Premarital Sex Norms: Cultural Interpretations in an Australian Aboriginal Community." *Ethos* 15, no. 2 (1987): 226–234.

Buringh, Eltjo, and Jan Luiten Van Zanden. "Charting the 'Rise of the West': Manuscripts and Printed Books in Europe: Long-Term Perspective from the Sixth Through Eighteenth Centuries." *The Journal of Economic History* 69, no. 2 (2009): 409–445.

Burton, Frances D. *Fire: The Spark That Ignited Human Evolution.* Albuquerque: University of New Mexico Press, 2009.

Burunat, Enrique. "Love Is the Cause of Human Evolution." *Advances in Anthropology* 4 (2014): 99–116.

Buss, David M. *The Handbook of Evolutionary Psychology.* New York: John Wiley & Sons, 2005.

Calvin, William H. "Hand-Ax Heaven: The Ambitious Ape's Guide to a Bigger Brain," in *The Ascent of Mind: Ice Age Climates and the Evolution of Intelligence*, chapter 8. New York: Bantam Books, 1990.

Cantor, Norman F. *The Civilization of the Middle Ages: A Completely Revised and Expanded Edition of Medieval History, the Life and Death of a Civilization*. New York: HarperCollins, 1994.

Carmody, Rachel N., and Richard W. Wrangham. "The Energetic Significance of Cooking." *Journal of Human Evolution* 57 (2009): 379–391.

Carneiro, Robert L. "A Theory of the Origin of the State." *Science* 169, no. 3947 (1970): 733–738.

Carroll, Lewis. *Alice's Adventures in Wonderland and Through the Looking Glass*. London: The Folio Society, 1961.

Casals, Pablo, and Josep M. Corredor. *Conversations with Casals*. New York: E.P. Dutton, 1957.

Casson, Lionel. *Travel in the Ancient World*. Baltimore: Johns Hopkins University Press, 1994.

Center for the Advancement of the Steady State Economy. *Discover the Steady State Economy*: http://steadystate.org (accessed June 6, 2014).

Cerpa, Juan Antonio. "Altamira, un calvario para Marcelino Sanz de Sautuola." *Red Española de Historia y Arqueología*. http://www.historiayarqueologia.com/profiles/blog/show?id=3814916%3ABlogPost%3A295493&commentId=3814916%3AComment%3A295461&xg_source=activity (accessed February 6, 2014).

Ceruzzi, Paul. *Computing: A Concise History*. Cambridge: MIT Press, 2012.

César, González Sainz, and Roberto Cacho Toca. "Paleolithic Cave Arts in Cantabria." *MUSE Digital Archiving Frontiers*. http://www.muse.or.jp/spain/eng/cantabria/cantabria_top.html (accessed February 10, 2014).

Chagnon, Napoleon. *Yanomamö: The Fierce People*. New York: Holt, Rinehart and Winston, 1968.

Chandler, Alfred D., and Bruce Mazlish, eds. *Leviathan: Multinational Corporations and the New Global History*. Cambridge: Cambridge University Press, 2005.

Chapais, Bernard. *Primeval Kinship: How Pair-Bonding Gave Birth to Human Society*. Cambridge: Harvard University Press, 2008.

Charles, J. A. "Early Arsenical Bronzes: A Metallurgical View." *American Journal of Archaeology* 71, no. 1 (1967): 21–26.

Chase, Kenneth. *Firearms: A Global History to 1700*. Cambridge, UK: Cambridge University Press, 2003.

Chazine, Jean-Michel. "Rock Art, Burials, and Habitations: Caves in East Kalimantan." *Asian Perspective* 44, no. 1 (2005): 219–230.

Childe, V. Gordon. "Chapter V: The Neolithic Revolution," in *Man Makes Himself.* Oxford: Oxford University Press, 1936.

Chu, Wei. "A Functional Approach to Paleolithic Open-Air Habitation Structures." *World Archaeology* 41, no. 3 (2009): 348–362.

———. "The Use of Dwellings During the Middle Paleolithic in Northern Europe." *Paleoanthropology Society Meeting Abstracts* 13–14 (2010).

Clark, J. Desmond, and J. W. K. Harris. "Fire and its Roles in Early Hominid Lifeways." *The African Archaeological Review* 3 (1985): 3–27.

Clayton, Brian. "The Incredible Shrinking Computer: Innovative Technology, Less Hardware." *Peer to Peer* 27, no. 4 (2011): 32–35.

Cohen, Avner. "Israel and Chemical/Biological Weapons: History, Deterrence, and Arms Control." *The Nonproliferation Review* 8, no. 33 (2001): 27–53.

Cohen, Joel. *How Many People Can the Earth Support?* New York: Norton, 1995.

Cohen, Joel E., and David Tilman. "Biosphere 2 and Biodiversity: The Lessons So Far." *Science* 274, no. 5290 (1996): 1150–1151.

Cohen, Mark N. *The Food Crisis in Prehistory: Overpopulation and the Origins of Agriculture.* New Haven: Yale University Press, 1977.

Collier, Bruce, and James MacLachlan. *Charles Babbage and the Engines of Perfection.* Oxford: Oxford University Press, 1998.

Collier, Paul. *Exodus: Immigration and Multiculturalism in the 21st Century.* London: Allen Lane, 2013.

Colomer, Josep M. "On Building the American and the European Empires." *London School of Economics and Political Science 'Europe in Question' Discussion Paper Series* 6 (2009): 1–29.

Computer History Museum. "A Brief History." *The Babbage Engine.* http://www.computerhistory.org/babbage/history/ (accessed June 12, 2014).

Conard, N. J., M. Malina, and S. C. Munzel. "New Flutes Document the Earliest Musical Tradition in Southwestern Germany." *Nature* 460, no. 7256 (2009): 737–740.

Connor, Walker. "A Nation Is a Nation, Is a State, Is an Ethnic Group, Is a..." *Ethnic and Racial Studies* 1, no. 4 (1978): 379–388.

Currier, Richard L. "Canine Teeth and Lethal Weapons: Was the Fabrication of Wooden Spears and Digging Sticks by Human Ancestors Responsible for the Evolution of Bipedal Locomotion?" Available at: http://www.richardlcurrier.com/articles/canine-teeth-and-lethal-weapons.html.

Cutress, Ian. "Intel Readying 15-core Xeon E7 v2." *AnandTech*, February 11, 2014. http://www.anandtech.com/show/7753/intel-readying-15core-xeon-e7-v2 (accessed June 6, 2014).

Dart, Raymond A. "Australopithecus Africanus: The Man-Ape of South Africa." *Nature* 115 (1925): 195–199.

Darwin, Charles. *The Descent of Man.* New York: Penguin Books, 2007.

Davidson, Iain, and William Noble. "The Archaeology of Perception: Traces of Depiction and Language." *Current Anthropology* 30, no. 2 (1989): 125–156.

Davis, Simon J. M. "The Age Profile of Gazelles Predated by Ancient Man in Israel: Possible Evidence for a Shift from Seasonality to Sedentism in the Natufian." *Paléorient* 9 (1983): 55–62.

Davis, Simon J. M. "Why Domesticate Food Animals? Some Zoo-Archaeological Evidence from the Levant." *Journal of Archaeological Science* 32 (2005): 1408–1416.

Dawkins, Richard. "Afterward," *in The Handbook of Evolutionary Psychology.* New York: John Wiley & Sons, 2005.

Dediu, Dan, and Stephen C. Levinson. "On the Antiquity of Language: The Reinterpretation of Neandertal Linguistic Capacities and Its Consequences." *Frontiers in Psychology* 4, no. 397 (2013): 1–17.

Defoe, Daniel. *The Life and Strange Surprizing Adventures of Robinson Crusoe, of York, Mariner.* London: W. Taylor, 1719.

DeLong, Bradford J. "The Reality of Economic Growth: History and Prospect," in Jeffrey Williamson et al., eds., *Globalization in Historical Perspective.* Appleton, Wisconsin: Lawrence University, 2000: 119–150.

Demay, Laëtitia, Stéphane Péan, and Marylène Patou-Mathis. "Mammoths Used as Food and Building Resources by Neanderthals: Zooarchaeological Study Applied to Layer 4, Molodova I (Ukraine)." *Quaternary International* 276–277 (2012): 212–226.

deMenocal, Peter B. "Cultural Responses to Climate Change During the Late Holocene." *Science* 292 (2001): 667–673.

Denham, Tim. "Early Agriculture and Plant Domestication in New Guinea and Island Southeast Asia." *Current Anthropology* 52, no. S4 (2011): S379–S395.

Dennell, Robin. "Dispersal and Colonisation, Long and Short Chronologies: How Continuous is the Early Pleistocene Record for Hominids Outside East Africa?" *Journal of Human Evolution* 45, no. 6 (2003): 421–440.

———. *The Paleolithic Settlement of Asia.* Cambridge, UK: Cambridge University Press, 2009.

d'Errico, Francesco, Christopher Henshilwood, Graeme Lawson, Marian Vanhaeren, Anne-Marie Tillier, Marie Soressi, Frédérique Bresson, et al. "Archaeological Evidence for the Emergence of Language, Symbolism, and Music—An Alternative Multidisciplinary Perspective." *Journal of World Prehistory* 17, no. 1 (2003): 1–70.

Deutscher, Guy. *Through the Language Glass: Why the World Looks Different In Other Languages*. New York: Metropolitan Books, 2010.

Devine, Warren D. "From Shafts to Wires: Historical Perspective on Electrification." *The Journal of Economic History* 43, no. 2 (1983): 347–372.

De Waal, Frans. "Bonobo Sex and Society." *Scientific American* 272, no. 3 (1995): 82–88, March, 1995.

De Waal, Frans, and Frans Lanting. *Bonobo: The Forgotten Ape*. Berkeley, CA: University of California Press, 1997.

Dewsbury, Donald A. "Patterns of Copulatory Behavior in Male Mammals." *The Quarterly Review of Biology* 47, no. 1 (March 1972): 133.

Diamond, Jared. *Guns, Germs, and Steel: The Fates of Human Societies*. New York: W. W. Norton, 1997.

———. *The World Until Yesterday*. New York: Viking, 2012.

Diehl, Michael W. "Architecture as a Material Correlate of Mobility Strategies: Some Implications for Archeological Interpretation." *Cross Cultural Research* 26, no. 1–4 (1992): 1–35.

Di Fiore, Anthony, and Drew Rendall. "Evolution of Social Organization: A Reappraisal for Primates by Using Phylogenetic Methods." *Proceedings of the National Academy of Sciences* 91, no. 21 (1994): 9941–9945.

Dirzo, Rodolfo, Hillary S. Young, Mauro Galetti, Gerardo Ceballos, Nick J. B. Isaac, and Ben Collen. "Defaunation in the Anthropocene." *Science* 345, no. 6195 (2014): 401–406.

Dodson, John, Xiaoqiang Li, Nan Sun, Pia Atahan, Xinying Zhou, Hanbin Liu, Keliang Zhao, Songmei Hu, and Zemeng Yang. "Use of Coal in the Bronze Age in China." *The Holocene* 24, no. 5 (2014): 525–530.

Dohrn-van Rossum, Gerhard, and Thomas Dunlap, trans. *History of the Hour: Clocks and Modern Temporal Orders*. Chicago: University of Chicago Press, 1996.

Domínguez-Rodrigo, Manuel. "Hunting and Scavenging by Early Humans: The State of the Debate." *Journal of World Prehistory* 16, no. 1 (2002): 1–54.

Douka, Katerina, Christopher A. Bergman, Robert E. M. Hedges, Frank P. Wesselingh, and Thomas F. G. Higham. "Chronology of Ksar Akil (Lebanon) and Implications for the Colonization of Europe by Anatomically Modern Humans." *PLOS ONE* 8, no. 9 (2013): e72931.

Duchin, Linda E. "The Evolution of Articulate Speech: Comparative Anatomy of the Oral Cavity in *Pan* and *Homo*." *Journal of Human Evolution* 19, no. 6–7 (1990): 687–697.

Dunbar, Robin I. M. "Neocortex Size as a Constraint on Group Size in Primates." *Journal of Human Evolution* 20 (1992): 469–493.

Dunbar, Robin, Clive Gamble, and John Gowlett, eds. *Social Brain, Distributed Mind*. Oxford: Oxford University Press for the British Academy, 2010.

Ecola, Lisa and Martin Wachs. "Exploring the Relationship between Travel Demand and Economic Growth." *Washington: Federal Highway Administration*. http://www.fhwa.dot.gov/policy/otps/pubs/vmt_gdp (accessed June 24, 2014).

Eldredge, Niles. *Life in the Balance: Humanity and the Biodiversity Crisis*. Princeton, NJ: Princeton University Press, 1998.

Elwin, Verrier, *The Kingdom of the Young*. London: Oxford University Press, 1968.

Eriksson, Anders, Lia Betti, Andrew D. Friend, Stephen J. Lycett, Joy S. Singarayer, Noreen von Cramon-Taubadel, Paul J. Valdes, Francois Balloux, and Andrea Manica. "Late Pleistocene Climate Change and the Global Expansion of Anatomically Modern Humans." *Proceedings of the National Academy of Sciences of the United States of America* 109, no. 40 (2012): 16089–16094.

Estebaranz, L., and A. Perez-Perez. *"Buccal Dental Microwear Signals in the Gracile Australopithecines A. anamensis, A. afarensis, and A. africanus: Adaptations to Open Environments with Climatic Shift."* 2010 Annual Meeting Paleoanthropology Society, St. Louis MO, April 2010.

Fagan, Brian. *Cro-Magnon: How the Ice Age Gave Birth to the First Modern Humans*. New York: Bloomsbury Press, 2010.

Febvre, Lucien, and Henri-Jean Martin. *The Coming of the Book: The Impact of Printing 1450–1800*. London: New Left Books, 1976.

Feliks, John. "The Golden Flute of Geissenklösterle: Mathematical Evidence for A Continuity Of Human Intelligence As Opposed To Evolutionary Change Through Time." *Aplimat-Journal of Applied Mathematics* 4, no. 4 (2011): 157–162.

———. "The Graphics of Bilzingsleben: Sophistication and Subtlety in the Mind of Homo Erectus." *Proceedings of the 15th Congress of the International Union of Prehistoric and Protohistoric Sciences*. http://www-personal.umich.edu/~feliks/graphics-of-bilzingsleben/index.html (accessed February 2, 2014).

Fildes, Jonathan. "Campaign Builds to Construct Babbage Analytical Engine." *BBC News*, October 14, 2010. http://www.bbc.co.uk/news/technology-11530905 (accessed June 12, 2014).

Finlayson, Clive, Darren A. Fa, Francisco Jiménez Espejo, Jóse S. Carrión, Geraldine Finlayson, Francisco Giles Pacheco, Joaquín Rodríguez Vidal, Chris Stringer, and Francisco Martínez Ruiz. "Gorham's Cave, Gibraltar—The Persistence of a Neanderthal Population." *Quaternary International* 181 (2008): 64–71.

Fish, Jennifer L., and Charles A. Lockwood. "Dietary Constraints on Encephalization in Primates." *American Journal of Physical Anthropology* 120, no. 2 (2003): 171–181.

Flannery, Kent. "The Origins of Agriculture." *Annual Review of Anthropology* 2 (1973): 271–310.

Flinn, Mark V., David C. Geary, and Carol V. Ward. "Ecological Dominance, Social Competition, and Coalitionary Arms Races: Why Humans Evolved Extraordinary Intelligence." *Evolution and Human Behavior* 26 (2005): 10–46.

Foley, Robert. "Adaptive Radiations and Dispersals in Hominin Evolutionary Ecology." *Evolutionary Anthropology, Supplement* 1 (2002): 32–37.

Foley, Robert, and Marta Lahr. "Mode 3 Technologies and the Evolution of Modern Humans." *Cambridge Archaeological Journal* 7, no. 1 (1997): 3–36.

Fong, Wen, W. Robert Bagley, Jenny F. So, and Maxwell K. Hearn, eds. *The Great Bronze Age of China: An Exhibition from the People's Republic of China.* New York: Metropolitan Museum of Art, 1980.

Fontana, Luigi, Jennifer L. Shew, John O. Holloszy, and Dennis T. Villareal. "Low Bone Mass in Subjects On a Long-Term Raw Vegetarian Diet." *Archives of Internal Medicine* 165 (2005): 684–689.

Ford, Clellan S., and Frank A. Beach. *Patterns of Sexual Behavior.* New York: Harper Torchbooks, 1951.

Froehle, Andrew W., and Steven E. Churchill. "Energetic Competition Between Neandertals and Anatomically Modern Humans." *PaleoAnthropology* (2009): 96–116.

Fuentes, Agustin. "Re-evaluating Primate Monogamy." *American Anthropologist* 100, no. 4 (1998): 890–907.

Galloway, Robert L. *A History of Coal Mining in Great Britain.* London: MacMillan and Co., 1882.

Georgano, Nick. *Cars Early and Vintage 1886–1930.* New York, Crescent Books, 1990.

Gibbons, Ann. "Stunning Skull Gives a Fresh Portrait of Early Humans." *Science* 342, no. 6156 (2013): 297–298.

Gibran, Kahlil, *The Prophet*. New York: Alfred A. Knopf, 1923.

Gilby, Ian C. "Meat Sharing Among the Gombe Chimpanzees: Harassment and Reciprocal Exchange." *Animal Behavior* 71 (2006): 953–863.

Gilby, Ian C., Lynn E. Eberly, Lilian Pintea, and Anne E. Pusey. "Ecological and Social Influences on the Hunting Behaviour of Wild Chimpanzees, *Pan Troglodytes Schweinfurthii*." *Animal Behaviour* 72 (2006): 169–180.

Gilligan, Ian. "Neanderthal Extinction and Modern Human Behaviour: The Role of Climate Change and Clothing." *World Archaeology* 39, no. 4 (2007): 499–514.

———. "The Prehistoric Development of Clothing: Archaeological Implications of a Thermal Model." *Journal of Archaeological Method and Theory* 17 (2010): 15–80.

Golitko, Mark, and Lawrence H. Keeley. "Beating Ploughshares Back into Swords: Warfare in the *Linearbandkeramik*." *Antiquity* 81 (2007): 332–342.

Good, Kenneth. *Into the Heart: One Man's Pursuit of Love and Knowledge Among the Yanomami*. New York: Simon & Schuster, 1991.

Goodall, Jane. *In the Shadow of Man*. Boston: Houghton Mifflin Company, 1971.

Goren-Inbar, Naama, Nira Alperson, Mordechai E. Kislev, Orit Simchoni, Yoel Melamed, Adi Ben-Nun, and Ella Werker. "Evidence of Hominin Control of Fire at Gesher Benot Ya'aqov, Israel." *Science* 304, no. 5671 (2004): 725–727.

Görlitz, Dominique. "Pre-Egyptian Reed Boat *Abora 2* Crosses the Mediterranean Sea." *Migration & Diffusion* 3, no. 12 (2002): 44–61.

Gowlett, John A. J. "The Early Settlement of Northern Europe: Fire History in the Context of Climate Change and the Social Brain." *Human Paleontology and Prehistory* 5 (2006): 299–310.

———. "Out in the Cold." *Nature* 413, no. 92 (2001): 33–34.

Gowlett, John A. J., J. W. K. Harris, D. Walton, and B. A. Wood. "Early Archaeological Sites, Hominid Remains, and Traces of Fire From Chesowanja, Kenya." *Nature* 294, no. 5837 (1981): 125–129.

Graber, Robert B., Randall R. Skelton, Ralph M. Rowlett, Ronald Kephart, and Susan Love Brown. *Meeting Anthropology, Phase to Phase: Growing Up, Spreading Out, Crowding In, Switching On*. Durham, NC: Carolina Academic Press, 2000.

Graham-Cumming, John. "Let's Build Babbage's Ultimate Mechanical Computer." *New Scientist*, December 23, 2010. http://www.newscientist.com/article/mg20827915.500-lets-build-babbages-ultimate-mechanical-computer.html#.U5njTGcg-Uk (accessed June 12, 2014).

Greenhill, Basil. *The Evolution of the Wooden Ship*. Caldwell, NJ: The Black-Burn Press, 2009.

Greenspan, Stanley, and Stuart Shanker. *The First Idea: How Symbols, Language, and Intelligence Evolved from Our Primate Ancestors to Modern Humans*. Cambridge, MA: Da Capo Press, 2006.

Gregory, J. W. "Edward Suess." *Science* 39, no. 1017 (1914): 933–935.

Groves, Colin. "The What, Why, and How of Primate Taxonomy." *International Journal of Primatology* 25, no. 5 (2004): 1105–1126.

Groves, Colin, and Jordi Sabater-Pi. "From Ape's Nest to Human Fx-Point." *Man* 20, no. 1 (1985): 22–47.

Guangwei, He. "China's Dirty Pollution Secret: The Boom Poisoned Its Soil and Crops." *Tainted Harvest: An e360 Special Report Part I*, June 30, 2014. http://e360.yale.edu/feature/chinas_dirty_pollution_secret_the_boom_poisoned_its_soil_and_crops/2782/ (accessed July 14, 2014).

———. "In China's Heartland, A Toxic Trail Leads from Factories to Fields to Food." *Tainted Harvest: An e360 Special Report Part II*. http://e360.yale.edu/feature/chinas_toxic_trail_leads_from_factories_to_food/2784/ (accessed June 28, 2014).

———. "The Soil Pollution Crisis in China: A Cleanup Presents Daunting Challenge." *Tainted Harvest: An e360 Special Report Part III*. http://e360.yale.edu/feature/the_soil_pollution_crisis_in_china_a_cleanup_presents_daunting_challenge/2786/ (accessed June 14, 2014).

Gullapalli, Sravani, and Michael Wong. "Nanotechnology: A Guide to Nano-Objects." *Chemical Engineering Progress* 107, no. 5 (2011): 28–32.

Haile-Selassie, Yohannes, Gen Suwa, and Tim D. White. "Late Miocene Teeth From Middle Awash, Ethiopia, and Early Hominid Dental Evolution." *Science* 303, no. 5663 (2004): 1503–1505.

Hansen, James, Makiko Sato, and Reto Ruedy. "*Global Temperature Update Through 2013.*" Goddard Institute for Space Studies, January 21, 2014.

Hansen, Karen T. "The World in Dress: Anthropological Perspectives in Clothing, Fashion, and Culture." *Annual Review of Anthropology* 33 (2004): 369–392.

Hardy, Bruce L., and Gary T. Garufi. "Identification of Woodworking on Stone Tools Through Residue and Use-Wear Analysis: Experimental Results." *Journal of Archaeological Science* 25 (1998): 177–184.

Hardy, Karen, Stephen Buckley, Matthew J. Collins, Almudena Estalrrich, Don Brothwell, Les Copeland, Antonio García-Tabernero, et al. "Neandertal Medics? Evidence for Food, Cooking, and Medicinal Plants Entrapped in Dental Calculus." *Naturwissenschaften* 99, no. 8 (2012): 617–626.

Harmand, Sonia, Jason E. Lewis, Craig S. Feibel, Christopher J. Lepre, Sandrine Prat, Arnaud Lenoble, Xavier Boës, Rhonda L. Quinn, Michel Brenet, Adrian Arroyo, et al., "3.3-million-year-old Stone Tools From Lomekwi 3, West Turkana, Kenya." *Nature* 521 (2015): 310–315.

Harrod, J. "Deciphering Later Acheulian Period Marking Motifs (LAmrk): Impressions of the Later Acheulian Mind." *OriginsNet Publications.* http://originsnet.org/publications.html#Deciphering LAmrk (accessed February 19, 2014).

Harvey, David. "Chapter 6: Time-Space Compression and the Postmodern Condition," in *The Global Transformations Reader: An Introduction to the Globalization Debate*, David Held and Anthony McGrew, eds. Stanford: Stanford University Press, 1999.

Harwood, Catherine. "Oral History Transcript: Frank Borman," *Johnson Space Center Oral History Archive,* April 13, 1999.

Hassan, Fekri. "The Gift of the Nile," in *Ancient Egypt*, David P. Silverman, ed. New York: Oxford University Press, 2003.

Hatley, Tom, and John Kappelman. "Bears, Pigs, and Plio-Pleistocene Hominids: A Case for the Exploitation of Belowground Food Resources." *Human Ecology* 4 (1980): 371–387.

Hauser, Marc D. "A Primate Dictionary? Decoding the Function and Meaning of Another Species' Vocalizations." *Cognitive Science* 24, no. 3 (2000): 445–475.

Hawkes, Kristen, and Nicholas Blurton Jones. "Human Age Structures, Paleodemography, and the Grandmother Hypothesis" *in Grandmotherhood: The Evolutionary Significance of the Second Half of Female Life.* New Brunswick, NJ: Rutgers University Press, 2005.

Hayden, Brian. "Nimrods, Piscators, Pluckers, and Planters: The Emergence of Food Production." *Journal of Anthropological Archaeology* 9, no. 1 (1990): 31–69.

Held, David, Anthony McGrew, David Goldblatt, and Jonathan Perraton, eds. *Global Transformations: Politics, Economics, and Culture.* Stanford: Stanford University Press, 1999.

Held, David, and Anthony McGrew, eds. *The Global Transformations Reader: An Introduction to the Globalization Debate.* Cambridge, MA: Polity Press, 2000.

Hernandez-Aguilar, R. Adriana, Jim Moore, and Travis Rayne Pickering. "Savanna Chimpanzees Use Tools to Harvest the Underground Storage Organs of Plants." *Proceedings of the National Academy of Sciences* 104, no. 49 (2007): 19210–19213.

Higham, Thomas, Laura Basell, Roger Jacobi, Rachel Wood, Christopher Bronk Ramsey, and Nicholas J. Conard. "Testing Models for the Begin-

nings of the Aurignacian and the Advent of Figurative Art and Music: The Radiocarbon Chronology of Geißenklösterle." *Journal of Human Evolution* 62, no. 6 (2012): 664–676.
Hilbert, Martin, and Priscila López. "The World's Technological Capacity to Store, Communicate, and Compute Information." *Science* 332, no. 6025 (2011): 60–65.
Hillman, Gordon C., and M. Stuart Davies "Measured Domestication Rates in Wild Wheats and Barley Under Primitive Implications." *Journal of World Prehistory* 4, no. 2 (1990): 157–222.
Hills, R. L., and A. J. Pacey. "The Measurement of Power in Early Steam-Driven Textile Mills." *Technology and Culture* 13, no. 1 (1972): 25–43.
Hirst, K. Kris. "Geißenklösterle (Germany): Aurignacian Site in the Swabian Jura of Germany." *About.com Archaeology*. http://archaeology.about.com/od/gterms/qt/Geissenklosterle-Germany.htm (accessed January 30, 2014).
———. "Molodova I (Ukraine)." *About.com Archeology*. http://archaeology.about.com/od/mterms/g/molodova.htm (accessed January 23, 2014).
Hobsbawm, Eric. *Nations and Nationalism Since 1780: Programme, Myth, Reality*. Cambridge, MA: Cambridge University Press, 1990.
Hodges, Henry. *Technology in the Ancient World*. New York: Alfred A. Knopf, 1974.
Hoebel, E. Adamson. *The Cheyennes: Indians of the Great Plains*. New York: Holt, Rinehart and Winston, 1960.
Hole, Frank. "Agricultural Sustainability in the Semi-Arid Near East." *Climate of the Past* 3 (2007): 193–203.
———. "A Reassessment of the Neolithic Revolution." *Paléorient* 10, no. 2 (1984): 49–60.
Holloway, Ralph L. "Tools and Teeth: Some Speculations Regarding Canine Reduction." *American Anthropologist* 69, no. 1 (1967): 63–67.
Hong, Sungmin, Jean-Pierre Candelone, Clair C. Patterson, and Claude F. Boutron."History of Ancient Copper Smelting Pollution During Roman and Medieval Times Recorded in Greenland Ice." *Science* 272, no. 5259 (1996): 246–249.
Hopkins, Anthony G., ed. *Global History: Interactions between the Universal and the Local*. Basingstoke, UK: Palgrave Macmillan, 2006.
Houghton, Richard A. "Why Are Estimates of the Terrestrial Carbon Balance So Different?" *Global Change Biology* 9, no. 4 (2003): 500–509.
Howard, William R. "Palaeoclimatology: A Warm Future in the Past." *Nature* 388 (1997): 418–419.

Hoyle, Fred, and Chandra Wickramasinghe. "On the Cause of Ice-Ages." *Cambridge-Conference Network*, July 1999. http://abob.libs.uga.edu/bobk/ccc/ce120799.html (accessed June 13, 2014).

Hu, Yaowu, Songmei Hu, Weilin Wang, Xiaohong Wu, Fiona B. Marshall, Xianglong Chen, Liangliang Hou, and Changsui Wang. "Earliest Evidence for Commensal Processes of Cat Domestication." *Proceedings of the National Academic of Sciences of the United States of America* 111, no. 1 (2014): 116–120.

Huffman, O. Frank, John De Vos, Aart W. Berkhout, and Fachroel Aziz. "Provenience Reassessment of the 1931–1933 Ngandong Homo Erectus (Java), Confirmation of the Bone-bed Origin Reported by the Discoverers." *PaleoAnthropology* (2010): 1-60.

Hultsman, John, and William Harper. "The Problem of Leisure Reconsidered." *Journal of American Culture* 16, no. 1 (2004): 47–54.

Hunt, Kevin D. "The Evolution of Human Bipedality: Ecology and Functional Morphology." *Journal of Human Evolution* 26 (1994): 183–202.

Hutchinson, Peter. "Magazine Growth in the Nineteenth Century." *A Publisher's History of American Magazines,* 2008.

Imbrie, John, A. Berger, E. A. Boyle, S. C. Clemens, A. Duffy, W. R. Howard, G. Kukla, J. Kutzbach, D. G. Martinson, A. Mcintyre et al. "On the Structure and Origin of Major Glaciation Cycles: 2. The 100,000-year cycle." *Paleoceanography* 8, no. 6 (1993): 699–735.

International Technology Roadmap for Semiconductors. "Overall Technology Roadmap Characteristics." *2010 Update.* http://www.itrs.net/Links/2010ITRS/Home2010.htm (accessed June 17, 2014).

Jablonski, Nina G., and George Chaplin. "Origin of Habitual Terrestrial Bipedalism in the Ancestor of the *Hominidae*." *Journal of Human Evolution* 24 (1993): 259–280.

James Martin Center for Nonproliferation Studies. "Chemical and Biological Weapons: Possession and Programs Past and Present." *Chemical & Biological Weapons Resource Page.* Monterey Institute of International Studies, 2008. http://cns.miis.edu/cbw/possess.htm (accessed June 16, 2014).

James, Steven R. "Hominid Use of Fire in the Lower and Middle Pleistocene: A Review of the Evidence." *Current Anthropology* 30, no. 1 (1989): 1–11.

Johanson, Donald and Blake Edgar. *From Lucy to Language.* New York, Simon & Schuster, 1996.

Jones, Eric L. *Cultures Merging: A Historical and Economic Critique of Culture.* Princeton: Princeton University Press, 2006.

Jones, Nate. "The Department of Defense List of 32 'Accidents Involving Nuclear Weapons.'" *Unredacted: The National Security Archive, Unedited and Uncensored*, October 9, 2013. http://nsarchive.wordpress.com/2013/10/09/document-friday-narrative-summaries-of-accidents-involving-nuclear-weapons/ (accessed June 26, 2014).

Keeley, Lawrence H. "War Before Civilization—15 Years On," in *The Evolution of Violence*. New York: Springer, 2014.

Kelly, Jack. *Gunpowder: Alchemy, Bombards, & Pyrotechnics: The History of the Explosive That Changed the World*. New York: Basic Books, 2004.

Kirkland, Joel. "Global Emissions Predicted to Grow through 2035." *Scientific American* (2010).

Kislev, Mordechai E., D. Nadel, and I. Carmi. "Epipalaeolithic (19,000 BP) Cereal and Fruit Diet at Ohalo II, Sea of Galilee, Israel." *Review of Palaeobotany and Palynology* 73, no. 1–4 (1992): 161–166.

Kitchen, Martin. *A History of Modern Germany: 1800 to the Present*. New York: John Wiley & Sons, 2011.

Kittler, Ralph, Manfred Kayser, and Mark Stoneking. "Molecular Evolution of *Pediculus Humanus* and the Origin of Clothing." *Current Biology* 13 (2003): 1414–1417.

Klein, Richard G. "Anatomy, Behaviour, and Modern Human Origins." *Journal of World Prehistory* 9, no. 2 (1995): 167–98.

———. "The Archaeology of Modern Human Origins." *Evolutionary Anthropology* 1 (1992): 5–14.

Knight, Chris, Camilla Power, and Ian Watts. "The Human Symbolic Revolution: A Darwinian Account." *Cambridge Archaeological Journal* 5, no. 1 (1995): 75–114.

Kortlandt, Adriaan. "How Might Early Hominids Have Defended Themselves Against Large Predators and Food Competitors?" *Journal of Human Evolution* 9, no. 2: 79–94.

Kremer, Michael. "Population Growth and Technological Change: One Million B.C. to 1990." *The Quarterly Journal of Economics* 108, no. 3 (1993): 681–716.

Kristensen, Hans M., and Robert S. Norris."Nuclear Warhead Stockpiles and Transparency," in *Global Fissile Material Report 2013: Increasing Transparency of Nuclear Warhead and Fissile Material Stocks as a Step Toward Disarmament*. Princeton: International Panel on Fissile Materials.

———. "Russian Nuclear Forces." *Bulletin of the Atomic Scientists* 69, no. (2013): 71–81.

———. "US Nuclear Forces." *Bulletin of the Atomic Scientists* 70, no. 11 (2014): 85–93.

Kuhn, Steven L., Mary C. Stiner, Erksin Güleç, Ismail Őzer, Hakan Yılmaz, Ismail Baykara, Ayşen Açıkkol, Paul Goldberg, Kenneth Martínez Molina, Engin Űnay, and Fadime Suata-Alpaslan. "The Early Upper Paleolithic Occupations at Űçalğızlı Cave (Hatay, Turkey)." *Journal of Human Evolution* 56 (2009): 87–113.

Kuijt, Ian. "Life in Neolithic Farming Communities: An Introduction," in *Life in Neolithic Farming Communities: Social Organization, Identity, and Differentiation,* Ian Kuijt, ed. New York: Plenum Publishers, 2000.

———. "Negotiating Equality Through Ritual: A Consideration of Late Natufian and Prepottery Neolithic A Period Mortuary Practices." *Journal of Anthropological Archaeology* 15 (1996): 313–336.

Kuijt, Ian, and Bill Finlayson. "Evidence for Food Storage and Predomestication Granaries 11,000 Years Ago in the Jordan Valley." *Proceedings of the National Academy of Sciences of the United States of America* 106, no. 27 (2009): 10966–10970.

Lallanilla, Marc. "World's Oldest Harbor Discovered in Egypt." *LiveScience* April 16, 2013.

Landes, David S. *Revolution in Time: Clocks and the Making of the Modern World.* Cambridge, MA: Belknap Press, 2000.

Laurance, William. "China's Appetite for Wood Takes a Heavy Toll on Forests." *Yale Environment 360*, November 17, 2011. http://e360.yale.edu/feature/chinas_appetite_for_wood_takes_a_heavy_toll_on_forests/2465/ (accessed June 28, 2014).

Lawler, Andrew. "Report of Oldest Boat Hints at Early Trade Routes." *Science* 296, no. 5574 (2002): 1791–1792.

Lawler, Richard R. "Monomorphism, Male-Male Competition, and Mechanisms of Sexual Dimorphism." *Journal of Human Evolution* 57 (2009): 321–325.

Leakey, Meave G., Craig S. Feibel, Ian McDougall, and Alan Walker. "New Four-Million-Year-Old Hominid Species from Kanapoi and Allia Bay, Kenya." *Nature* 376 (2002): 565–571.

Leo, Natalie P., and Stephen C. Barker. "Unravelling the Evolution of the Head Lice and Body Lice of Humans." *Parasitology Research* 98 (2005): 44–47.

Leonard, William R. "Food for Thought: Dietary Change was a Driving Force in Human Evolution." *Scientific American* 287, no. 6 (2002): 106–115.

Lessa, William A., *Ulithi: A Micronesian Design for Living.* New York: Holt, Rinehart and Winston, 1966.

Lewis, Bill, "The Gaslight Era: A Revolution in Lighting." *About.com/ Lighting*.http://lighting.about.com/od/Fixtures/a/The-Gaslight-Era.htm (accessed May 30, 2014).

Lewis, M. Paul, Gary F. Simons, and Charles D. Fennig, eds. *Ethnologue: Languages of the World, Seventeenth Edition*. Dallas: SIL International, 2004

Lieberman, Daniel E., Brandeis M. McBratney, and Gail Krovitz. "The Evolution and Development of Cranial Form in Homo Sapiens." *Proceedings of the National Academy of Sciences* 99, no. 3 (2002): 1134–1139.

Lieberman, Daniel E., Gail E. Krovitz, Franklin W. Yates, Maureen Devlin, and Marisa St. Claire. "Effects of Food Processing on Masticatory Strain and Craniofacial Growth in a Retrognathic Face." *Journal of Human Evolution* 46 (2004): 655–677.

Lieberman, Philip, Jeffrey T. Laitman, Joy S. Reidenberg, and Patrick J. Gannon. "The Anatomy, Physiology, Acoustics and Perception of Speech: Essential Elements in Analysis of the Evolution of Human Speech." *Journal of Human Evolution* 23 (1992): 447–467.

Lindsey, Rebecca. "Tropical Deforestation." *National Aeronautics and Space Administration, Earth Observatory*, March 30, 2007. http://earthobservatory.nasa.gov/Features/Deforestation/ (accessed June 27, 2014).

Lordkipanidze, David, Marcia S. Ponce de León, Ann Margvelashvili, Yoel Rak, G. Philip Rightmire, Abesalom Vekua, and Christoph P. E. Zollikofer. "A Complete Skull from Dmanisi, Georgia, and the Evolutionary Biology of Early *Homo*." *Science* 342, no. 6156 (2013): 326–331.

Lovejoy, C. Owen. "The Origin of Man." *Science (New Series)* 211, no. 4480 (1982): 341–350.

Lu, Caitlin. *Matteo Ricci and the Jesuit Mission in China 1583-1610*. Boston: The Concord Review.

Lucas, Adam Robert. "Industrial Milling in the Ancient and Medieval Worlds: A Survey of the Evidence for an Industrial Revolution in Medieval Europe." *Technology and Culture* 46, no. 1 (2005): 1–30.

Luengen, Hans B., Michael Peters, and Peter Schmöle. "Ironmaking in Western Europe." *Association for Iron & Steel Technology 2011 Proceedings* 1 (2011): 387–400.

Mcbrearty, Sally, and Alison S. Brooks. "The Revolution That Wasn't: A New Interpretation of the Origin of Modern Human Behavior." *Journal of Human Evolution* 39, no. 5 (2000): 453–563.

McCallum, Malcolm L. "Amphibian Decline or Extinction? Current Declines Dwarf Background Extinction Rate." *Journal of Herpetology* 41, no. 3 (2007): 483–491.

McCorriston, Joy, and Frank Hole. "The Ecology of Seasonal Stress and the Origins of Agriculture in the Near East." *American Anthropologist* 93, no. 1 (1991): 46–69.

Macdonald, David, ed. *Primates.* Oxford, England: Equinox, 1984.

McGrew, William C. *Chimpanzee Material Culture: Implications for Human Evolution.* Cambridge, UK: Cambridge University Press, 1992.

McKibben, Bill. *Deep Economy: Economics as if the World Mattered.* Oxford: OneWorld Publications, 2007.

McPherron, Shannon Patrick. "Handaxes as a Measure of the Mental Capabilities of Early Hominids." *Journal of Archaeological Science* 27 (2000): 655–663.

Maher, Lisa A., Tobias Richter, and Jay T. Stock. "The Pre-Natufian Epipaleolithic: Long-term Behavioral Trends in the Levant." *Evolutionary Anthropology* 21 (2012): 69–81.

Maisels, Charles K. "The Institutions of Urbanism," in *The Emergence of Civilization: From Hunting and Gathering to Agriculture, Cities, and the State in the Near East.* London: Routledge, 1990.

———. "The Interactive Evolution of Alphabetic Script," in *The Emergence of Civilization: From Hunting and Gathering to Agriculture, Cities, and the State in the Near East.* London: Routledge, 1990.

Malinowski, Bronislaw. *Magic, Science, and Religion.* Garden City, NY: Doubleday & Company, 1954.

———. *The Sexual Life of Savages in North-Western Melanesia.* New York: Harcourt, Brace and Company, 1929.

Mania, Dietrich, and Ursula Mania. "Deliberate Engravings on Bone Artefacts of Homo Erectus." *Rock Art Research* 5 (1988): 91–107.

———. "The Natural and Socio-Cultural Environment of Homo Erectus at Bilzingsleben, Germany," in *The Hominid Individual in Context, Archaeological Investigations of Lower and Middle Palaeolithic Landscapes, Locales and Artefacts,* Clive Gamble and Martin Porr, eds. New York: Routledge, 98–114.

Mann, Alan, and Mark Weiss. "Hominoid Phylogeny and Taxonomy: A Consideration of the Molecular and Fossil Evidence in a Historical Perspective." *Molecular Phylogenetics and Evolution* 5, no. 1 (1996): 169–181.

Mann, Charles C. *1491: New Revelations of the Americas Before Columbus.* New York: Vintage Books, 2006.

Manning, Patrick. *Navigating World History: Historians Create a Global Past.* New York: Palgrave Macmillan, 2003.

Marland, G., T. A. Boden, and R. J. Andres. "Global, Regional, and National Fossil Fuel CO2 Emissions," in *Trends: A Compendium of Data on Global*

Change. Oak Ridge, Tennessee: Carbon Dioxide Information Analysis Center, Oak Ridge National Laboratory, US Department of Energy, 2012.

Marlowe, Frank W. "Hunting and Gathering: The Human Sexual Division of Foraging Labor." *Cross-Cultural Research* 41, no. 2 (2007): 170–195.

Martin, R. M., ed. *State of the World's Forests 2012*. Rome: United Nations Food and Agriculture Organization, 2012.

Martínez, I., M. Rosa, J.-L. Arsuaga, P. Jarabo, R. Quam, C. Lorenzo, A. Gracia, J.-M. Carretero, J.-M. Bermúdez de Castro, and E. Carbonell. "Auditory Capacities in Middle Pleistocene Humans from the Sierra de Atapuerca in Spain." *Proceedings of the National Academy of Sciences of the United States of America* 10.1 no. 27 (2004): 9976–9981.

Martínez, I., M. Rosa, R. Quam, P. Jarabo, C. Lorenzo, A. Bonmatí, A. Gómez-Olivencia, A. Gracia, and J. L. Arsuaga. "Communicative Capacities in Middle Pleistocene Humans from the Sierra de Atapuerca in Spain." *Quaternary International* 295 (2013): 94–101.

Martínez, Maria del Carmen Rodríguez, Ponciano Ortíz Ceballos, Michael D. Coe, Richard A. Diehl, Stephen D. Houston, Karl A. Taube, and Alfredo Delgado Calderón. "Oldest Writing in the New World." *Science* 313, no. 5793 (2006): 1610–1614.

Mastny, Lisa. "Traveling Light: New Paths for International Tourism." *Worldwatch Paper 159*, December 2001.

Mead, Margaret. *Culture and Commitment: A Study of the Generation Gap*. London: The Bodley Head, 1970.

Mellars, Paul. "Cognitive Changes and the Emergence of Modern Humans in Europe." *Cambridge Archaeological Journal* 1, no. 1 (1991): 63–76.

Mendoza, Sally P., Deeann M. Reeder, and William A. Mason. "Nature of Proximate Mechanisms Underlying Primate Social Systems: Simplicity and Redundancy." *Evolutionary Anthropology, Supplement* 1 (2002): 112–116.

Mgeladzea, Ana, David Lordkipanidze, Marie-Hélène Moncelb, Jackie Desprieeb, Rusudan Chagelishvilia, Medea Nioradzea, and Giorgi Nioradzea. "Hominin Occupations at the Dmanisi Site, Georgia, Southern Caucasus: Raw Materials and Technical Behaviours of Europe's First Hominins." *Journal of Human Evolution* 60, no. 5 (2011): 571–596.

Milton, Katharine. "Diet and Primate Evolution." *Scientific American* 269 (1993): 86–93.

———. "A Hypothesis to Explain the Role of Meat-Eating in Human Evolution." *Evolutionary Anthropology* 8 (1999): 11–21.

Mitani, John C. and David P. Watts. "Correlates of Territorial Boundary Patrol Behavior in Wild Chimpanzees." *Animal Behavior* 70 (2005): 1079–1086.

———. "Demographic Influences on the Hunting Behavior of Chimpanzees." *American Journal of Physical Anthropology* 109 (1999): 439–454.

Mitani, John C., David P. Watts, and Sylvia J. Amsler. "Lethal Intergroup Aggression Leads to Territorial Expansion in Wild Chimpanzees." *Current Biology* 20, no. 12 (2010): 507–508.

Moore, Gordon E. "Cramming More Components onto Integrated Circuits." *Electronics* 38, no. 8 (1965): 114–117.

———. "Cramming More Components onto Integrated Circuits." *Proceedings of the IEEE* 86(1): 82–85 (1998). Available at: http://www.cs.utexas.edu/users/fussell/courses/cs352h/papers/moore.pdf (accessed June 17, 2014).

Muller, Martin N., and John C. Mitani. "Conflict and Cooperation in Wild Chimpanzees." *Advances in the Study of Behavior* 35 (2005): 275–331.

Müller, Werner, and Clemens Pasda. "Site Formation and Faunal Remains of the Middle Pleistocene Site Bilzingsleben." *Quartär* 58 (2011): 25–49.

Münzel, S., F. Seeberger, and W. Hein. "The Geissenklösterle Flute—Discovery, Experiments, Reconstruction," in *The Archaeology of Sound: Origin and Organisation*. Rahden/Westfalen: Verlag Marie Leidorf, 2002, 107–118.

Murdock, George P. *Social Structure*. New York: MacMillan Company, 1960.

Murnane, William J. "Three Kingdoms and Thirty-Four Dynasties," in *Ancient Egypt*, Daniel Silverman, ed. New York: Oxford University Press, 20–57.

National Aeronautics and Space Administration. Goddard Institute for Space Studies.*GISS Surface Temperature Analysis (GISTEMP)*. Available at: http://data.giss.nasa.gov/gistemp/(accessed June 19, 2014).

Needham, Joseph, "Military Technology: the Gunpowder Epic," in *Science and Civilisation in China, Chemistry and Chemical Technology (vol. 5, part 2)*. Cambridge, UK: Cambridge University Press,1986.

Newman, Russell W. "Why Man Is Such a Sweaty and Thirsty Naked Animal: A Speculative Review." *Human Biology* 42 (1970): 12–27.

Newton-Fisher, Nicholas E. "Chimpanzee Hunting Behavior," in *Handbook of Paleoanthropology*, eds. Winfried Henke, Ian Tattersall, and Thorolf Hardt. New York: Springer, 2007.

Niépce House Museum. "The Pyrelophore." *Other Inventions*. http://www.niepce.com/pagus/pagus-other.html (accessed May 30, 2014).

Noble, William, and Iain Davidson. "The Evolutionary Emergence of Modern Human Behavior: Language and its Archaeology." *Man (New Series)* 26 (1991): 223–253.

Norris, Robert S., William M. Arkin, Hans M. Kristensen, and Joshua Handler. "Israeli Nuclear Forces" *Bulletin of the Atomic Scientists* 58, no. 5 (2002): 72–75.

North, J. D., "Wallingford, Richard (c.1292–1336)," in *The Oxford Dictionary of National Biography*. Oxford: Oxford University Press, 2004.

Oates, Joan, Augusta McMahon, Philip Karsgaard, Salam Al Quntar, and Jason Ur. "Early Mesopotamian Urbanism: A New View from the North." *Antiquity* 81, no. 313 (2007): 585–600.

O'Brien, Patrick. Review of *Global History: Interactions between the Universal and the Local*, ed. Anthony G. Hopkins. Basingstoke: Palgrave Macmillan, 2006.

Ogawa, Hideshi, Gen'ich Idani, Jim Moore, Lilian Pintea, and Adriana Hernandez-Aguilar. "Sleeping Parties and Nest Distribution of Chimpanzees in the Savanna Woodland, Ugalla, Tanzania." *International Journal of Primatology* 28 (2007): 1397–1412.

Okołów, Czesław, ed. *Białowieża National Park: Know It, Understand It, Protect It.* Białowieża, Poland: Białowieski Park Narodowy, 2009.

O'Neill, Brian C., Michael Dalton, Regina Fuchs, Leiwen Jiang, Shonali Pachauri, and Katarina Zigova. "Global Demographic Trends and Future Carbon Emissions." *Proceedings of the National Academy of Sciences of the United States of America* 107, no. 41 (2010): 17521–17526.

Organ, Chris, Charles L. Nunn, Zarin Machanda, and Richard W. Wrangham. "Phylogenetic Rate Shifts in Feeding Time During the Evolution of Homo." *Proceedings of the National Academy of Sciences* 108, no. 5 (2011): 1455–1459.

Osborne, Colin P., and David J. Beerling. "Nature's Green Revolution: The Remarkable Evolutionary Rise of C4 Plants." *Philosophical Transactions of the Royal Society B: Biological Sciences* 361, no. 1465 (2006): 173–194.

Pagel, Mark, and Walter Bodmer. "A Naked Ape Would Have Fewer Parasites." *Proceedings of the Royal Society of London B (Supplement)* 270 (2003): S117–S119.

Parfitt, Simon A., Nick M. Ashton, Simon G. Lewis, Richard L. Abel, G. Russell Coope, Mike H. Field, Rowena Gale, et al. "Early Pleistocene Human Occupation at the Edge of the Boreal Zone in Northwest Europe." *Nature* 466 (2010): 229–233.

Parpola, Asko. "Study of the Indus Script." *Paper, International Conference of Eastern Studies*, Tokyo, May 19, 2005.

———. "Towards Further Understanding of the Indus Script." *Proceedings of SCRIPTA 2008*, Seoul, October 2008.

Partain, Gary. “Bilzingsleben: Providing a New View of the Lower Paleolithic.” *Yahoo Voices* Dec 14, 2009. http://voices.yahoo.com/bilzingsleben-providing-view-lower-paleolithic-5056010.html (accessed January 12, 2014).

Patterson, Claire C. “Native Copper, Silver, and Gold Accessible to Early Metallurgists.” *American Antiquity* 36, no. 3 (1970): 286–321.

Pennisi, Elizabeth. “Did Cooked Tubers Spur the Evolution of Big Brains?” *Science* 283, no. 5410 (1999): 2004–2005.

Peresania, Marco, Ivana Fiore, Monica Gala, Matteo Romandini, and Antonio Tagliacozzo. “Late Neandertals and the Intentional Removal of Feathers as Evidenced from Bird Bone Taphonomy at Fumane Cave 44 ky B.P., Italy.” *Proceedings of the National Academy of Sciences of the United States of America* 108, no. 10 (2011): 3888–3893.

Perles, Catherine. “Hearth and Home in the Old Stone Age.” *Natural History* 90, no. 10 (1981): 38.

Peterson, Ivars. “The Incredible Shrinking Computer.” *Science News* 123, no. 24 (1983): 378–380.

Petigura, Eric A., Andrew W. Howard, and Geoffrey W. Marcy. “Prevalence of Earth-Size Planets Orbiting Sun-like Stars.” *Proceedings of the National Academy of Sciences of the United States of America* 110, no. 48 (2013): 19273–19278.

Petit, John-Robert, J. Jouzel, D. Raynaud, N. I. Barkov, J. M. Barnola, I. Basile, M. Bender, J. Chappellaz, M. Davisk, G. Delaygue, et al. “Climate and Atmospheric History of the Past 420,000 Years from the Vostok Ice Core, Antarctica.” *Nature* 399 (1999): 429–436.

Pike, A.W.G., D.L. Hoffmann, M. García-Diez, P.B. Pettitt, J. Alcolea, R. De Balbín, C. González-Sainz, C. de las Heras, J.A. Lasheras, R. Montes, and J. Zilhão. “U-Series Dating of Paleolithic Art in 11 Caves in Spain.” *Science* 336 (2012): 1409–1413.

Pimentel, David, and Anne Wilson. “World Population, Agriculture, and Malnutrition.” *World Watch Magazine* 17, no. 5 (2004): 22–25.

Pimm, Stuart, Peter Raven, Alan Peterson, Çağan H. Şekercioğlu, and Paul R. Ehrlich. “Human Impacts on the Rates of Recent, Present, and Future Bird Extinctions.” *Proceedings of the National Academy of Sciences of the United States of America* 103, no. 29 (2006): 10941–10946.

Pirzada, Syed M. “Ivy Bridge-EX Arrives—Intel Xeon E7 v2 Released with 20 New SKU Lineup.” *WCCFTech.com*, Feb. 19, (accessed June 24, 2014).

Pol, K., M. Debret, V. Masson-Delmotte, E. Capron, O. Cattani, G. Dreyfus, S. Falourd, S. Johnsen, J. Jouzel, A. Landais, B. Minster, and B.

Stenni. "Links between MIS 11 Millennial to Sub-Millennial Climate Variability and Long Term Trends as revealed by New High Resolution EPICA Dome C Deuterium Data – A Comparison with the Holocene." *Climate of the Past* 7 (2011): 437–450.

Poole, Robert. *Earthrise: How Man First Saw the Earth*. New Haven: Yale University Press, 2010.

Potter, Jack M., May N. Diaz, and George M. Foster. *Peasant Society: A Reader*. Boston: Little, Brown and Company, 1967.

Potts, Richard. "Environmental Hypotheses of Hominin Evolution." *Yearbook of Physical Anthropology* 41 (1998): 93–138.

Preece, R. C., John A. J. Gowlett, Simon A. Parfitt, D. R. Bridgland, and S. G. Lewis. "Humans in the Hoxnian: Habitat, Context, and Fire Use at Beeches Pit, West Stow, Suffolk, UK." *Journal of Quaternary Science* 21 (2006): 485–496.

Price, T. Douglas, and Ofer Bar-Yosef. "The Origins of Agriculture: New Data, New Ideas." *Current Anthropology* 52, no. S4 (2011): S163–S174.

Proença, Vânia, and Henrique Miguel Pereira. "Comparing Extinction Rates: Past, Present, and Future," in *Encyclopedia of Biodiversity*, vol. 2, 167–176.

Pruetz, J. "Use of Caves by Savanna Chimpanzees (*Pan Troglodytes Verus*) in the Tomboronkoto Region of Southeastern Senegal." *Pan Africa News* 8, no. 2 (2001): 26–28.

Pruetz, Jill D., and Paco Bertolani. "Savanna Chimpanzees, *Pan Troglodytes Verus*, Hunt with Tools." *Current Biology* 17, no. 5 (2007): 412–417.

Raask, Erich. *Mineral Impurities in Coal Combustion: Behavior, Problems, and Remedial Measures*. New York: Hemisphere Publishing Corporation, 1985.

Ragir, Sonia. "Diet and Food Preparation: Rethinking Early Hominid Behavior." *Evolutionary Anthropology* 9, no. 4 (2000): 153–155.

Reed David L., Jessica E. Light, Julie M. Allen, and Jeremy J. Kirchman. "Pair of Lice Lost or Parasites Regained: The Evolutionary History of Anthropoid Primate Lice." *BMC Biology* 5 (2007): 7.

Reed, Kaye E. "Early Hominid Evolution and Ecological Change through the African Plio-Pleistocene." *Journal of Human Evolution* 32 (1997): 289–322.

Relman, David A., Eileen R. Choffnes, and Alison Mack. *Infectious Disease Movement in a Borderless World: Workshop Summary*. Washington, DC: National Academies Press, 2010.

Renfrew, Colin. *The Emergence of Civilisation: The Cyclades and the Aegean in the Third Millennium BC*. Oxford: Oxbow Books, 1972.

Reyes, Alberto V., Anders E. Carlson, Brian L. Beard, Robert G. Hatfield, Joseph S. Stoner, Kelsey Winsor, Bethany Welke, and David J. Ullman.

"South Greenland Ice-Sheet Collapse During Marine Isotope Stage 11." *Nature* 510 (2014): 525–528.

Riall, Lucy. *The Italian Risorgimento: State, Society, and National Unification.* London: Routledge, 1994.

Richerson, Peter J., Robert Boyd, and Robert L. Bettinger. "Was Agriculture Impossible during the Pleistocene but Mandatory during the Holocene?" *American Antiquity* 66, no. 3 (2001): 387–411.

Ridley, Matt. "Humans: Why They Triumphed." *Wall Street Journal*, May 22, 2010.

Riehl, Simone, Mohsen Zeidi, and Nicholas J. Conard. "Emergence of Agriculture in the Foothills of the Zagros Mountains of Iran." *Science* 341, no. 6141 (2013): 65–67.

Rindos, David. *The Origins of Agriculture: An Evolutionary Perspective.* New York: Academic Press, 1984.

Roach, Neil T., Madhusudhan Venkadesan, Michael J. Rainbow, and Daniel E. Lieberman. "Elastic Energy Storage in the Shoulder and the Evolution of High-Speed Throwing in Homo." *Nature* 498, (2013): 483–486.

Robinson, Andrew,. "Decoding Antiquity: Eight Scripts That Still Can't Be Read." *New Scientist* 2710, May 27, 2009.

Rodman, Peter S., and Henry M. McHenry. "Bioenergetics and the Origin of Hominid Bipedalism." *American Journal of Physical Anthropology* 52 (1980): 103–106.

Roebroeks Wil. "The Human Colonisation of Europe: Where Are We?" *Journal of Quaternary Science* 21 (2006): 425–435.

Roebroeks, Wil, and Paola Villa. "On the Earliest Evidence for Habitual Use of Fire in Europe." *Proceedings of the National Academy of Sciences* 108, no. 3 (2011): 5209–5214.

Roebroeks, Wil, Nicholas J. Conard, and Thijs van Kolfschoten. "Dense Forests, Cold Steppes, and the Palaeolithic Settlement of Northern Europe." *Current Anthropology* 33 (1992): 551–86.

Roebroeks, Wil, Mark J. Siera, Trine Kellberg Nielsen, Dimitri De Loecker, Josep Maria Parés, Charles E. S. Arps, and Herman J. Mücher. "Use of Red Ochre by Early Neandertals." *Proceedings of the National Academy of Sciences of the United States of America* 109, no. 6 (2102): 1889–1894.

Roeder, Philip G. *Where Nation-States Come From: Institutional Change in the Age of Nationalism.* Princeton: Princeton University Press, 2007.

Rosena, Arlene M., and Isabel Rivera-Collazo. "Climate Change, Adaptive Cycles, and the Persistence of Foraging Economies during the late Pleistocene/Holocene Transition in the Levant." *Proceedings of the National Academic of Sciences of the United States of America* 109, no. 10 (2012): 3640–3645.

Ross, Philip E. "When Will We Have Unmanned Commercial Airliners?" *IEEE Spectrum*, November 29, 2011. http://spectrum.ieee.org/aerospace/aviation/when-will-we-have-unmanned-commercial-airliners/0 (accessed June 12, 2014).

Rowlett, Ralph M. "Letter: Did the Use of Fire for Cooking Lead to a Diet Change That Resulted in the Expansion of Brain Size in *Homo erectus* from That of *Australopithecus Africanus*?" *Science* 283 no. 5410 (1999): 2005.

Royer, Dana L., Robert A. Berner, Isabel P. Montañez, Neil J. Tabor, and David J. Beerling. "CO2 as a Primary Driver of Phanerozoic Climate." *GSA Today* 14, no. 3 (2004): 4–10.

Ryan, Peter G., Charles J. Moore, Jan A. van Franeker, and Coleen L. Moloney. "Monitoring the Abundance of Plastic Debris in the Marine Environment." *Philosophical Transactions of the Royal Society B: Biological Sciences* 364, no. 1526 (2009): 1999–2012.

Samarth, Nitin. "The Incredible Shrinking Computer." *Penn State News*. http://news.psu.edu/story/141560/1999/05/01/research/incredible-shrinking-computer (accessed June 21, 2014).

Saraswat, Krishna. "Trends in Integrated Circuits Technology." *Stanford University, EE311/Trends*. http://www.google.com/url?sa=t&rct=-j&q=&esrc=s&source=web&cd=1&ved=0CB0QFjAA&url=http%3A%2F%2Fwww.stanford.edu%2Fclass%2Fee311%2FNOTES%2F-TrendsSlides.pdf&ei=RCWnU_O2IJPjoATd94D4Bw&usg=AFQ-jCNGUsBM2PT3G1d7laao1cgoXdc3yOQ&sig2=HlpraPlhV-7LR3-9T9tqJcA (accessed June 22, 2014).

Saturno, William A., David Stuart, and Boris Beltrán. "Early Maya Writing at San Bartolo, Guatemala." *Science* 31, 1 no. 5765 (2006): 1281–1283.

Scharf, Caleb A. "The Fastest Spacecraft Ever?" *Scientific American,* February 25, 2013. http://blogs.scientificamerican.com/life-unbounded/2013/02/25/the-fastest-spacecraft-ever/ (accessed June 12, 2014).

Schlesier, Karl H. "More on the 'Venus' Figurines." *Current Anthropology* 42, no. 3 (2001): 410.

Schlosser, Eric *Command and Control: Nuclear Weapons, the Damascus Accident, and the Illusion of Safety.* New York: Penguin Press, 2013.

Schmid, Peter. "Functional Interpretation of the Laetoli Footprints" in *From Biped to Strider: The Emergence of Modern Human Walking, Running, and Resource Transport,* Jeffrey Meldrum and Charles E. Hilton, eds. New York: Kluwer Academic/Plenum Publishers.

Schmid, Randolph E. "Ancient Shells May Be Oldest Jewelry." *LiveScience,* June 22, 2006. http://www.livescience.com/842-ancient-shells-oldest-jewelry.html (accessed October 22, 2013).

Schopenhauer, Arthur. *Die Welt als Wille und Vorstellung.* New York: Koneman, 1998.

Service, Elman R. "The Cheyenne of the North American Plains," in *Profiles in Ethnology,* 3rd ed. New York: Harper & Row, 1978.

———. "The Nootka of British Columbia," in *Profiles in Ethnology,* 3rd ed. New York: Harper & Row, 1978.

Shaw, Ian. "The Settled World," in *Ancient Egypt,* Daniel Silverman, ed. New York: Oxford University Press, 68–79.

Sherratt, Andrew. "Climatic Cycles and Behavioural Revolutions: The Emergence of Modern Humans and the Beginning of Farming." *Antiquity* 71 (1997): 271–287.

Shipman, Pat. *The Invaders: How Humans and Their Dogs Drove Neanderthals to Extinction.* Cambridge, Massachusetts: Harvard University Press, 2015.

Shostak, Marjorie, *Nisa: The Life and Words of a ¡Kung Woman.* New York: Vintage Books, 1983.

Shurkin, Joel. *Broken Genius: The Rise and Fall of William Shockley, Creator of the Electronic Age.* London: Macmillan, 2006.

Sica, Edgardo. "International Tourism: A Driving Force for Economic Growth of Commonwealth Countries." *The Commonwealth Finance Ministers Meeting 2007.* https://www.academia.edu/982748/International_tourism_a_driving_force_for_economic_growth_of_Commonwealth_countries (accessed June 25, 2014).

Simmons, Alan. "Mediterranean Island Voyages." *Science* 338, no. 6109 (2012): 895–897.

Skjelsbaek, Kjell. "The Growth of International Nongovernmental Organization in the Twentieth Century." *International Organization* 25, no. 3 (1971): 420–442.

Slurink, Pouwel. "Ecological Dominance and the Final Sprint in Hominid Evolution." *Human Evolution* 8, no. 4 (1993): 265–273.

Smart, Jeffery K. "History of Chemical and Biological Warfare: An American Perspective," *in* US Army Medical Department, AMEDD Center and School, *Medical Aspects of Chemical and Biological Warfare.* Washington: The Borden Institute, 2008.

Smith, A. H. V. "Provenance of Coals from Roman Sites in England and Wales." *Britannia* 28 (1997): 297–324.

Soffer, Olga. "Recovering Perishable Technologies through Use Wear on Tools: Preliminary Evidence for Upper Paleolithic Weaving and Net Making." *Current Anthropology* 45 (2004): 407–413.

Solem, Børge and Trond Austheim. "Statistics Concerning the Transatlantic Crossing." *Norway-Heritage: Hands Across the Sea*, April 16, 2004.

Spoor, F., M. G. Leakey, P. N. Gathogo, F. H. Brown, S. C. Anton, I. McDougall, C. Kiarie, F. K. Manthi, and L. N. Leakey. "Implications of New Early *Homo* Fossils from Ileret, East of Lake Turkana, Kenya." *Nature* 448 (2007): 688–691.

Stahl, Ann B. "Hominid Dietary Selection Before Fire." *Current Anthropology* 25. no. 2 (1984): 151–168.

Stanford, Craig. "Chimpanzee Hunting Behavior and Human Evolution." *American Scientist*, May-June, 1995. http://www.americanscientist.org/issues/feature/chimpanzee-hunting-behavior-and-human-evolution/1 (accessed February 1, 2013).

———. *Upright: The Evolutionary Key to Becoming Human*. Boston: Houghton Mifflin Company, 2003.

Stewart, John. *Evolution's Arrow: The Direction of Evolution and the Future of Humanity*. Canberra: Chapman Press, 2000.

———. "The Meaning of Life in a Developing Universe." *Foundations of Science* 15, no. 4 (2010): 395-409.

Stockwell, Foster. *Westerners in China: A History of Exploration and Trade, Ancient Times Through the Present*. Jefferson, NC: Mcfarland & Co., 2002.

Stoimenov, Miodrag, Branislav Popkonstantinović, Ljubomir Miladinović, and Dragan Petrović. "Evolution of Clock Escapement Mechanisms." *FME Transactions* 40 (2012): 17–23.

Subramanian, Sushma. "Fact or Fiction: Raw Veggies Are Healthier than Cooked Ones." *Scientific American,* March 31, 2009.

Suwa, Gen, Reiko T. Kono, Scott W. Simpson, Berhane Asfaw, C. Owen Lovejoy, and Tim D. White. "Paleobiological Implications of the *Ardipithecus Ramidus* Dentition." *Science* 326, no. 5949 (2009): 94–99.

Swade, Doron D., 2005, The Construction of Charles Babbage's Difference Engine No. 2. *IEEE Annals of the History of Computing*, July-September 2005, 70–88.

Swisher, Carl C., III, W. J. Rink, S. C. Antón, H. P. Schwarcz, Garniss H. Curtis, A. Suprijo, Widiasmoro. "Latest Homo Erectus of Java: Potential Contemporaneity with Homo Sapiens in Southeast Asia." *Science* 274, no. 5294 (1996): 1870–1874.

Teitelbaum, Michael S. and Jay M. Winter. "Bye-bye, Baby." *New York Times Book Review*, April 4, 2014. http://www.nytimes.com/2014/04/05/opinion/sunday/bye-bye-baby.html?_r=0 (accessed June 8, 2014).

Testart, Alain, Richard G. Forbis, Brian Hayden, Tim Ingold, Stephen M. Perlman, David L. Pokotylo, Peter Rowley-Conwy, and David E. Stuart. "The Significance of Food Storage among Hunter-Gatherers: Residence Patterns, Population Densities, and Social Inequalities." *Current Anthropology* 23, no. 5 (1982): 523–537.

Thieme, Hartmut. "Lower Palaeolithic Hunting Spears from Germany." *Nature* 385, no. 6619 (1997): 807–810.

Thirgood, J. V. "The Historical Significance of Oak, in: *Oak Symposium Proceedings, 1971 August 16-20*: 1–18. Upper Darby, PA.: US Department of Agriculture, Forest Service, Northeastern Forest Experiment Station, 1971.

Thomas, Donald E. *Diesel: Technology and Society in Industrial Germany*. Tuscaloosa: University of Alabama Press, 1987.

Tobias, Phillip V. "Recent Studies on Sterkfontein and Makapansgat and their Bearing on Hominid Phylogeny in Africa." *South African Archaeological Society, Goodwin Series* 10, no. 2 (1974): 5–11.

Tomasello, Michael. "Primate Cognition." *Cognitive Science* 24, no. 3 (2000): 351–361.

Torres, Abel Mendez. "A Nearby Super-Earth with the Right Temperature but Extreme Seasons." *Planetary Habitability Laboratory, University of Puerto Rico at Arecibo*. June 25, 2014. http://phl.upr.edu/press-releases/Gliese832 (accessed July 12, 2014).

Toups, Melissa A., Andrew Kitchen, Jessica E. Light, and David L. Reed. "Origin of Clothing Lice Indicates Early Clothing Use by Anatomically Modern Humans in Africa." *Molecular Biology and Evolution* 28 (2011): 29–32.

Trinkaus, Erik. "Anatomical Evidence for the Antiquity of Human Footwear Use." *Journal of Archaeological Science* 32, no. 10 (2005): 1515–1526.

Trinkaus, Eric, and Pat Shipman. *The Neandertals: Changing the Image of Mankind*. New York: Alfred A. Knopf, 1993.

Tsukahara, Takahiro. "Lions Eat Chimpanzees: The First Evidence of Predation by Lions on Wild Chimpanzees." *American Journal of Primatology* 29, no. 1 (2005): 1–11.

Twomey, Terrence. "The Cognitive Implications of Controlled Fire Use by Early Humans." *Cambridge Archaeological Journal* 23, no. 1 (2013): 113–128.

Tyrrell, Toby, John G. Shepherd, and Stephanie Castle. "The Long-term Legacy of Fossil Fuels." *Tellus B* 59 (2007): 664–672.

United Nations, Dept. of Economic and Social Affairs. *World Population Prospects: The 2012 Revision*. http://esa.un.org/wpp/Excel-Data/population.htm (accessed June 23, 2014).

———. Net International Migration. *International Migration Report 2013*, pp. 11–17. http://www.un.org/en/development/desa/population/publications/migration/migration-report-2013.shtml (accessed June 20, 2014).

United States Department of Agriculture. "2007 Census of Agriculture," in *United States Summary and State Data* 1, no. 51 (2009): 1–639.

United States Department of Defense. "*Narrative Summaries of Accidents Involving U.S. Nuclear Weapons, 1950–1980.*" http://nsarchive.files.wordpress.com/2010/04/635.pdf (accessed June 26, 2014).

Van Derbeken, Jaxton, Demian Bulwa, and Erin Allday. "SF Plane Crash: Crew Tried to Abort Landing." *San Francisco Chronicle*, July 8, 2013. http://www.sfchronicle.com/multimedia/item/boeing-777-crashes-at-sfo-22447.php (accessed June 24, 2014).

Vanhaeren, Marian, Francesco d'Errico, Chris Stringer, Sarah L. James, Jonathan A. Todd, and Henk K. Mienis. "Middle Paleolithic Shell Beads in Israel and Algeria." *Science* 312, no. 5781 (2006): 1785–1788.

Verhaegen, Marc and Pierre-François Puech. "Hominid Lifestyle and Diet Reconsidered: Paleo-Environmental and Comparative Data." *Human Evolution* 15 (2000): 151–162.

Videan, Elaine N. and W.C. McGrew. "Bipedality in Chimpanzee (*Pan Troglodytes*) and Bonobo (*Pan Paniscus*): Testing Hypotheses on the Evolution of Bipedalism." *American Journal of Physical Anthropology* 118, no. 2 (2002): 184–190.

Vigne, Jean-Denis, François Briois, Antoine Zazzo, George Willcox, Thomas Cucchi, Stéphanie Thiébault, Isabelle Carrère, Yodrik Franel, Régis Touquet, Chloé Martin, Christophe Moreau, Clothilde Comby, and Jean Guilaine. "First Wave of Cultivators Spread to Cyprus at Least 10,600 y Ago." *Proceedings of the National Academic of Sciences of the United States of America, PNAS Early Edition*. (May 7, 2012): 1–5.

Villa, Paola. *Terra Amata and The Middle Pleistocene Archaeological Record of Southern France*. Berkeley: University of California Press, 1983.

Wagner, Donald B. "Chemistry and Chemical Technology, Part II: Ferrous Metallurgy," in *Science and Civilization in China*, ed. Joseph Needham. Cambridge: Cambridge University Press, 2008.

Wade, Nicholas. "Chimps, Too, Wage War and Annex Rival Territory." *New York Times*, June 21, 2010.

Wake, David B., and Vance T. Vredenburg. "Are We in the Midst of the Sixth Mass Extinction? A View from the World of Amphibians." *Proceed-*

ings of the National Academy of Sciences of the United States of America 105, no. 1 (2008): 11466–11473.

Wales, Nathan. "A Fresh Perspective on Neandertal Clothing: Inferring Pleistocene Attire Using Modern Analogues." *2010 Annual Meeting, Paleoanthropology Society Abstracts.* St. Louis, Missouri, 13-14 April 2010.

———. "Modeling Neanderthal Clothing Using Ethnographic Analogues." *Journal of Human Evolution* 63, no. 6 (2012): 781–95.

Wallis, David A. "History of Angle Measurement," in *Pharaohs to Geoinformatics.* Cairo: FIG Working Week, 2005.

Ward, Cheryl "Boatbuilding in Ancient Egypt." In *The Philosophy of Shipbuilding,* Frederick M. Hocker and Cheryl A. Ward eds. College Station: Texas A&M University Press, 2004.

———. "Boat-Building and Its Social Context in Early Egypt: Interpretations from the First Dynasty Boat-Grave Cemetery at Abydos." *Antiquity* 80 (2003): 118–129.

———. "Sewn Planked Boats from Early Dynastic Abydos, Egypt," in *Ship Archaeology of the Ancient and Medieval World,* ed. C. Beltrame, 2003.

Washburn, Sherwood and R. Ciochon. "Canine Teeth: Notes on Controversies in the Study of Human Evolution." *American Anthropologist* 76, no. 4 (1974): 765–784.

Watts, Anthony. "NASA and Multi-Year Arctic Ice and Historical Context." *Watts Up With That,* March 1, 2012. http://wattsupwiththat.com/2012/03/01/nasa-and-multi-year-arctic-ice-and-historical-context/ (accessed June 27, 2014).

Webb, John and Marian Domanski. "Fire and Stone." *Science* 325, no. 5942 (2009): 829–821.

Weber, Johannes. "Strassburg, 1605: The Origins of the Newspaper in Europe." *German History* 24(3) (2006): 387–412.

Weik, M. H., ed. "Computers with Names Starting with E through H." *A Survey of Domestic Electronic Digital Computing Systems, US Department of Commerce, Office of Technical Services.* http://ed-thelen.org/comp-hist/BRL-e-h.html (accessed June 15, 2014).

Weinstein-Evron, Mina and Shimon Ilani. "Provenance of Ochre in the Natufian Layers of El-Wad Cave, Mount Carmel, Israel." *Journal of Archaeological Science* 21, no. 4 (1994): 461–467.

Weisman, Alan. *The World Without Us.* New York: St. Martin's / Thomas Dunne Books, 2007.

Welsh, Jennifer. "Man Entered the Kitchen 1.9 Million Years Ago." *LiveScience,* August 22, 2011. http://www.livescience.com/15688-man-cooking-homo-erectus.html (accessed May 22, 2013).

Wenbin, Huang, and Sun Xiufang. "Tropical Hardwood Flows in China: Case Studies of Rosewood and Okoumé." *Forest Trends*, December 2013. http://www.forest-trends.org/publication_details.php?publicationID=4138 (accessed June 28, 2014).

Wertime, Theodore A. *The Coming of the Age of Steel.* Chicago: University of Chicago Press, 1962.

Wheeler, P. E. "The Evolution of Bipedality and Loss of Functional Body Hair in Humans." *Journal of Human Evolution* 13 (1984): 91–98.

———. "The Influence of the Loss of Functional Body Hair on the Water Budgets of Early Hominids." *Journal of Human Evolution* 23 (1992): 379–388.

White, Mark J."Things to Do in Doggerland When You're Dead : Surviving OIS3 at the Northwestern-most Fringe of Middle Palaeolithic Europe." *World Archaeology* 38, no. 4 (2006): 547–575.

White, Randall. "Personal Ornaments from the Grotte du Renne at Arcy-sur-Cure." *Athena Review* 2, no. 4 (2001): 41–46. Abridged version at: http://www.athenapub.com/8white1.htm (accessed January 29, 2014).

White, Tim D., Berhane Asfaw, Yonas Beyene, Yohannes Haile-Selassie, C. Owen Lovejoy, Gen Suwa, and Giday WoldeGabriel. "*Ardipithecus Ramidus* and the Paleobiology of Early Hominids." *Science* 326, no. 5949 (2009): 64,75–86.

Whiten, Andrew. "Primate Culture And Social Learning." *Cognitive Science* 24, no. 3 (2000): 477–508.

Whiten, Andrew, Jane Goodall, W. C. McGrew, T. Nishida, V. Reynolds, Y. Sugiyama, C. E. G. Tutin, R. W. Wrangham, and Christophe Boesch. "Cultures in Chimpanzees." *Nature* 399 (1999): 682–685.

———. "Charting Cultural Variation in Chimpanzees." *Behaviour* 138, no. 11–12 (2001): 1481–1516.

Whiten, Andrew, and Christophe Boesch. "The Cultures of Chimpanzees." *Scientific American* 284, no. 1 (2001).

Wikipedia The Free Encyclopedia, 2014, *List of Nobel Laureates.* Available at: http://en.wikipedia.org/wiki/List_of_Nobel_laureates (accessed June 19, 2014).

Wildman, Derek E., Monica Uddin, Guozhen Liu, Lawrence I. Grossman, and Morris Goodman. "Implications of Natural Selection in Shaping 99.4% Nonsynonymous DNA Identity between Humans and Chimpanzees: Enlarging Genus Homo." *Proceedings of the National Academy of Sciences of the United States of America* 100, no. 12 (2003): 7181–7188.

Wilford, John N. "Fossil Skeleton from Africa Predates Lucy." *New York Times,* October 1, 2009.

Willcox, George. "The Roots of Cultivation in Southwestern Asia." *Science* 341, no. 6141 (2013): 39–40.

Willcox, George, Ramon Buxo, and Linda Herveux. "Late Pleistocene and Early Holocene Climate and the Beginnings of Cultivation in Northern Syria." *The Holocene* 19, no. 1 (2009): 151–158.

Williams, Michael. *Deforesting the Earth: From Prehistory to Global Crisis.* Chicago: University of Chicago Press, 2003.

Willman, David, 2014." $40-Billion Missile Defense System Proves Unreliable." *Los Angeles Times,* June 15, 2014. http://www.latimes.com/nation/la-na-missile-defense-20140615-story.html#page=1 (accessed June 6, 2014).

Wills, Christopher. *Children of Prometheus: The Accelerating Pace of Human Evolution.* New York: Basic Books, 1999.

Wilson, David Sloan and Edward O. Wilson. "Rethinking the Theoretical Foundation of Sociobiology." *Quarterly Review of Biology* 82, no. 4 (2007): 327–348.

Wilson, Edward O. *The Diversity of Life.* Cambridge, MA: Belknap Press / Harvard University Press, 1992.

Wobber, Victoria, Brian Hare, and Richard Wrangham. "Great Apes Prefer Cooked Food." *Journal of Human Evolution* 55 (2008): 343–348.

World Nuclear Stockpile Report. Washington, DC: The Ploughshares Fund. http://www.ploughshares.org/world-nuclear-stockpile-report (accessed June 3, 2014).

Wrangham, Richard. *Catching Fire: How Cooking Made Us Human.* New York: Basic Books, 2009.

———. "The Significance of African Apes for Reconstructing Human Evolution," in *The Evolution of Human Behavior: Primate Models,* ed. W. Kinzey. State University of New York Press, 1987: 51–71.

Wrangham, Richard, James Holland Jones, Greg Laden, David Pilbeam, and Nancy Lou Conklin-Brittain. "The Raw and the Stolen: Cooking and the Ecology of Human Origins." *Current Anthropology* 40, no. 5 (1999): 567–594.

Wrangham, Richard and NancyLou Conklin-Brittain. "The Biological Significance of Cooking in Human Evolution." *Comparative Biochemistry and Physiology, Part A* 136 (2003): 35–46.

Wright, Ronald. *A Short History of Progress.* Cambridge, MA: Da Capo Press, 2005.

Wynn, Jonathan G., Matt Sponheimer, William H. Kimbel, Zeresenay Alemseged, Kaye Reed, Zelalem K. Bedaso, and Jessica N. Wilson. "Diet of Australopithecus Afarensis from the Pliocene Hadar Formation, Ethiopia." *Proceedings of the National Academy of Sciences of the United States of America* 110, no. 26 (2013).

Zeder, Melinda A. "Central Questions in the Domestication of Plants and Animals." *Evolutionary Anthropology* 15 (2006): 1 05–117.

———. "The Origins of Agriculture in the Near East." *Current Anthropology* 52, no. S4 (2011): S221–S235.

Zhua, Rixiang, Zhisheng An, Richard Potts, and Kenneth A. Hoffman. "Magnetostratigraphic Dating of Early Humans in China." *Earth-Science Reviews* 61 (2003): 341–359.

Zilhão, João, Diego E. Angelucci, Ernestina Badal-García, Francesco d'Errico, Floréal Daniel, Laure Dayet, Katerina Douka, et al. "Symbolic Use of Marine Shells and Mineral Pigments by Iberian Neandertals." *Proceedings of the National Academy of Sciences of the United States of America* 107, no. 3 (2010): 1023–1028.

Zimmer, Carl. "Hope for Frogs in Face of a Deadly Fungus." *New York Times*, July 9, 2014. http://www.nytimes.com/2014/07/09/science/hope-for-frogs-facing-a-deadly-fungus.html?ref=science&_r=1 (accessed July 12, 2014).

Zimmerman, Andreas, Johanna Hilpert, and Karl Peter Wendt. "Estimations of Population Density for Selected Periods Between the Neolithic and AD 1800." *Human Biology* 81, issue 2, article 13 (2009).

Index

*Figure numbers are given in italics and preceded by "fig." If a page number is given (*e.g., *"174 fig. 7.1"), the illustration referenced will be found on the page indicated. If a page number is not given (*e.g., *"fig. 3.2"), the illustration referenced will be found in the sixteen-page photo insert following page **170**. Page numbers followed by the letter* n *refer to information found in the notes. The number of the referenced note immediately follows the letter* n.

absence of evidence in paleontological research, 30–32, 88
Acheulean hand axe, xvi, 61, 62, 121, *fig. 3.2*, 313n6
adornment, 109, 122, 187: use of beads for, 109, 187, 310n2; use of feathers for, 122; use of shells for, 122
adze, stone, in Neolithic, 154
agriculture: co-evolution theory 315n1, competitive feasting theory, 142, 143, 315n1; decline of, as a way of life, 236–237; earliest evidence of, 143–145; hilly flanks theory, 142, 315n1; hospitable climate theory, 143, 315n1; in modern times, 236-237; oasis theory, 142, 315n1; origins of, xi, 141–145, 315n1; population pressure theory, 142, 315n1; role of language in facilitating, 141, 145–147
Aiello, Leslie, 69, 311n9, 312n4, 315n13
Akkadian civilization, 173, 183
Aldrin, Buzz, 266
alpha male, 10–11, 12, 15, 23
Altamira Cave, 113, 114, 183
analytical engine, 241, 318n1
anatomically modern humans, xvii, 109, 123, 134, 314n7
animal tracks, 119, 120: in fossil form, 29, 30
Anthony, David W., 167, 191, 317n7, 317n8
antimony, 208–209
Apollo 8 mission, 266
Ardipithecus ramidus, 37, 39
Armstrong, Neil, 266
arsenic, use in making bronze, 188
artificial ecosystem, 270
asteroids, impact with Earth: cause of mass extinctions, 281; threat to biosphere, 275, 280
atlatl. *See* spear throwers
atmospheric engine, 214–215, 216, 290
Aurignacians, 129, 314n7
Australian Aborigines, 159, 316n9
australopithecines, 40
Australopithecus afarensis, 32
Australopithecus africanus, xv, 56, 58
Australopithecus prometheus, 58
automation systems, 253–255

Babbage, Charles, 240, 318n1
barges, of ancient Egyptians, 182
Barker, Graeme, 139, 143, 315n2
Batrachochytrium dendrobatidis, 322n16
battery, wet cell, 225
BD. See *Batrachochytrium dendrobatidis*

beads, prehistoric, worn as adornment, 109, 187, 310n2. *See* adornment: use of beads for
beaver lodge, 85, *fig. 4.1*
Bednarik, Robert G., 180, 317n4, 317n6
Ben Gurion, David, 321n9
Benz, Karl, 220
Bergmann, Ernst David, 321n9
Bernifal Cave, France, 314n8
Białowieża Forest, Poland, 278–279, 321n14
Bilzingsleben, Germany, 90, 121, 310n1, 311n7
biodiversity, restoration of, after mass extinctions, 283
biosphere, 320n5; humans as masters of, ix, 300; loss of human contact with, 270–272; threats to health of, 275, 277, 278, 282, 293, 299
Biosphere 2, 270–272: atmospheric pollution in, 271; ecosystem collapse in, 282; naming of, 320n6
Biospherians, 271
bipedal locomotion, xv, xviii, 34, 36: advantages of, 44; antiquity of, 30, 88, 136; disadvantages of, 44, 47, 99; in early hominids, 32–36. *See also* bipedalism, evolution of
bipedalism, evolution of, 34, 36–40, 44, 77, 106, 306n14: postural feeding hypothesis, 38; provisioning hypothesis, 37–38; savannah-based hypothesis, 36–37; thermal loading hypothesis, 38; warning display hypothesis, 38. *See also* bipedal locomotion
birth rates, decline in, 292; in modern times, 293
bison, prehistoric paintings of, 114
bitumen, 181
Black Death, 291
black powder, 221
blast furnace, 211, 213, 277, *fig. 8.3*
Blumenbach, Johann Friedrich, 77
boat-building, in early civilizations, 173
Bontius, 77
Boorstin, Daniel, 201, 315n1
Borman, Frank, 266, 320n1
Boulton, Matthew, 216
Brace, C. Loring, 41, 306n16
brain
—size: of beavers, 85; of chimpanzees, xv, 68; of early hominids, xv, xvi, 61, 68; of emerging humans, xvi, 61, 71, 102; and group size, 98–99; of *Homo floresiensis,* 111, 312n4; of modern humans, xv, 68, 69; of Neandertals, 77, 103.
—underdevelopment of, in human newborn, 98–101. *See also* human brain, massive size of
Brain, Charles K., 58, 307n1
branching logic, 241
Breuil, Abbé Henri, 115
Breviary, Catholic, 204
bronze,186–190: advantages over iron, 189–190; in stimulating early seafaring, 188–189. *See also* metal-working
Bulletin of Atomic Scientists, 273
Bullfinch, Thomas, 55
Burton, Frances, 307n2, 308n7
Bushmen, ¡Kung, 17–18, 101; fission-fusion society of, 133
Butler, Edward, 219

California condor, 283
canine teeth, reduction in size of, 38–41, 305n13, 306n14
cannons, 222
Cantabria, Spain, 112, 113, 123
carbon dioxide: atmospheric concentration of, 287–288; effect on global temperatures, 284, 285

fig. 10.4, 286, *fig. 10.5;* global emissions of, 284–290, 322n17
carbon nanotubes, 249
Carnot engine, 220
Carnot, Nicolas, 220
carpenters, Neolithic, 169
Carroll, Lewis, 315n14
Cartailhac, Émile, 115
carts, two-wheeled, 192
catapults, 222
cave-like dwellings, of hominids, 78–80
cave man, 63, 76–79
cave paintings, xvii, 112–117
CBS News, reporting US presidential election of 1952, 244–245
Celtic cultures, 251
Chabot, grotto of, 314n8
chain pump, 214
charcoal: for smelting, 211, 212, 213; use in gunpowder, 221; wood consumed in production of, 211–212
chariots, 192–193
chastity, 160–161
Châtelperronians, 314n7
Chauvet Cave, France, 116
Chesowanja, Kenya, 59
Cheyenne Indians: Arrow Renewal ceremony of, 116, 129–132; atonement for murder, 130; Sacred Arrows of, 130, 131; sexual repression among, 161; tribal solidarity of, 130, 132; warlike culture of, 164; Warrior Society of, 132
childbirth, hominid difficulty of, 99, 311n11
children
—burden of raising, in hunting and gathering societies, 101, 155–156; sexual freedom of, among hunter-gatherers, 157, 158–160; as source of support to aging parents, 13, 157, 160, 229, 232, 292
—economic significance of: in agricultural societies, 155–157; to hunters and gatherers, 155–156; in industrial societies, 231, 233, 235, 292; in traditional Greek island society, 229
chimpanzees: hammer and anvil technique used by, 25, 27; handling of fire, 308n7; hunting and warfare among, 22–23; of Ngogo, 23; preference for cooked food, 309n12; tool-making among, xii, 24–25; weapons made by, 25
Chinese civilization, 178–179
Chinese writing, 185, *fig. 7.7*
circadian rhythm, 66, 308
city-states, 170–171
civilization, birth of, xviii, 165-166
Clacton-on-Sea, England, 31, 310n1
class system: of *Linearbandkeramik* people, 154–155; of Nootka, 152
Clement, Joseph, 241
clock face, 205
clocks: European clockmakers, 200, 201, 203–207, 261; mechanical, 203, 204, 205–207. *See also* escapement, pendulum
clothing, fabrication of, x, xviii: by emerging humans, 92–943; foreshadowed by apes, 87; by modern humans, 92–98; by Neandertals, 94–95, 311n8, 311n9, *fig. 4.3*; non-tailored, 97; by Plains Cree, 94, *fig. 4.2*; role in protecting human infants, 101–102; by Tlingit women, 94
clothing, tailored: of Plains Cree, 94, *fig. 4.2*; of Cro-Magnons, 95, 97, 311n10; of Paleolithic people,

clothing, tailored (continued) 311n10
clubs, used as weapons, 40
coal: gas, 219; mining of, 212–214; use as fuel, 212; use for smelting, 212–213
Cohen, Joel E., 272, 320n8
coke (fuel), 213, 219
Collingwood, Charles, 244
colonization, extraterrestrial: of the "gas giants," 268; of Gliese 832 c, 269, 270, 320n4; of Mars, 268; of Mercury, 267; of the moon, 267; of Venus, 267
colony collapse, in bees and bats, 282
Colossi of Memnon, 182
communication, social: in nomadic hunting and gathering societies, 257; in sedentary agricultural societies, 258; technology of, xi; in urban industrial societies, 258
composite bow, 193
computers, 107, 137, 226, 232: as storage devices, 249–253; automation by means of, 253–255; integrated circuits in, 246–247; mechanical, 240–242; miniaturization of, 248–249; vacuum tube, 241, 243–245, 247, 319n5
conquistadores, 194
consort pairs, 11, 48, 49, 310n5
cooking: advantages of, 67–68; breakdown of cellulose and toxins in plant foods, 67–68; breakdown of collagen in meat, 67; disinfecting meat by means of, 67; earliest evidence of, 66, 72; enabling development of massive hominid brain, 68–71
corn. *See* maize
co-wives, jealousy among, 13
cradles of civilization, 166, 171, 173, 179
craft specialization, development of, 169–170
crankshaft. *See* rotary motion
Cretaceous Period, 280–281
Cro-Magnon people, 78, 111, 112, 118, 120, 313n6
Cronkite, Walter, 244
cultural fusion: in commerce, 297–298; in culinary traditions, 297; as a global process, 297; in music, 297; in science, 298
cultural inertia, 232
cuneiform writing, 126, 183, 184, 250
curanderas, 135
customs and traditions, in non-human species, 3, 26–28

Dadiwan, China, 185
Daimler, Gottlieb, 220
Dart, Raymond, 55–59, 77
Darwin, Charles, xix, 1, 36, 41, 56, 76, 186, 306n15
Dawson, Charles, 57
de Lumley, Henri, 90
de Mortillet, Gabriel, 114
de Sautuola. *See* Sanz de Sautuola
Dead Sea Scrolls, 78
Defoe, Daniel, 29
deforestation, 212, 213, 277–279: in China, 277; in Fertile Crescent, 277; in Mississippi River Valley, 277–278; of rain forests, 278
dental weaponry, xix, 33, 44
Devonian Period, 279
Diamond, Jared, 167, 316n1
diesel engine, 220
Diesel, Rudolf, 220
difference engine, 240–241
digestive organs, reduction in size among hominids, 70–71
digging sticks, 25, 41, 44, 45
digital communication, xviii
digital soapbox, 259

digital technology, 140, 232, 239, 240–265
diminished male rivalry hypothesis, 38–39, 305n12
division of labor, sexual: in early hominids, 44–48; in Greek Island society, 228; in hunting and gathering societies, 231
divorce. *See* marriage and divorce
DNA: advances in analysis of, leading to reclassification of primates, xiii; attempts to recreate extinct species through sequencing of, 283; as mechanism of inheritance, 136; in nanotechnology, 249; role in biological evolution, 136; role in construction of nests and shelters by *Hominidae*, 87–88; role in hominid capacity for symbolic communication, 110
domestic technology. *See* mechanization of the household
domestication of plants and animals, xviii. *See also* agriculture
dominance: displays of, 17; social, 15
doomsday machines, 273, 274
Dubois, Eugène, 74–75, 77, 121: in Sumatra, 74; in Java, 74
Dunbar, Robin, 198, 316n9
Dunbar's number, 197–198
Dutch East India Company, 205
dwellings, fabrication of, x, xviii, 84–92

early hominids, xv–xvii: anatomical difference from emerging humans, 59–60; brain, size of, xv, 68; discovery by Raymond Dart, 56–59; eventual disappearance of, xvi; evidence of dwellings of, 89; evolution of bipedalism among, 30–39; evolution of sexual-ecological symbiosis among, 45–48; habitation exclusively in Africa, 75; loss of weapons-grade canines among, 40–44; remains of, found in caves, 64; as scavengers, 67; size of digestive organs, 70–71; size of teeth and jaws, 71; sleeping in trees by, 33, 60, 62, 64, 72, 75
Egyptian civilization, 174
Einstein, Albert, 266
Eisenhower, Dwight D., 244
Elbe River, Germany, 162, 172
Eldredge, Niles, 282, 322n15
electric generators, 224, 225
electric motors, 224–226
electric power, 225–227: first availability of, 225; use in air conditioning, 226, 228; use in artificial lighting, 225, 226; use in communication, 225, 226, 227; use in elevators, 225, 226; use in mechanization of the household, 230–231; use in refrigeration, 225, 226, 228; use in transportation, 225
electromagnetic induction, 224, 225
electromechanical relays, 243
emerging humans, xv–xvii: anatomical characteristics of, 59–62; control of fire by, 64–66, 72, 79–80; different species of, 61–62, 307n4; discovery of by Dubois, 74–75; dwellings of, 81, 91; evidence of seafaring by, 180–181; invention of cooking by, 66–69; loss of body hair among, 72–74; migration out of Africa, 76, 102; reduction in size of digestive organs among, 70–71; use of clothing by, 92–94, 101; use of symbols by, 109, 121. *See also Homo erectus*
empires, in ancient civilizations, 194, 196–197

employment, 231: psychological hazards of, 238
employment society, 237–239
endocast, 55, 56
ENIAC (Electronic Numerical Integrator And Computer), 243, 245, 247, 319n4, 319n5
Erlitou, China, 178
escapement, 205: verge and foliot, 206, *fig. 8.1*; anchor and deadbeat, 206, *fig. 8.1*
estrus, 8–9, 12, 48
ethnic groups, 105, 110, 111
ethnic identity, xviii, 104, 105, 165, 304n3: and shared symbols, 129–132
Euphrates River, 171, 181
Eurasian steppes, 191: nomads of, 190–192, 194
evolution: cultural, 104, 135–137; vs. biological, 111, 136
exogamy, 17, 19–21: female, 20; village, 20
expensive tissue hypothesis, 69–71, 309n14
extinction, rates of, 279, 282, 322n15

facial hair, 303n1
Fagan, Brian, 106
families: extended, 18; nuclear, 18
Faraday, Michael, 224, 225, 240
farms, decline in number of: in the modern world, 233; in the U.S., 236–237, 318n9
Federation of American Scientists, 273
Feliks, John, 122, 314n11
female breast: enlargement at puberty, xix; as a sexual signal, 51–52; sexual significance of in humans, 50–53
female sex organs, concealing of, 51–52
Fertile Crescent, 141, 142–145, 149, 174, *fig. 7.1*
fetal hair. *See* lanugo
figurines, Paleolithic, 109. *See also* Venus figurines
fire lances, 221
fire: attraction to, xix; control of, x, xviii, 63–65; for inhabiting caves, 65; for promoting spread of grasses, 64; for protection from predators, 64; sleeping with, 72; for staying up late, 65–66; use in hunting, 65; use in tool-making, 65. *See also* cooking
firearms, 179, 221–224
firewood, 212, 213, 231
fission, as a process in human societies, 16, 19, 147–148, 178, 263–264, 296–297, 299, 323n23
fission-fusion society, 16–19, 28, 104, 132–134, 147: among chimpanzees, 132; among Inuit Eskimos, 133; among San Bushmen, 133
Five Dynasties and Ten Kingdoms Period, China, 317n3
Flores Island, Indonesia, 111, 312n4
flutes, prehistoric, 110, 129
foliot. *See* escapement: verge and foliot
Font-de-Gaume Cave, France, 314n8
food-sharing, 45–46: between sex partners, 48
foramen magnum, 34, 57
Forbidden City, China, 201
fossil fuels, 226: consumption of, global, 284–285; replacement of, by sustainable energy sources, 290
free time, problem of, 254
friendship, bonds of, 14
fundamental transformation. *See* metamorphosis
fungus infection, in bat colonies. *See* *Batrachochytrium dendrobatidis*
fusion: in the birth of civilizations, 197–200; in human history, 262–263; as a process in human

societies, 138, 165; in the rise of nation-states, 295

Galileo Galilei, 206
Gaslight Era, 219
gasoline engine, 219, 220
Gibran, Kahlil, 81
Gitano people. *See* Romani
global civilization, birth of, xi, 262, 264, 295–299
global climate change, 283–290
Goodall, Jane, 24, 31
Graber, Robert Bates, 41, 306n17, 314n9
Graham-Cumming, John, 241, 318n1
granaries, Neolithic, 148, 150
Grand Canal, China, 179
grasping reflex, 46
grasslands, prehistoric expansion of, 76, 309n17
Gravettian, 129, 314n7
Great Wall of China, 179
Great Western, 218
greenhouse effect, on Venus, 267
Greenland Ice Sheet, collapse of, 287, 323n19
group size in primates, 4–5
group solidarity, 3–5, 19
Groves, Colin P., 85, 86, 87–88, 310n4
gunpowder, 179: invention by Chinese, 221
guns: barrels, 222–223; hand-held, 222; naval, 223; rifling of, 223
Gutenberg, Johannes, 208–209
gyres, 276

hamadryas baboon, multilayered social structure of, 303n2
Harappa, Indus Valley, 175, 177, 185
harem system, 10–11
Herxheim, Germany, 163
Heyerdahl, Thor, 182
hieroglyphics, 126, 174: in Ancient Egypt, 184–185, *fig. 7.5*, 250; in Mayan writing, 186, *fig. 7.8*, 250
Hollerith, Herman, 242
homeland: among primates, 3–5; defense of, 4
Hominidae: xiii; classification by Linnaeus, 76; definition of, xiii; forms once believed to exist, 76–77, *fig. 3.4*
hominids, use of term, xii–xiv, 304n2
Hominina, xiv
Homininae, xiv
Hominini, xiv
hominins, xiv
Homo (genus), xiv
Homo antecessor, 307n4
Homo erectus, xvi, 62: antiquity of, 308n5; discovery by Eugène Dubois, 74–75; migration into northern latitudes, 93, 102; migration out of Africa, 93; size of brain, xvi; spears made by, 43; survival into Upper Paleolithic, 312n3; use of clothing and shelter by, 89; use of fire by, 59. *See also* emerging humans
Homo ergaster, 61
Homo floresiensis, 111, 312n4, 307
Homo habilis, 61
Homo heidelbergensis, 121, 314n10, 307: hyoid bone of, 127
Homo rhodesiensis, 307
Homo rudolfensis, 307
Homo sapiens neanderthalensis, 77
Homo sapiens, xvi, 77
Homo troglodytes, 76
horizon of digital history, 250–253
horse: domestication of, 145, 190–191, 315n4; invention of bit and bridle, for controlling, 191; use in warfare, 192–194. *See also* chariots

howitzer, 223
human anatomy, influence of tree-dwelling ancestry on, 2–3
human brain, massive size of, xv, 98, 309n15
human evolution, three phases of, xiv–xvii. *See also* anatomically modern humans, bipedalism, early hominids, emerging humans
hunting and gathering way of life: development of among early hominids, 45; material culture of, 82–83
Huygens, Christiaan, 206
hydraulic fracturing, 290
hyoid bone, 126–127

ice ages: conditions during most recent, 139, 287–289; recurrence of, 287, 288–289, 322–323n20, 323n21
Imo (Japanese macaque), 26
incense clocks, 204, 205
incest, 158, 235: prohibition or taboo against, 21
inclusionary religions, 296
Indus River, 171
Indus script, 185, *fig. 7.6*, 250
Indus Valley civilization, 175–178, 176 *fig. 7.2*, 185
industrial revolution, xvii
infanticide, in non-human primates, 10–11, 12, 23
information, recording of: fidelity of digital media, 252; problems with pre-digital media, 251–252
inheritance: of rank and status, 152–155; of wealth and property, 151, 153–155
ink, oil-based, invention by Gutenberg, 209
integrated circuit, 246–249: 15-Core Xeon E7, 247, 319n4
Intel Corporation, 246
intercontinental ballistic missiles, 273, 275
interglacial periods, 146, 287
internal combustion engine, 219, 220, 273
International Business Machines Company (IBM), 242
international organizations, growth of, 298
International Technology Roadmap for Semiconductors, 319n7
Inuit Eskimos, 17: fission-fusion society of, 133; sharing of sexual partners among, 13
Ios Island, Greece, 227–230
iron and steel: smelting of, 211, 277; wrought iron, 211. *See also*: metal working
irrigation systems, 173
isotopic signature, 154–155
Israel, nuclear weapons program of, 321n9

James, Steven R., 307n2
Japanese macaque, 26
Java Man, 75
javelins, prehistoric, 32, 43, 118, 311n7, 312n1
Jesuits, in China, 201–203
Jesus of Nazareth, 295–296
Jiahu, China, 185
Joordens, Josephine, 121

kangaroo, locomotion and posture of, 305n5
kaross, 101
Kibale National Park, Uganda, 23
kinship group, 21, 235
knee joints, locking, 33, 35
Koobi Fora, Kenya, 59
Koshima Island, Japan, 26
Kroeber, A. L., 305n3

La Calevie Cave, France, 314n8
La Dame à la Capuche. *See* Lady with the Hood
La Mouthe Cave, 314n8
La Pasiega Cave, Spain, 123–125, 183
Lady with the Hood, 118
Laetoli footprints, 29–31, 304n1
Lake Tanganyika, East Africa, 24
language: among Neandertals, 128; origins of, 128; role in social bonding, 128
lanugo, 309n15
Lartet, Louis, 113
Lascaux Cave, France, 116, 183
last glacial maximum, 146
lathes, 207
Leakey, Louis, 89
leather, preparation from hides, 92
leisure time. *See* free time
Les Combarelles Cave, France, 314n8
Les Eyzies Caves, France, 113, 183, 314n8
lethal weapons: for defense against predators, 223; by early hominids, 32–34, 39–45, 53, 60, 306n14; by emerging humans, 65, 79; fabrication and use of, xviii, xix, 22, 25, 44, 68; of metal, 165, 175, 187–190, 211; in Neolithic, 169; by precision machinery, 221–233; use by children, 155. *See also* nuclear weapons
Lévi-Strauss, Claude, 308n10
lice: in hominids, 74; three types of, 95–98. *See also* louse
lineage. *See* kinship group
Linearbandkeramik people, 162–164, 172: deforestation by, 277; mass executions by, 163–164
Linnaeus, Carolus, xiii, 76
living in one place. *See* sedentary lifestyle
loess soils, 163, 172
long houses, of Nootka, 152
longbow, 193
longhorn crazy ant, 321n7
Longshan script, China, 250
Los Angeles: air quality, 277; smog alerts, 321n11
Lothal, India, 175, 177
louse: of chimpanzee, 96; of gorilla, 96; of human body, 96–97; of human head, 96; of human pubic area, 96
Lovell, James, 266, 320n2
Lower Paleolithic, 112, 313n6
Lucy, xv, 32–33
lunar calendar, of *Homo erectus*, 121

machine gun, M-61, 223
machine tools, 201, 207, 226
magazines, printed, 210
Magdalenians, 112, 115, 116, 123, 125, 126, 314n7
maize, 144, 315n3
Makapansgat Cave, South Africa, 58
male dominance and sexual access, 49–50
Malinowski, Bronisław, 117, 314n9, 316n7
mammalian body plan, radical redesign of, x, 34–39
mammoth bone dwellings, 91
Mania, Dietrich, 90, 121
Mania, Ursula, 90, 121
Marine Isotope Stages, 287, 322n18
market towns, 165
marriage, and divorce: arranged, 232, 235; among hunter-gatherers, 13, 157–162; in hunting and gathering societies, 157–158, 234–235; in modern society, 232–235, 262. *See also* exogamy, monogamy, polygamy
marrying for love, 233–234
Marsoulas Cave, France, 314n8
Mas d'Azil Cave, France, 118

mass extinctions, ix, 279–283, 281 *fig. 10.3*. *See also* sixth mass extinction
material wealth, pursuit of, 149–151
maternal bond: in primates, 5–7, 46–47, 52; in humans, 52
maternal burden, in hominids, 45–48
Maybach, Wilhelm, 220
Mead, Margaret, 240
mechanization of the household, 227–232
Medicine Arrow Lodge. *See* Cheyenne Indians
megafauna, extinction of, 103, 133–134
melatonin, 65, 308n9
Memphis, Egypt, 196
Menes, 196
Mesolithic, 313n6
Mesopotamia, 173
metal-working: of bronze, 188–190; of copper: in the Middle East, 187-188; by Europeans, 276; of gold, silver, and copper, in pre-Columbian America, 187, 310n3. *See also* iron and steel
metamorphoses, x, xvii–xviii: 139–140, 200, 261, 265, 299, 300: vs. revolutions, xvii–xviii
metamorphosis, x, xi: definition of, xvii
Mickelson, Sig, 244
microchip, 246
microfilm, 252
microwear, 92
Middle Paleolithic, 112, 311
migration, out of Africa, xvi
milling machines, 207
Ming dynasty, China, 202
miniaturization of computer components, 247–249: quantum effects of, 248–249
MIS. *See* Marine Isotope Stages
modern human brain, size of, xvi
modern humans, xv, xvii
Mohenjo-daro, Indus Valley, 175, 177, 185
Moldova I, Ukraine, 91
Mongol invasions, 291
monogamy, 9, 12, 13, 306n18
Moore, Gordon E., 246, 319n3
Moore's Law, 246–247, 248, 249
Mousterian flake tools, 313n6
movable type, 208–209
multi-male–multi-female system, 11
music: in prehistoric cultures, 128–129; as symbolic communication, 110
musket: flintlock, 222; matchlock, 222
mutations, genetic, 136, 137

nakedness, development in hominids, xix, 72–74
nanorods, of gold and zinc oxide, 249
nanotechnology, 249
narrative story: possible role in invention of agriculture, 141, 146; power of, 134–135
Nasser, Gamal Abdel, 321n9
nation-states: 105: in ancient civilizations, 194, 196; mythical independence of, 294–295; number of, in modern times, 294, 323n23, 323–324n24; rarity until modern times, 294
Natufians, 148–150, 194, 289
nature of technology. *See* technology: definitions of
Neander Valley, Germany, 77
Neandertals, xvi, 77, 103: burial of dead among, 122–123; extinction of, 313n5; use of symbolism by, 122
necklaces, prehistoric, 122
Neolithic revolution, xvii
Neolithic, 112, 145, 313n6
nest site, 86
nest-building: by apes, 85–88, 310–311n5; by beavers, 85; by birds, 84–85; by rodents, 85; by social insects, 84
New Guinea Highlands, 167–168, 171

New Horizons Mission, 269
Newcomen, Thomas, 214, 215
newspapers, 210
Niaux Cave, France, 115
Niépce, Nicéphore, 219, 318n5
Nile River, Egypt, 171, 174–175
Nobel Prizes: awarded to international teams, 298, 324n25; in Physics, of 1956, 245
nomadic way of life, 82, 93, 147–148, 150–153, 168, 262
Nootka (Northwest Coast tribe), 151–153
nuclear family: human, 50, 53; evolution of, 53
nuclear nations, 273
nuclear weapons, 273–276: stockpiles of, global, 321n10; total explosive power of, 273–274

Obermaier, Hugo, 123
OIS. *See* Marine Isotope Stages
old stone age. *See* Paleolithic
Oldowan pebble tools, xv, 61, 313n6
Olduvai Gorge, Tanzania, 89
onagers, 192
oracle bone writing, 185
oral traditions, 135, 146–147
Ordovician Period, 279
Ørsted, Hans Christian, 224
Otto, Nikolaus, 219, 220
overpopulation, global, 290–294, 291 *fig. 10.6*
Oxygen Isotope Stages. *See* Marine Isotope Stages

Pacific Trash Vortex, 276
Paleolithic art, 112–117: cultural significance of, 116
Paleolithic, definition of, 112, 313n6
Pan troglodytes, 77
Paranthropus robustus, 59
Paratrechina longicornis. See longhorn crazy ant
passage of time, concept of, 135
Pech Merle Cave, 116
Pediculus humanus capitis, 96
Pediculus humanus humanus, 96
Pediculus schaeffi, 96
Peking Man, 78
pelvis: in early hominids, 33; obstacle to birth of large-brained infants, 98–102; restructuring of, 35
pendulum, 206
percussion tool, elephant bone, 121
Peres, Modesto, 113
Peres, Shimon, 321n9
Permian Period, 279
personal adornment. *See* adornment
petroglyphs, 110, 312n2, 119, 123–126: of la Pasiega, 123–125, *figs. 5.5, 5.6*
petroleum industry, 220
photocopies, 252
Pickard, James, 215
Piltdown Man, 57
piston and cylinder: in internal combustion engines, 218–220; in steam engines, 214–216
Pithecanthropus erectus, 74
planet of the hominids, 292–294
planets, existence of outside of solar system, 320n3
plank boats, 182
platen, 209
Platte River, North America, 130
polar ice sheet: destruction of inhabited areas, 288; effects on agriculture, 288; effects on global climate and environments, 289; possible return of, 289, 323n21
pollution: atmospheric, 276–277; from industrial processes, 276; oceanic, 276
polygamy, 12
polygyny, 12–13
Pongidae, xiii

population explosion. *See* overpopulation: global
population increase, of hominids, global, 104, 135, 291–292, 291 *fig. 10.6*
positive feedback loop, 195
postural feeding hypothesis, 38, 305n11
potato-washing, 26
potlatch, 153
precision instruments, xi
precision machinery: 140, 200, 201–239, 240, 253, 261–262
precision machining: 203–204, 207–210, 223, 225
pre-Columbian civilizations, 186, 187
predators: in prehistoric Africa, 60; defense against, 223–224
predatory hunting, in primates, 47
predatory species, 45
prehistorical congress, Lisbon, Portugal, 1880, 114–115
prehistory, as compared to history, 250–251
premarital chastity, 233
premature birth, of human infant, 98–102
primates: customs and traditions among, 26; exogamy in, 19–21; friendships among, 13–14; group size among, 4–5; hunting and warfare among, 21–24; maternal burden of, 6–7; motherhood among, 5–7; nocturnal species of, 308n8; sexual relationships among, 7–13; social hierarchies among, 14–16; taxonomic order of, xiii; terrestrial, sleeping in trees by, 307n3; types of relationships among, 5; use of tools and weapons among, 24–26
printing press, 209–210
promiscuity, 11, 13
property, communal vs. personal, 4
provisioning hypothesis, 37, 305n8
Pthirus gorillae, 96
Pthirus pubis, 96
Pueblo Indians, of American Southwest, 78
punched cards, 241, 242

Qing dynasty, China, 178, 179
quadrupedal species, xiv, 37, 51, 57, 63, 99, 139, 283, 304n2

railroads, 217, 218, 260
raw food diet, disadvantages of, 68, 309n13
reason and emotion, in non-human species, 318n8
reciprocating engines, xi, 218, 226
Red Flag Act, 219
red ochre, symbolic use of, 123
reed boats, of ancient Egyptians, 181–182
reforestation: in China, 278; in Europe, 279
Remington Rand Corporation, 243–244
Renaissance, 208, 209, 210
replacement birth rate, 292
Rhine River, Germany, 162, 172
Ricci, Father Matteo, 201–203, 206
rifle, M-16, 223
river valleys: habitat of Aurignacians, 129; habitat of early hominids, 75; habitat of *Linearbandkeramik* people, 154–155, 162–164; location of first civilizations, 171–178; settlement by early agricultural people, 172
rivers: deltas of, 172; role in development of civilizations, 171–179; role in transportation and trade, 173; seasonal flooding of, 172
robots. *See* automation systems
Roman alphabet, 126

Roman citizenship, benefits of, 296
Romani, 78
rotary motion: crank-and-flywheel design, 215–216; crankshaft, 218; vs. oscillating motion, 215, 216

Sabater Pi, Jordi, 85, 86, 87–88, 310n4
sailing ships, 166, 212, 217
saltpeter, use in gunpowder, 221
sandglasses, 204, 205
Sanz de Sautuola: Marcelino, 113–115, *fig. 5.1*; Maria, 113, 115, *fig. 5.1*
savannah-based hypothesis, 36
sawmills, 216, 217
Schöningen javelins, 43, 65, 93, 310n1, 311n7
Schöningen, Germany, 32, 43
Schopenhauer, Arthur, 303n1
screw press, 209
seafaring: in ancient Egypt, 174–175; in ancient Mesopotamia, 175, 181; antiquity of, 180–181; by emerging humans, 180; in Upper Paleolithic, 180–181
sedentary way of life, 148, 150, 151–154
settlements, permanent, 140, 172. *See also* sedentary way of life
sexual and maternal bonds, integration of, 50–53
sexual behavior: continuous among human females, 48, 52; increased duration of, among male hominids, 49; receptivity to, among female hominids, 49
sexual freedom, in modern society, 233
sexual partners, sharing of, among Inuit, 13
sexual permissiveness: in childhood, 157, 158–160; in adolescence, 158
sexual relationships: in goats and sheep, 7; in bonobos, 8–9; in orangutans, 9; in titi monkeys, 7
sexual repression, 161–162
shamans, 135
Shang dynasty, China, 185
Shanidar, Iraq, 123
shelter: fabrication of, by great apes, 85–88; fabrication of, by other animals, 84–85; fabrication of, by prehistoric hominids, 88–91; lack of paleontological remains of, 89–92; role in protecting human infants, 101–102. *See also* nest-building
Sherratt, Andrew, 145, 316n5
Shockley, William, 245
Sillen, Andrew, 58
sinew-backed bow. *See* composite bow
single-person household, 230
Sistine Chapel of Paleolithic Art. *See* Altamira Cave
Sixteen Kingdoms Period, China, 317n3
sixth mass extinction, 279, 282, 283
smell, sense of, reduction of in primates, 119
Smoot-Hawley Tariff Act, 294
social media, 256–259. *See also* communication: social
Solar Probe Plus, 269
Solutreans, 123, 125, 314n7
Song, Liao, Jin, and Western Dynasties Period, China, 317n3
song: in gibbons, 2; in humans, 128
spear throwers, 109, 118, 312n1
spears: and digging sticks, x, xv, 42, 53; role in promoting bipedal locomotion, 42; as throwing weapons, 43; use as weapons by early hominids, 40–44; wooden, 25, 31
status symbols, 122, 310n3
staying up late, 65–66
steam engine, 213–218, 226. *See also* steamships, railroads
steamboat, 277

steamships, 217–218, 260
Stevenson, Adlai, 244
stone age, 83–84, 186: artifacts missing from, 81–83
stone, artifacts made from, 83
stones, as throwing weapons, 40
storytellers, 135
strangers, fear and avoidance of, before civilization, 167-169
stratified societies. *See* class system
Su Song (Chinese civil servant), 202
Suess, Eduard, 320n5
Sumerian civilization, 173, 183
sun and planet gear, 216
sundials, 204
Swartkrans Cave, South Africa, 58
sweat glands, 73
sweating, advantages of, 73
Sweet Medicine. *See* Cheyenne Indians: Sacred Arrows of
symbolic communication, xviii, 104, 107–135, 137–138, 289: and agriculture, 145–147; among insects, 108; among Neandertals, 109; oldest evidence of, 120–123; vocalizations, 109
symbols, x–xi, 104, 106, 108–113, 120–124: and ethnic identity, 129–132, 199, 262; verbal, x, 126–128; visual, x, 126, 175, 183–186. *See also* symbolic communication
Systema Naturae, 76

Tabulating Machine Company, 242
Talheim, Germany, 163
tar sands, 290
Tau Ceti, 269
Taung Child, 56–57
technologies of interaction, xi, 166, 168, 173, 195, 260–261, 264, 295, 304n3, 315–316n4. *See also* horse, seafaring, transportation, writing
technology: of agriculture, 140–145, 165–166, 168; in ancient China, 178–179, 203; in chimpanzees, 24–25; of clothing and shelter, 88–95, 101-104, 139; definition of, xii-xiii; of fire, 62–74, 79–80, 139; in the house and home, 227–231; of information processing, 240–250; of mechanical clocks, 204–207; of movable type, 208–210; of musical instruments, 128; of precision machinery, 140, 200, 203–204; of spears and digging sticks, 40–45, 53, 139; of symbolic communication, 107–109, 140. *See also* electric power, metal working, reciprocating engine, rotary motion
termite fishing, 24
Terra Amata, France, 90, 311n8
Texas Instruments Corporation, 245
The Raw and the Cooked, 66, 308n10
thermal loading hypothesis, 38, 305n9
thermonuclear annihilation, 274, 275
thermonuclear weapons. *See* nuclear weapons
Thomsen, Christian Jürgensen, 186
Tigris River, Mesopotamia, 171
Tilman, G. David, 272, 321n8
Timor Island, Indonesia, 180, 317n5
tin: use in bronze, 188–189; use in movable type, 208–209
toes, 35–36: grasping, 35; of early hominids, 33; opposable, 35
tools for thought, 107–108
tourism, growth of, international, 261
town clock, 210
transistor, 245: transistor radio, 246
transportation, technologies of, xi, 170, 179, 187, 189, 195, 217, 221, 260, 284, 296. *See also*

boat-building, seafaring, wheel: invention of
Triassic Period, 279
Trobriand Islands, Papua New Guinea, 117

underemployment, resulting from automation, 254–255
unemployment, resulting from automation, 254–255
United Nations World Tourism Organization, 261
UNIVAC (Universal Automatic Computer), 243–245, 247
Upper Paleolithic, definition of, 112, 313n6
upright posture, xv, xviii. *See also* bipedalism, bipedal locomotion
urban centers, 170
urban civilization: achievements of, 200; advantages of, to citizenry, 195; characteristics of, 197; emergence of, 194–197
urbanization: global phenomenon of, 237; recent trend toward, 236–237
US census, problems with, 242

vacuum tube, 240. *See also* computers: vacuum tube
Van Derbeken, Jaxon, 319n6
vellus hair, 73
Velocycle, 219
Venus figurines, 117–118, *fig. 5.3*, 311n10
Vilanova y Piera, Juan, 114
visual tracking, in pursuit of game, 119–120
voluntary associations, 18
Vostok Station, ice cores taken at, 286
wagons, four-wheeled, 192
Wales, Nathan, 94–95
walking on hind legs, by apes, 40
Wanli (Chinese emperor, Ming dynasty), 201, 206
warfare, 221–222: among chimpanzees, 22–23; among hunters and gatherers, 164; among Nootka, 164–165; organized, 155, 162–165, 221
warning display hypothesis, 38, 305n10
Warring States Period, China, 317n3
washing machines, 228, 230, 231
water clocks, 202–203, 204–205
waterwheels, 215, 217
Watt, James, 215–217, 223
weapons: biological, 33, 39, 41; use by early hominids, 32, 34, 40, 42, 44, 53, 60; use by emerging humans, 43, 65; use by non-human primates, 3, 22, 25. *See also* lethal weapons
weavers, Neolithic, 169
weaving and sewing, in Upper Paleolithic, 311n10
wheat-washing, 26
wheel, invention of, 191, *fig. 7.9*
Wheeler, Peter, 69
Wilkinson, John, 216, 223
Wills, Christopher, 308n5
Wilson, Edward O., 282, 322n15
windmills, 217
Wonderwerk Cave, South Africa, 59, 183, 309n11
woodblock printing, 208
wooden artifacts, prehistoric remains of, 31–32, 42–43, 65, 91, 93, 310n1
work, replacement of, by digital technologies, 239, 253–255
Wrangham, Richard, 63, 66, 67, 308n6, 309n15
writing, development of, 170

Yangtze River, 171, 178
Yąnomamö, 161–162, 198, 303n1, 316n10
yaodongs, 78
Yellow River, 171, 178
Zhezong (Chinese emperor, Song dynasty), 202
Zuse, Konrad, 242

Made in the USA
Monee, IL
09 September 2021

77526366R00238